Printreading

based on the

2011 NEC®
National Electrical Code

AMERICAN TECHNICAL PUBLISHERS
ORLAND PARK, ILLINOIS 60467-5756

R. T. Miller

The author and publisher are grateful to the following companies, organizations, and individuals for providing software, photographs, information, and technical assistance.

ABB Power T&D Company, Inc.
Barclay and Associates
Carlon, an Indian Head Company
Carrier Corporation
Cooper Bussmann
Cooper Crouse-Hinds
Fluke Corporation
General Electric Company
Hubbell Incorporated (Delaware), Wiring Device-Kellems
Rockwell Automation, Allen-Bradley Company, Inc.
Rodger A. Brooks, Architect
Square D Company

American Technical Publishers, Inc., Editorial Staff

Editor in Chief:
Jonathan F. Gosse
Vice President—Production:
Peter A. Zurlis
Art Manager:
James M. Clarke
Copy Editor:
Talia J. Lambarki
Cover Design:
Jennifer M. Hines
Illustration/Layout:
Nicole D. Bigos

1 2 3 4 5 6 7 8 9 – 11 – 9 8 7 6 5 4 3 2 1

Printed in the United States of America

ISBN 978-0-8269-1569-6

This book is printed on recycled paper.

atp made in USA

Contents

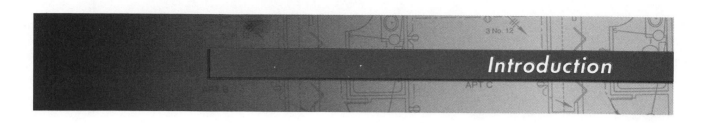

PREFACE

Printreading Based on the 2011 National Electrical Code® reflects many of the important changes that appear in the 2011 National Electrical Code®. The changes are presented as they pertain to Onc-Family Dwellings, Multifamily Dwellings, Commercial Locations, Industrial Locations, and Hazardous Locations. In addition to changes to the text and illustrations, the Trade Competency Tests and two Final Tests have been thoroughly updated.

Printreading Based on the 2011 National Electrical Code® is designed to enable the student to learn electrical printreading and become familiar with applicable sections of the 2011 National Electrical Code®. Complete references to the 2011 NEC® are presented throughout the text. Trade Competency Tests are included at the end of each chapter to help students check their understanding of the text material and the NEC®. Two Final Tests are included in Chapter 7 of the text. Answers with NEC® validation for the Trade Competency Tests and Final Tests are included in the *Printreading Based on the 2011 National Electrical Code® Answer Key*.

NATIONAL ELECTRICAL CODE®

The National Electrical Code® is sponsored and controlled by the National Fire Protection Association, Inc., (NFPA). The primary function of the NEC® is to safeguard people and property against electrical hazards. It is mandatory that the *Printreading Based on the 2011 National Electrical Code®* be used only in conjunction with the 2011 NEC®. Copies of the 2011 NEC® (NFPA No. 70) can be ordered directly from its publisher:

> National Fire Protection Association, Inc.
> 1 Batterymarch Park
> Quincy, MA 02169

WATTS AND VOLT-AMPERES

In general, within the NEC®, the term *watts* (W) has been superseded by the term *volt-amperes* (VA) for the computation of loads. However, references to nameplate ratings still reflect the term *watts* on certain loads.

CALCULATIONS

When total wattage or VA is to be divided by phase-to-phase (3ϕ) voltage times 1.732, the following values may be substituted:

> for 208 V × 1.732, use 360
> for 230 V × 1.723, use 398
> for 240 V × 1.723, use 416
> for 440 V × 1.723, use 762
> for 460 V × 1.723, use 797
> for 480 V × 1.723, use 831

MANDATORY USE OF SHALL

Section 90.5(A) states that mandatory rules use the word shall or shall not. Always refer to the NEC® for mandatory rules.

INFORMATIONAL NOTES

Informational Notes (Notes) are explanatory or provide additional information. Notes are not mandatory and do not contain any mandatory provisions.

FRACTIONS OF AN AMP

Section 220.5(B) states that for calculations, fractions less than 0.5 may be dropped.

THW Cu

Unless otherwise specified, copper conductors are sized on THW per Table 310.15(B)(16).

TRADE COMPETENCY TESTS AND FINAL TESTS

Printreading Based on the 2011 National Electrical Code® contains 23 Trade Competency Tests that should be completed after studying the corresponding chapter. The two Final Tests include questions that cover the content of the entire text. The Final Tests should be taken after completing the text and all of the Trade Competency Tests.

Topic Definitions
are provided at the beginning
of each chapter

**Related Code References and
Definitions** are indicated in black type

Large Detailed Illustrations
organize and simplify topics for
easy learning

Applicable Code Requirements
within the figures are indicated in red type;
letters correspond to specific printreading
applications

Electrical Floor Plans
provide examples of typical
installations in the field

**Complete References to the
2011 NEC®** are presented throughout
the text

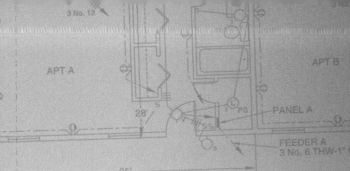

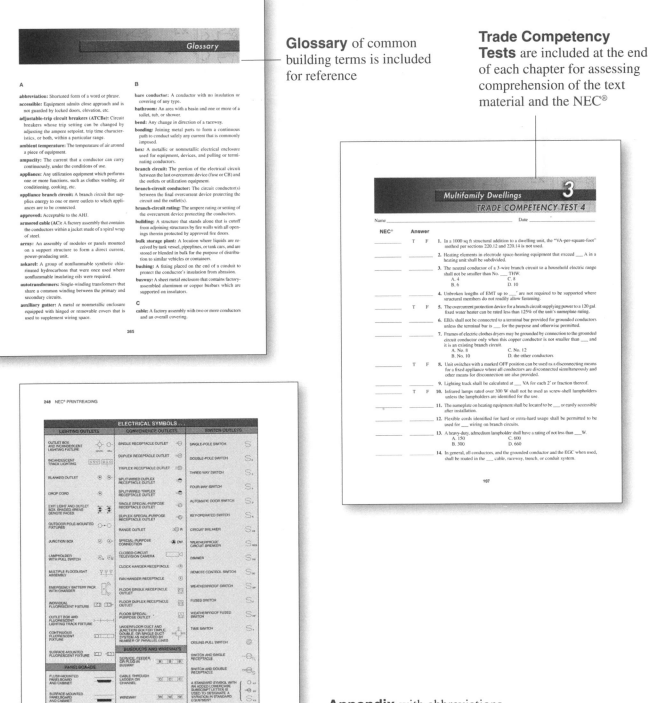

Glossary of common building terms is included for reference

Trade Competency Tests are included at the end of each chapter for assessing comprehension of the text material and the NEC®

Appendix with abbreviations, formulas, and symbols is included for quick and convenient reference

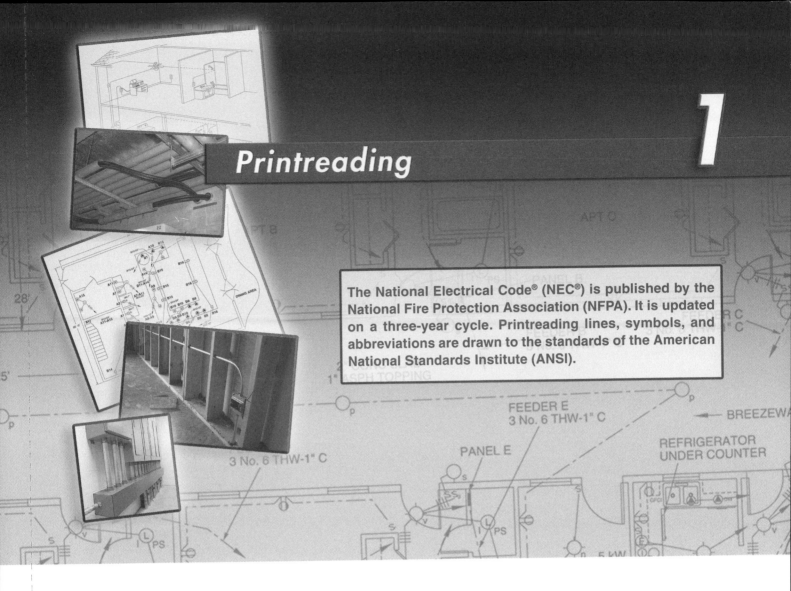

The National Electrical Code® (NEC®) is published by the National Fire Protection Association (NFPA). It is updated on a three-year cycle. Printreading lines, symbols, and abbreviations are drawn to the standards of the American National Standards Institute (ANSI).

THE NATIONAL ELECTRICAL CODE®

The National Electrical Code® (NEC®) is one of the most widely used and recognized consensus standards in the world today. It is a true consensus standard because members from throughout the electrical industry contribute to its development. The NEC® is updated and revised every three years to reflect current trends in the electrical industry. The purpose of the NEC® is the practical safeguarding of persons and property from hazards arising from the use of electricity.

NEC® Process

The National Fire Protection Association, Inc. (NFPA) sponsors the development of the NEC®. The NFPA publishes guidelines in the Regulations Governing Committee Projects for the procedures for all of the standards it publishes. For the NEC®, the procedures call for essentially a five-step process:

1. The Call for Proposals
2. Report on Proposals (ROP)
3. Report on Comments (ROC)
4. The Technical Report Session of the NFPA Annual Meeting
5. Standards Council Consideration and Issuance

The Call for Proposals. The first step in the Code process involves issuing a public notice that the NEC® revision process has begun and asking for interested parties to submit proposals for revising the Document. The notice is placed in appropriate publications such as the *NFPA News*, the *U.S. Federal Register*, and the American National Standards Institute's *Standards Action*. This step provides approximately twenty weeks for respondents to submit their proposals.

Report on Proposals. The second step in the Code process is the receipt of proposals. Anyone can submit a proposal to change the NEC® provided it contains the required information. **See Figure 1-1.**

<table>
<tr><td colspan="1">

NEC® PROPOSALS

NEC® proposals shall contain:

1. Identification of the submitter.
   (name and organization or company affiliation)

2. Identification of the specific Code section and type of revision that is proposed.

3. A statement of the problem and substantiation for why the change is necessary.

4. The actual wording of the revised text or the wording to be deleted.

</td></tr>
</table>

Figure 1-1. NEC® proposals shall contain the required information in order to be considered.

The key to successful Code proposals is proper substantiation for the proposed changes. The time window to submit proposals is very small. For example, the last date to submit changes for the 2011 NEC® was November 2008. After the last date to receive proposals, code panels meet to discuss and vote on each of the proposed changes. These votes are recorded and published in the Report on Proposals.

Report on Comments. Once the Report on Proposals is published, everyone has the opportunity to submit a comment on each of the proposed changes whether the Code-Making Panel (CMP) voted to accept or reject them. A closing date for comments is published, and after the closing date the Code-Making Panel meets to vote on each of the comments.

The Report on Comments contains all comments and the votes of the respective code panels. If a proposal or comment is not accepted by the Code-Making Panel, the statement must state the reason the proposal or comment was not accepted. New proposals cannot be submitted during this stage of the Code cycle. The Code-Making Panel can only take action on proposals that have received adequate public review during the proposal stage.

NFPA Annual Meeting. Once the Code-Making Panels have reviewed the proposals and comments, the next step is to present the changes at the NFPA Annual Meeting. Floor action on individual proposals can occur during this meeting. Usually there are very few of these actions on the floor of the NFPA Annual Meeting because most members choose to support the actions of the Code panels during the initial stages of the Code process. Beginning with the 2008 NEC®, the NFPA at the their annual meeting modified the method by which floor actions can be

implemented. The revised process permits votes only in cases where a Notice of Intent to Make a Motion (NIT-MAM) has been properly filed.

Voting on floor actions at the NFPA Annual Meeting is limited to NFPA members only. A simple majority vote is required for a floor action to pass. Actions that occur on the floor of the NFPA Annual Meeting are still subject to review by the NFPA Standards Council.

Standards Council Issuance. The Standards Council has the responsibility for overseeing all of the codes and standards developed for the NFPA. The NEC® Correlating Committee works directly under the Standards Council. The Correlating Committee steers the panels through the process, ensuring that each proposal and comment received is handled according to an established operating procedure. Once the process is complete, the Standards Council reviews the entire process and actually issues the document for publication.

The NEC® is a legal document designed to be adopted by local and/or state governmental bodies. Local jurisdictions may choose to adopt the Code in its entirety, with specific additions or exceptions, or they may choose not to adopt the Code at all.

Code-Making Panel Membership. Each Code proposal or comment is reviewed by the representatives of various segments of the electrical industry. There are 20 Code-Making Panels to cover the articles in the NEC®. The members of each Code panel represent labor, manufacturing, electrical utilities, electrical inspectors, contractor associations, and testing laboratories.

NFPA Operating Procedures for the National Electrical Code® Committee state that no interest group shall comprise more than one-third of the total voting panel membership. Membership is designated as either principal or alternate. Alternate members assume the participation and voting rights only when the principal member is not present.

The official vote on panel proposals and comments occurs on a written ballot after the meetings conclude. Although each proposal is voted on during the Code-Making Panel discussions, the official vote on each proposal occurs when the Code-Making Panel member returns a written ballot which was mailed after the conclusion of the Code-Making Panel meetings. All members who vote against a panel action must state the reasons for doing so, and their comments are recirculated to each panel member.

Using the NEC®

The NEC® is available in soft cover, loose-leaf, and electronic versions. It may be ordered directly from its publisher:

National Fire Protection Association, Inc.
1 Batterymarch Park
Quincy, MA 02169

While the NEC® is not difficult to use, it does have its own method of organization which must be understood for ease of use.

Revisions and Extracted Text. Beginning with the 2008 NEC®, vertical bars are no longer used to identify code changes. Shading behind the revised text is now used to identify changes. The Code also contains extracted text that has been taken from other NFPA documents identified with brackets. The number, title and edition of the NFPA document from which the extract is taken appears at the beginning of the article in which the extract has been used. For example, in Article 517, extracted text from NFPA 99-2005 *Standard for Health Care Facilities* appears throughout the article. These editorial markings are important for helping to trace the development of Code requirements. Rules for formatting the NEC® are found in Chapter 2 of the NFPA *National Electrical Code©* Style Manual. **See Figure 1-2.**

Outline Format. The NEC® is arranged in a simple outline format. **See Figure 1-3.** Chapters are divided into articles. With the exception of Article 90, these are three-digit numbers. For example, Article 210 covers Branch Circuits. The Contents in the front matter of the Code contains a list of the chapters with the article numbers covered in each chapter.

Section numbers are designated by a period following the article number. For example, 210.8 indicates Article 210, Section 8. Section numbers often proceed in a normal numbering sequence, but sometimes sections are not used in numerical sequence. This occurs when sections have been deleted from the NEC® or perhaps if the Code panel wanted to leave room for future Code sections.

Subsections are part of the sections and are designated by uppercase letters set off in parentheses. For example, 210.8(A) is a subsection of 210.8. Subsections are further broken down by numbers to indicate parts of subsections. For example, 210.8(A)(4) is a part of subsection 210.8(A). In a few instances, the NEC® has further subsections of subsections. In most cases, the format remains in this number-letter-number sequence. The only time the NEC® deviates from this fashion is when sections do not contain any subsections, but do contain numbered items. For example, 230.43(4) is a numbered item of Section 230.43.

Exceptions. Exceptions permit alternate methods to the general (main) rule. Exceptions are permitted in lieu of the main rule when all of the provisions of the exception are met. For example, 210.8(A)(5) requires that, in general, all receptacles in unfinished basements of dwelling units shall be GFCI-protected. There is an exception to this section, however, that permits receptacles that supply only a permanently installed fire alarm or burglar alarm system to be installed without GFCI protection. The reason for the exception is to ensure that these life-safety and security systems remain in operation and are not inadvertently de-energized or rendered inoperable in the event of a GFCI tripping. Similarly, in 210.8(B)(4), there is an exception to the requirement of GFCI protection for outdoor locations for electric snow-melting, deicing, or pipeline and vessel heating equipment. Where exceptions contain qualifying provisions or conditions, all of the conditions of the exception must be met in order to apply the exception. In this case, the electric snow-melting, deicing, or pipeline and vessel heating equipment must be not readily accessible and must be supplied from a dedicated branch circuit.

NEC® REVISIONS AND EXTRACTED TEXT	
SHADED TEXT	Shading behind text is used to indicate revisions from the previous issue of the Code. For example, see Article 100 where the definition of equipment had been revised from the previous edition of the NEC®.
[EXTRACTED TEXT]	Brackets are used to denote extracted text from another document that appears in the Code. For example, see 517.35(A) which was extracted from the *Standard for Health Care Facilities*, NPFA 99-2005, Section 4.4.1.1.4.

Figure 1-2. NEC® editorial symbols are useful tools for studying the latest NEC® changes.

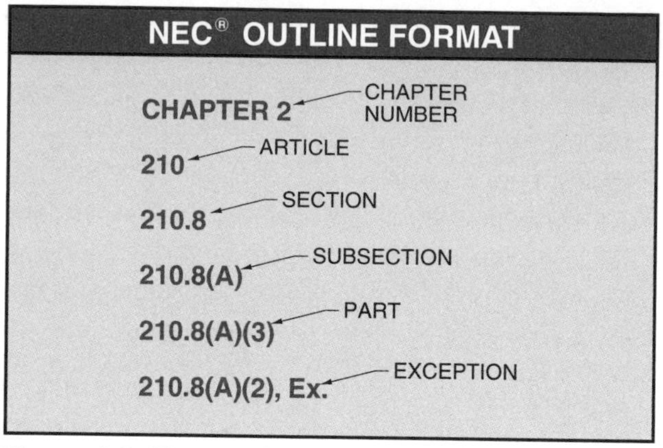

NEC® OUTLINE FORMAT

CHAPTER 2 —— CHAPTER NUMBER

210 —— ARTICLE

210.8 —— SECTION

210.8(A) —— SUBSECTION

210.8(A)(3) —— PART

210.8(A)(2), Ex. —— EXCEPTION

Figure 1-3. The NEC® is arranged in a simple outline format.

Because the NEC® is organized in an outline format, exceptions are applied only to the section they follow or as noted. For example, in 210.8(B) there is an exception to the requirements of GFCI protection in other than dwelling units for electric snow-melting, deicing, or pipeline and vessel heating equipment. Note that the exception only applies to (3) rooftops and (4) outdoors. It cannot be applied except in those two specific cases.

Applying the NEC®

The National Electrical Code® is adopted for use by local (village, town, city, etc.), county or parish, and state authorities. These authorities are generally represented by building officials who issue permits for jobs and make periodic inspections of work in progress. Additionally, these authorities certify that the completed building meets all applicable codes before occupancy occurs.

Purpose and Intent – 90.1. The two-fold purpose of the NEC® is to protect people and property from the dangers associated with the use of electricity. This two-fold purpose should be kept in mind whether applying the NEC® or submitting new proposals to modify the NEC®.

The NEC® is not intended to be used as an instruction manual per 90.1(C) for untrained persons. The rules are stated without any reasons for their existence. For example, 210.8(A)(1) states that all 15 A and 20 A receptacles installed in dwelling bathrooms shall have

GFCI protection. In order to understand the reason for the rule, the Report on Proposals would have to be read to determine the substantiation when the Code panel first accepted the change.

The NEC® is also not a design specification per 90.1(C). It contains the rules and necessary provisions for electrical systems, but leaves the design and layout of the electrical systems to others.

Scope – 90.2. Before applying the NEC®, the scope should be reviewed to determine if the electrical installation falls within the jurisdiction of the NEC®. Installations which are not covered are given in 90.2(B). Often, the distinction between what is covered and what is not covered is very fine and can lead to some difficult interpretations. For example, generally installations for electrical utilities are not covered if they involve the metering, generation, transmission, or distribution of electric energy. Those installations, however, that are used by an electric utility, but are not an integral part of the electrical energy transmission or distribution are covered. Another example is watercraft. Installations on ships are not covered, but installations on floating buildings are.

Code Arrangement – 90.3. The arrangement of the NEC® is specified in 90.3. The NEC® contains an Introduction and nine chapters. The first four chapters apply generally, unless they are modified by the latter chapters. For example, 300.14 requires that, in general, a minimum of 6″ of free conductor be provided at each outlet for splices or terminations. This applies throughout the NEC® unless modified.

Chapter 8 covers communications systems. None of the NEC® provisions apply unless they are directly referenced. Chapter 8 is independent of the other chapters because it covers communication circuits. These circuits are part of the premises wiring, but are treated separately because they do not, in general, pose a threat of electrical shock to persons. None of the NEC® provisions therefore apply, unless they are directly referenced in Chapter 8.

Chapter 9 contains tables and examples which are referenced throughout the entire NEC®. The Appendix, which appears at the end of the Code, does not contain any mandatory provisions, but it does contain information that may be helpful in designing or installing electrical installations. For example, Appendix C contains many useful tables for determining the maximum number of conductors permitted in a raceway.

A comprehensive Index follows Chapter 9. The Index is copyrighted separately from the NEC®. The alphabetical entries are boldfaced with Code numbers giving the location of the entry as it applies to particular subjects. Page numbers are not referenced in the Index.

Enforcement – 90.4. All applications of the NEC® are ultimately subject to approval by the authority having jurisdiction (AHJ). **See Figure 1-4.** The NEC® is designed to be used as a legal document. Any interpretations and approval of equipment rest with the authority having jurisdiction. In many cases, the AHJ is a local or municipal building official.

AUTHORITY HAVING JURISDICTION

Duties and responsibilities:

1. Make interpretations of NEC® rules.

2. Approve equipment and materials.

3. Grant special permission.

4. Waive specific requirements.

5. Permit alternate methods.

Figure 1-4. The AHJ is responsible for enforcement of the NEC® per 90.4.

Some localities permit the use of independent third party electrical inspection agencies. In other cases, the AHJ may be a governmental official, local fire marshal, or an insurance official with jurisdiction over the installation. In any case, whoever the AHJ is, they can permit alternate methods or even waive specific requirements if they believe the installation is equally effective at meeting the intended objective.

Mandatory Rules and Explanatory Material – 90.5. The use of the word "shall" indicates mandatory rules of the Code. For example, per 220.12, lighting loads for occupancies shall be figured using the values in Table 220.12. Mandatory provisions of the code can also be expressed as "shall not."

Informational Notes (Notes) are used throughout the NEC® to explain information that is contained in the Code. For the purposes of enforcement, Notes are not mandatory and do not contain any mandatory provisions. For example, 220.12 states that lighting loads shall not be less than the unit load per square foot in Table 220.12. The Notes clarify that the unit values are based on minimum load conditions and 100% power factor, and may not be sufficient for the installation.

In addition to Informational Notes, Informative Annexes are included for information only and are not enforceable requirements of the Code.

Future Use – 90.8. Sections 90.1(B) and 90.8(A), taken together, indicate that installations completed in accordance with the NEC® merely meet a minimum established standard. The installations may not be adequate for future use, and while it is not required, allowance for future expansion is generally in the best interest of all concerned. Essentially, the NEC® is a minimum standard. The designer or installer can choose to provide for future use if desired. Installations in accordance with the NEC® are not necessarily efficient and may not meet the future needs and demands for electricity. In other words, the NEC® only contains requirements that are necessary for safety. Section 90.1(B) clearly states that following the provisions of the NEC® will only result in an installation that is "essentially free from hazard." The NEC® rules and requirements are the minimum standards. Following the NEC® provisions does not ensure an electrical installation that will be suitable for future use or expansion.

PRINTS

A *print* is a detailed plan of a building drawn orthographically using a conventional representation of lines, and including abbreviations and symbols. **See Figure 1-5.** Prints were formerly known as blueprints because the method used to produce them yielded white lines on a blue background. Today, most prints are black lines on white paper. Conventional representation of lines refers to prescribed line types as detailed in American National Standards Institute (ANSI) Standard Y14.2M, *Line Conventions and Lettering.*

By drawing lines to ANSI standards, an "alphabet of lines" is established and easily read. **See Appendix.** For example, a line that is always drawn with a series of dashes of equal length is immediately identified as a hidden line. Lines used on drawings include object lines, hidden lines, centerlines, dimension lines, extension lines, leaders, break lines, and cutting plane lines. Other types of lines are also used on drawings. See ANSI Standard Y14.2M.

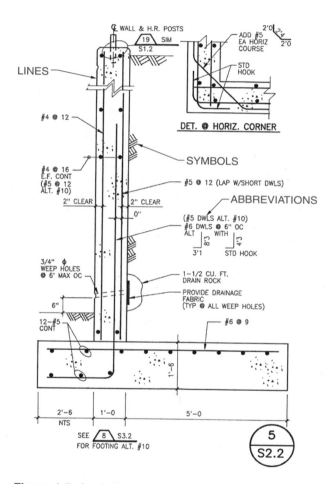

Figure 1-5. A print is a detailed plan of a building drawn orthographically using a conventional representation of lines, and including abbreviations and symbols.

Prints are first drawn as working drawings and copies are made to produce the prints. Working drawings may be drawn with pencil or ink and drafting instruments, although the common practice today is to use a computer software program such as CAD (computer-assisted drafting). Some computer software programs are available which automatically locate electrical receptacles per the NEC®. **See Figure 1-6.**

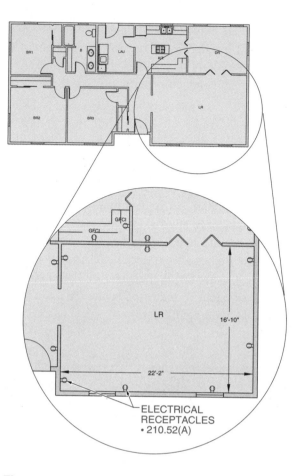

Figure 1-6. Some computer software programs automatically locate electrical receptacles per the NEC®.

Shape Description

Each print of a set of working drawings represents a part of a building drawn as an orthographic projection. This method of drawing allows the individual parts of the building to be shown on a flat plane in their true shape. A substantial number of individual prints may be required to clearly show the shape and size of all components. The most essential plans required to show a building include plot plans, floor plans, elevations, section views, and detail views. Of these, the floor plan is the primary plan in that it gives the most information about the building.

Plot Plans. A *plot plan* is an orthographic drawing that shows the location and orientation of the structure on the lot and the size of the lot. **See Figure 1-7.** A legal survey of the property made by a licensed surveyor establishes the corners of the lot in relation to official points of measurement in the vicinity. The compass direction and the angles at the corners of the lot are included. The survey locates existing buildings, shows the contour of the land, large trees, and other physical features. Other information on the plot plan includes property dimensions, grade elevations, and the location of water, sewer, gas, and electrical utilities. A utility pole indicates that the electrical service is overhead. A lateral service is underground.

The *point of beginning* is a fixed point from which all measurements, both vertical and horizontal, are made to show the building lot. The point of beginning, in turn, is related to a bench mark or city datum point. These datum points may be concrete markers within sighting distance of the lot, a mark on a fire hydrant, the top of a sewer in the street, or other points established by the city as a local point of reference. The datum points are not only points from which the point of beginning is located by horizontal measurement, but are also the basic points for vertical dimensions.

Floor Plans. A *floor plan* is an orthographic drawing of a building as though cutting planes were made through it horizontally. **See Figure 1-8.** The cutting plane is generally taken 5'-0" above the floor being shown. Objects below the 5'-0" elevation are shown with object lines. Objects above the 5'-0" elevation are shown with hidden lines.

Floor plans are the most frequently used drawings in a set of plans. They are drawn to scale to show the size relationship of various components. A one-story building requires one floor plan. A one-story building with a basement requires two floor plans. A building with more than one story generally requires a floor plan for each level.

The location of all exterior walls and interior partitions are shown and dimensioned on the floor plans. Overall size dimensions and specific location dimensions are given for structural elements. Door and window openings are shown. Reference letters or numbers refer to door and window schedules showing the type and size of door or window required at each opening.

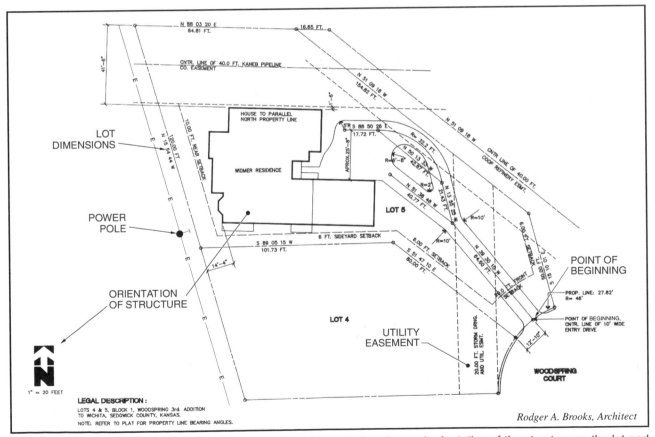

Rodger A. Brooks, Architect

Figure 1-7. A plot plan is an orthographic drawing that shows the location and orientation of the structure on the lot and the size of the lot.

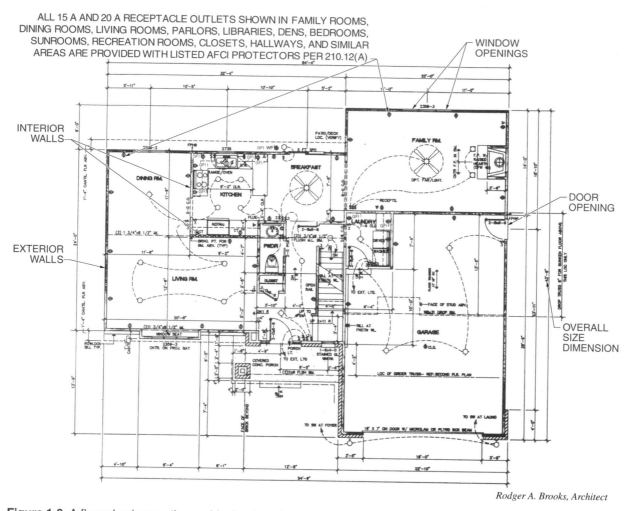

ALL 15 A AND 20 A RECEPTACLE OUTLETS SHOWN IN FAMILY ROOMS, DINING ROOMS, LIVING ROOMS, PARLORS, LIBRARIES, DENS, BEDROOMS, SUNROOMS, RECREATION ROOMS, CLOSETS, HALLWAYS, AND SIMILAR AREAS ARE PROVIDED WITH LISTED AFCI PROTECTORS PER 210.12(A)

Rodger A. Brooks, Architect

Figure 1-8. A floor plan is an orthographic drawing of a building as though cutting planes were made through it horizontally. (The floor plan is for graphic representation only.)

On smaller sets of prints, electrical information is shown directly on the floor plans. On larger sets of residential plans and on commercial plans, electrical information is usually shown on separate electrical plans. The general location of receptacles, lights, and switches is shown. The electrician determines the specific location in conformance to applicable NEC® articles. For example, per 210.52(A)(1), in kitchens, family rooms, dining rooms, living rooms, parlors, libraries, dens, sunrooms, bedrooms, recreation rooms, or similar rooms of dwelling units, receptacles shall be installed such that no point along the floor line, measured horizontally, shall be more than 6′ from a receptacle outlet.

Plumbing fixtures such as water closets, sinks, shower stalls, and bathtubs are shown on residential floor plans. Information pertaining to heating, ventilating, and air conditioning (HVAC) systems may also be shown on floor plans. Registers connecting to the system are located on the floor plan. On larger sets of plans, HVAC information is usually shown on mechanical plans.

Elevations. An *elevation* is an orthographic drawing showing vertical planes of a building. **See Figure 1-9.** Depending on building size, shape, and complexity, four or more elevations may be required to clearly show all exterior walls. Generally, however, four exterior views are sufficient. These are the front, rear, right side, and left side or North, South, East, and West. The North Elevation of a building is the elevation facing North, not the direction a person faces to see that side of the house. Elevations are identified and the scale is given on elevation drawings. Interior elevations are used to clarify locations of such items as cabinets, mantels, and bookcases.

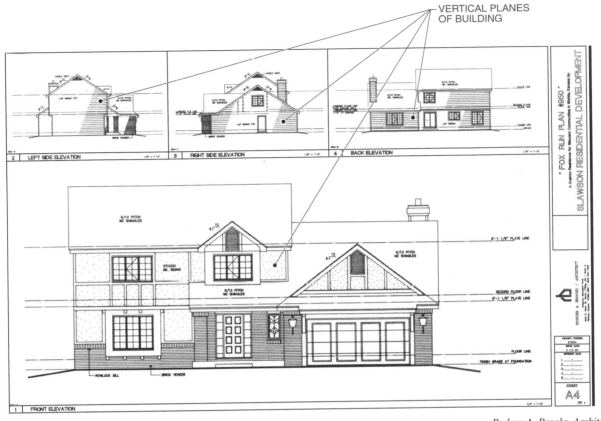

Figure 1-9. An elevation is an orthographic drawing showing vertical planes of a building.

Sections. A *section* is an orthographic drawing created by passing a cutting plane (either horizontal or vertical) through a portion of a building. **See Figure 1-10.** Section views are shown wherever an important part of a building is not apparent on the plan views or elevations views. Cutting planes are passed through the building at the most advantageous point to provide clear information.

The most common section view is taken through an outside wall. It shows information such as details on the foundation wall and footing, details of wall and floor framing, height of windows above the floor, eave construction, and roof construction.

Cutting planes for section views are shown on the floor plan. Direction arrows indicate line of sight and a code reference shows the sheet number on which the identified section is found.

Several section views may be required if the construction varies from one part of the building to another part of the building. In some instances, cutting planes are taken lengthwise or crosswise in a building. A *longitudinal section* is a section taken lengthwise. A *transverse section* is a section taken crosswise.

Details. A *detail* is a part of a plan, elevation, or section view drawn at a larger scale. **See Figure 1-11.** Often the plan, elevation, and section views do not show all the information the builder must have to construct the building. The basic set of prints may be drawn at a small scale such as $\frac{1}{8}''$ = 1' or $\frac{1}{4}''$ = 1' and not show small details clearly. When this occurs, a small part of the building is drawn at a larger scale. Details may be drawn of any part of the building that cannot be shown clearly and conveniently on the working drawings. The scale is always given for the detail.

Written Description

Working drawings show all walls, partitions, details, electrical devices and equipment, and other features of a building by means of lines, symbols, and conventional representations. With few exceptions, each part is located by precise dimensions. However, in order to develop accurate cost estimates and to quickly solve problems that may arise during the construction process, information is shown on the prints in the form of notations. Also, additional information is included in the specifications. Specifications may consist of a few sheets of paper or a book of several hundred pages, depending on the complexity of the project.

Written information is shown on prints in four ways:

1. Some information is found in the title block.

2. Descriptive titles are placed near a specific item and connected with a leader line terminated by an arrowhead or a dot.

3. Specific information that refers to only one situation may be placed near the situation.

4. General information applying to several sheets in the prints may be placed in any convenient space.

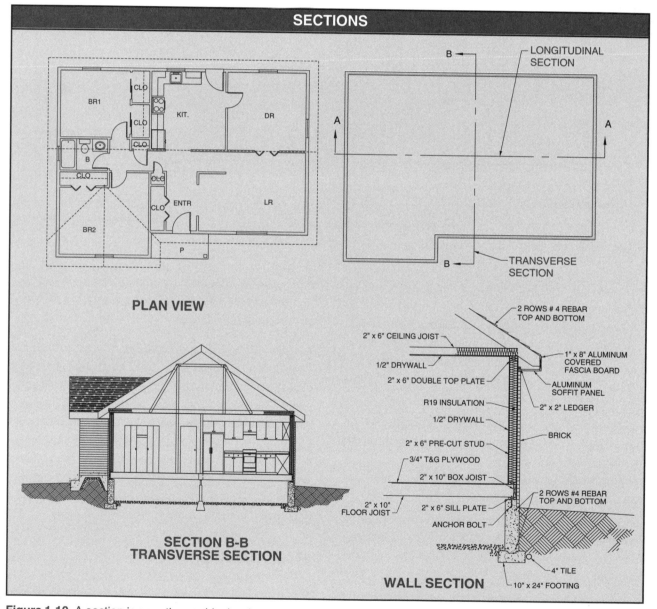

Figure 1-10. A section is an orthographic drawing created by passing a cutting plane (either horizontal or vertical) through a portion of a building.

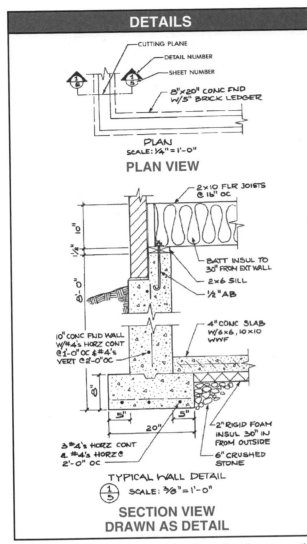

DETAILS

CUTTING PLANE
DETAIL NUMBER
SHEET NUMBER
8"x20" CONC FND
W/5" BRICK LEDGER

PLAN
SCALE: ¼" = 1'-0"
PLAN VIEW

2x10 FLR JOISTS
@ 16" OC

BATT INSUL TO
30" FROM EXT WALL
2x6 SILL
½" AB

10" CONC FND WALL
W/#4's HORZ CONT
@ 1'-0" OC & #4's
VERT @ 2'-0" OC

4" CONC SLAB
W/6x6, 10x10
WWF

5"   5"
20"

3 #4's HORZ CONT
& #4's HORZ @
2'-0" OC

2" RIGID FOAM
INSUL 30" IN
FROM OUTSIDE
6" CRUSHED
STONE

TYPICAL WALL DETAIL
1/5   SCALE: ⅜" = 1'-0"
SECTION VIEW
DRAWN AS DETAIL

Figure 1-11. A detail is a part of a plan, elevation, or section view drawn at a larger scale.

Specifications. *Specifications* are sheets of paper, gathered together into a pamphlet (or into several volumes for a large building) covering a number of subjects in detail. These subjects include information, in paragraph form, that describes the building to be constructed, the materials to be used, and the responsibilities of those involved.

The simplest type of construction job should have specifications. They are included with prints and a written contract to become the basis for agreement between the owner and the contractor.

Complete specifications are written to give detailed information about the job. This information cannot conveniently be shown on a set of prints because it is too lengthy. The general purpose of specifications includes the following:

- Specifications provide the general conditions giving the broad provisions of the contract and outline the responsibilities of the owner, architect, contractor, and subcontractors. Performance guarantees are included.
- Specifications supplement the working drawings with detailed technical information about the work to be done, specifying the material, equipment, and fixtures to be used.
- Specifications serve, with the contract agreement and the prints, as the legal basis for the transaction of erecting the building from start to finish.

Specifications are intended to supplement the set of prints with data that will spell out the job in a general sense. They also include the details of the job to be done by each individual trade. The set of prints and the specifications are intended to be in agreement. In the event of any discrepancy or conflict between the two, the specifications take precedence.

Specifications provide the contractor and subcontractors with detailed information for precise and competitive bidding when estimating the costs of labor and material. Subcontractors bid their work based upon codes referenced in the specifications. For example, electricians comply with the requirements of the National Electrical Code® when estimating cost for specific materials.

The building department of the town or city in which the building is to be constructed uses the specifications and the prints to determine their compliance with all applicable building codes and zoning ordinances to meet structural, fire, and health standards. Banks and loan agencies use the same information to help appraise the building to determine its value. When governmental agencies such as the Veterans Administration (VA) or the Federal Housing Administration (FHA) provide part of the financing, they require a copy of the specifications for their approval. Certain sections of the specifications must be written on forms supplied by the individual agency when either VA or FHA financing is to be used.

The Construction Specifications Institute (CSI) has developed standard procedures that are helpful in writing specifications, particularly in writing in the CSI MasterFormat™. Specification writing is divided into 50 technical divisions, each with several broad (smaller subheads) section headings. **See Figure 1-12.**

CSI MASTERFORMAT™

Procurement and Contracting Requirements Group
 Division 00—Procurement and Contracting
 Requirements
Specifications Group
 General Requirements Subgroup
  Division 01—General Requirements
 Facility Construction Subgroup
  Division 02—Existing Conditions
  Division 03—Concrete
  Division 04—Masonry
  Division 05—Metals
  Division 06—Woods, Plastics, and Composites
  Division 07—Thermal and Moisture Protection
  Division 08—Openings
  Division 09—Finishes
  Division 10—Specialties
  Division 11—Equipment
  Division 12—Furnishings
  Division 13—Special Construction
  Division 14—Conveying Equipment
  Division 15 to 19—Unassigned
 Facility Services Subgroup
  Division 20—Unassigned
  Division 21—Fire Suppression
  Division 22—Plumbing
  Division 23—Heating, Ventilating, and Air-Conditioning
  Division 24—Unassigned
  Division 25—Integrated Automation
  Division 26—Electrical
   260100—Operation and Maintenance of Electrical
   Systems
   260500—Common Work Results for Electrical

   260600—Schedules for Electrical
   260800—Commissioning of Electrical Systems
   260900—Instrumentation and Control for Electrical
   Systems
   261000—Medium-Voltage Electrical Distribution
   262000—Low-Voltage Electrical Transmission
   263000—Facility Electrical Power Generating and
   Storing Equipment
   264000—Electrical and Cathodic Protection
   265000—Lighting
  Division 27—Communications
  Division 28—Electronic Safety and Security
  Division 29—Unassigned
 Site and Infrastructure Subgroup
  Division 30—Unassigned
  Division 31—Earthwork
  Division 32—Exterior Improvements
  Division 33—Utilities
  Division 34—Transportation
  Division 35—Waterway and Marine Construction
  Divisions 36 to 39—Unassigned
 Process Equipment Subgroup
  Division 40—Process Integration
  Division 41—Material Processing and Handling Equipment
  Division 42—Process Heating, Cooling, and Drying Equipment
  Division 43—Process Gas and Liquid Handling, Purification,
   and Storage Equipment
  Division 44—Pollution Control Equipment
  Division 45—Industry-Specific Manufacturing Equipment
  Divisions 46 to 47—Unassigned
  Division 48—Electrical Power Generation
  Division 49—Unassigned

Figure 1-12. The CSI MasterFormat™ contains 50 divisions.

Title Block. The title block is the logical place to begin reading a set of prints. **See Figure 1-13.** Information in the title block includes the name of the project, project location, architect's name and office location, date on which plans were completed, initials of the drafter, number of the sheet, number of sheets in the set of prints, name of the sheet, and other information as determined by the architect. Title blocks are commonly located in the lower right-hand corner of the sheet.

Abbreviations

An *abbreviation* is the shortened form of a word or phrase. **See Figure 1-14.** Abbreviations are used throughout a set of plans to describe materials and processes while conserving space on the drawing sheets. They save drafting time while presenting information in a standard manner.

Uppercase letters are used for most abbreviations. All abbreviations that make an actual word end in a period in order to distinguish the abbreviation from the actual word. For example, IN. is the abbreviation for inch.

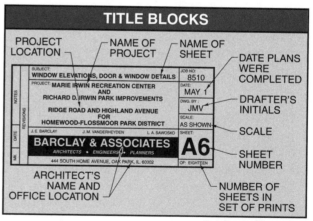

Barclay and Associates

Figure 1-13. The title block contains specific information related to the prints.

While many abbreviations are accepted by virtue of their common usage, only standardized abbreviations should be used on drawings. Standardized abbreviations eliminate confusion and misinterpretation. Although some different words use the same standardized abbreviations, an interpretation based upon the context in which they are used helps clarify the specific use. Examples of the same abbreviations for different words are A and V. A is the abbreviation for amp and area. V is the abbreviation for volts, valve, and vent. **See Appendix.**

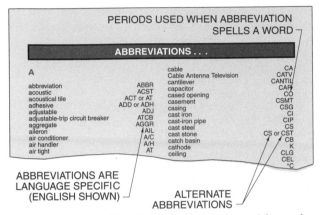

Figure 1-14. An abbreviation is the shortened form of a word or phrase.

Symbols

A symbol is a graphic representation of an object. **See Figure 1-15.** Symbols provide a uniform representation of building materials, fixtures, and structural parts that are easily

recognizable. Symbols are universal in that the symbol can be understood no matter what language a person speaks. Symbols are shown on plan views (plot, floor, framing, etc.), elevation views, sections, and details. **See Appendix.**

Architectural symbols show building materials such as wood, brick, concrete, glass, roofing, and structural steel. See Appendix. Electrical symbols show various types of outlets, receptacles, lights, wiring, signaling systems, etc. **See Appendix.**

PRINTREADING

Prints for the one-family dwelling include the plot plan, foundation and unfinished basement plan, floor plan, elevations, sections, and details. A door schedule and a window schedule are shown. Specifications are not included. The electrical plans are shown as part of the foundation and basement floor plans and the floor plan.

Plot Plan

The **Plot Plan** shows a one-family dwelling located on S. Juniper Dr. From the point of beginning on the NE corner (which is assigned an elevation of ±0′-0″), the property

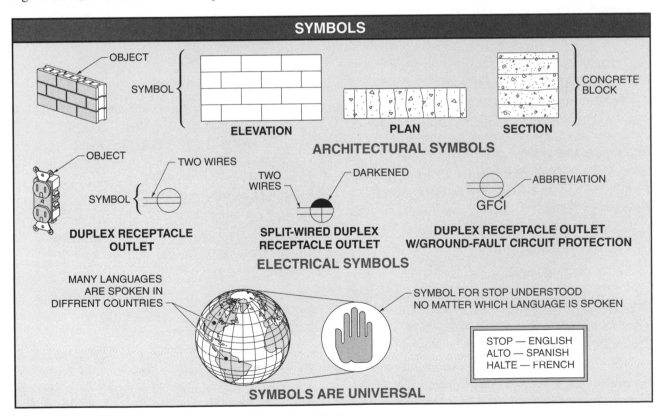

Figure 1-15. A symbol is a graphic representation of an object.

line parallels S. Juniper Dr. to the SE corner for 192'-6". From that point to the SW corner is 180'-0". The four corners of the property are at 90° angles. The lot contains 34,650 sq ft (192.5' × 180'= 34,650 sq ft). Contour lines are not shown as the lot is relatively flat.

The house is set back 58'-5" from S. Juniper Dr. and 63'-8" from the North property line. This is well beyond the 30'-0" minimum building line. The main floor is 2'-10" above the POB and the garage floor is 2'-0" above the POB to provide good drainage away from the house. A 4'-0" concrete sidewalk is shown from the blacktop drive to the front entrance.

The gas line is on the East side of the street. It is represented by the long-dashed line with the letter G interspersed at regular intervals. The transformer is located near the SW corner of the lot. It is represented by a symbol showing a square within a square. The transformer provides lateral service to the dwelling. The electrical utility line is shown along the West property line. It is located in a 10'-0" easement. An *easement* is a space set aside by agreement with property owners for use by utility companies to install and service their equipment.

Water and sewer utilities are not provided in the area of the county where the house is to be built. Water will be provided by a well to be drilled in the front yard. The well may not be located within 75'-0" of the proposed septic field. A 1000 gal. septic tank is located in the back yard. The area for the septic field is defined.

Foundation and Unfinished Basement Plan

The **Foundation and Unfinished Basement Plan** shows the overall dimensions and location dimensions for the unfinished basement and garage foundation. The elevation of the basement floor is 5'-10¾" below the POB, which is shown on the Plot Plan. The basement floor is a 4" concrete slab with a 6 mil vapor barrier over 4" of gravel fill. The garage is unexcavated.

Basement foundation walls are 10" thick and 7'-10" high from finished floor to the bottom of the 2" × 10" floor joists, which are 16" OC. All foundation walls are reinforced by two rows of #4 rebar top and bottom. See Details.

W 8 × 21 beams support the floor joists. These wide flange steel beams are 8" deep and weigh 21 lb per foot. The beams are supported by 8" wide × 8½" high beam pockets 4" deep and 3" diameter lally columns, which

are placed in 3' × 3' × 1' deep concrete footings. The lally columns are 17'-1¹¹⁄₁₆" OC from North to South and 14'-7" OC East to West.

Two windows are set in window wells in the South wall. The larger window well, in the SE corner, is designed for egress in the event of an emergency. The glass can be easily removed and the window well is large enough to allow a person to exit. The sizes of the windows are given in the Window Schedule.

The stairs have 13 risers (R) up to the main floor. The location of the stairs and stairwell is given on the Floor Plan.

A 200 A service is located on the West wall near the NW corner of the basement. At least one receptacle is required for one-family basements per 210.52(G). This 20 A, 125 V, 1ϕ wall receptacle with GFCI protection per 210.8(A)(5) is located on the West wall of the basement. All 125 V, 1ϕ, 15 A and 20 A receptacles installed in unfinished basements or dwelling units shall have GFCI protection unless they are supplying only a permanently installed fire alarm or burglar alarm system.

Lighting for the basement is provided by seven incandescent lamps. These lights have pull switches. One light is located in the stairwell and is controlled by three-way switches located at the top and bottom of the stairs per 210.70(A)(2)(c).

Floor Plan

The **Floor Plan** shows the overall dimensions and location dimensions for the main floor living area and the two-car garage. The main living area contains approximately 2226 sq ft (42 × 53 = 2226 sq ft). This figure is used to determine the lighting load of 6678 VA (2226 × 3 = 6678) per 220.12. When figuring the lighting load, the floor area for dwelling units does not include open porches, garages, or unused or unfinished areas. The floor area is figured using the outside dimensions.

The one-family dwelling is entered from an open Porch (P) through an A door. All doors on this plan are keyed to the Door Schedule by an encircled letter. For example, one A door is required. It is a 3'-0" × 6'-8", right-hand, Therma-Tru, Model # CC45 door. The porch has an incandescent lighting fixture on the wall to the left of the front door. The symbol for a wall-mounted incandescent lighting fixture is a circle with one leg. This lighting fixture, and the two garage front wall exterior lighting fixtures, are controlled by three-way switches near the front door and in the Garage (G) at the B door near the Dining Room (DR). S_3 is the symbol for a three-way switch. A three-way switch used with another three-way switch provides control from two locations. Three-way switches have no OFF-ON markings.

The weatherproof (WP) receptacle on the porch has ground-fault circuit interrupter (GFCI) protection. All outdoor 125 V, 1φ, 15 A and 20 A receptacles shall have ground-fault circuit interrupter protection for personnel per 210.8(A)(3) unless they are supplying electric snow-melting, deicing, or pipeline and vessel heating equipment and they are not readily accessible.

Two single-pole switches near door A control ceiling-mounted lighting fixtures in the entry area and the closet. S is the symbol for single-pole switches. The symbol for a ceiling-mounted incandescent lighting fixture is a circle with four legs.

The entry area contains a coat closet with a shelf. The shelf is represented by hidden lines because it is over 5′-0″ above the floor. The coat closet has a 6′-0″ × 6′-8″ bypass door. This model door is mirrored. See the M door in the Door Schedule. The entry area is separated from the Living Room (LR) by a kneewall of glass block (GLB). Details are referenced on this floor plan by an encircled number with a darkened arrow indicating the direction of sight. See Detail 3 for an elevation view of the kneewall.

An F door leads to the unfinished basement. This is a left-hand reverse door (LHR). See the Door Schedule. There are 13 risers (R) down to the basement. A ceiling-mounted incandescent lighting fixture is controlled by three-way switches located at the top of the stairs and at the foot of the stairs in the unfinished basement. See the Foundation and Unfinished Basement Plan.

The Living Room (LR) has an A window on the front (East) wall and a B window on the North wall. The Window Schedule gives the quantity, sash size, rough openings, and other information for all windows. **See the Window Schedule.**

Two ceiling-mounted incandescent lighting fixtures are shown in the Living Room. The B in one of the symbols indicates that the fixture is blanked out. This fixture can be used later when a lighting fixture is hung. The active ceiling-mounted incandescent lighting fixture is controlled by three-way switches located near the stairs and near the Pantry (PAN.) in the Kitchen (KIT.).

Receptacle outlets are placed so that no point along the floor line in any unbroken wall space is more than 6′-0″ measured horizontally from an outlet per 210.52(A)(1). These duplex receptacle outlets are shown by a circle with two legs. The split-wired duplex receptacle outlet on the East wall beneath Window A is split-wired as shown by the darkened area in the symbol. One receptacle is controlled by a wall-mounted switch and the other receptacle remains energized. A television outlet is shown on the North Wall. The symbol for a television outlet is a square with the letters TV inscribed.

The Kitchen (KIT.) and Dining Area (DR) are entered from the Living Room (LR), the Garage (G), the laundry room (LAU), or through sliding doors on the West side of the dwelling. See Code E of the Door Schedule. The Dining Area has a ceiling-mounted incandescent lighting fixture controlled by three-way switches located near the B door to the Garage and on the wall near the Pantry (P). Receptacle outlets are provided as required.

The Kitchen (KIT.) is L-shaped with an island counter and a Pantry (PAN.). The Pantry contains a ceiling lampholder with a pull switch (PS). The ceiling-mounted incandescent lighting fixture in the Kitchen is controlled by three-way switches on both sides of the Pantry door. See Code N of the Door Schedule. A receptacle outlet and a telephone jack are located in the Pantry wall near the GLB kneewall. This area is designed for a small desk. The symbol for a telephone jack is a triangle with a base line.

The island counter holds a 9.0 kW, 240 V, counter-mounted electric cooktop. This appliance is on a minimum branch circuit of 40 A per 210.19(A)(3). The 240 V range receptacle outlet is shown by a circle with three legs and an R. A receptacle outlet is shown on each end of the island counter. These receptacle outlets serve the countertop of the island per 210.52(C)(2); 210.52(C)(3); and 210.52(C)(5), Ex. Because these receptacles serve kitchen countertops, they have GFCI protection per 210.8(A)(6).

The South wall of the Kitchen shows base cabinets, wall cabinets, and the refrigerator. See Detail 1, South Wall Elevation. The receptacle outlet for the refrigerator does not have GFCI protection because it is a dedicated receptacle and does not serve the countertop. Two receptacle outlets along the South wall are intended for countertop use and have GFCI protection per 210.8(A)(6). One receptacle outlet does not have GFCI protection. This receptacle outlet is located in a wall cabinet, is dedicated to an under-the-cabinet light, is over 5½′ above the floor, and not intended to serve the countertop surface. This receptacle outlet does not count as part of the required countertop receptacles. Two soffit-mounted incandescent lighting fixtures are controlled by switches located in the West wall.

The West wall of the Kitchen shows base cabinets, wall cabinets, sink, and dishwasher. See Detail 2, West Kitchen Elevation. The sink is centered on a D window. See the Window Schedule. The dishwasher (DW) is located to the right of the sink. The symbol for a dishwasher is a triangle within a circle with one leg. The letters DW are placed beside the symbol. Two receptacle outlets with GFCI protection serve the countertop. One soffit-mounted incandescent lighting fixture is controlled by a switch located in the West wall. This is the same switch that controls the two soffit-mounted incandescent lighting fixtures on the South wall. Individual switches near the E door control three exterior wall-mounted incandescent lighting fixtures on the West wall of the dwelling. **See the Door Schedule.**

The Laundry (LAU) is conveniently located as a pass-through from Bedroom 1 (BR1) and the Kitchen. Plumbing economy is realized by the proximity of the Laundry, Baths 1 and 2, and the Kitchen. The K doors of the Laundry are pocket doors. See the Door Schedule. The ceiling-mounted incandescent lighting fixture is controlled by three-way switches near each door. A wall receptacle outlet is provided near the shelving on the West wall per 210.52(F). This outlet is not required to be protected by GFCI since it is not within 6′ of the outside edge of the sink. The washing machine (WM) and gas dryer (D) are located on the East wall next to the tubs. The appliance outlet is within 6′-0″ of the laundry equipment per 210.50(C) and must be GFCI-protected per 210.8 (A)(7).

Bathroom 1 (B1) has an I door and B2 has an H door. See the Door Schedule. B1 has a small bumpout to the rear of the house. GLB provides natural daylight in B1. B1 has a basin, toilet, and shower. B2 has a basin, toilet, and tub. Each of these meets the definition of a bathroom. See Article 100.

Bathrooms 1 and 2 (B1 and B2) have a ceiling-mounted incandescent lighting fixture, a fan, and a heater in their ceilings. The symbol for the fan is the letter F inside a square. The symbol for the heater is the letter H inside a square. All ceiling fixtures are controlled by switches near the bathroom doors. See the Door Schedule for I and H doors. All receptacle outlets in the bathrooms are GFCI-protected per 210.8(A)(1). At least one 20 A branch circuit shall be provided to supply the bathroom receptacle outlets per 210.11(C)(3). This circuit is not permitted to serve any other outlets. Incandescent track lighting is shown on the East wall of B1. The symbol for incandescent track lighting is a rectangle with inscribed circles and one leg. The bathtub (BT) in B2 has a dedicated receptacle for a whirlpool. This is on a separate 15 A circuit. **See the Panel Schedule.**

BR1, BR2, and BR3 have G doors and C windows. See the Door Schedule and the Window Schedule. A large walk-in closet (WIC) is entered through an I door from BR1. Dashed lines indicate shelving on three walls. A Scuttle (S) is shown in the ceiling with a Lampholder (L) and a Pull Switch (PS) to light the attic area per 210.70(A)(3). A ceiling-mounted incandescent lighting fixture for the walk-in closet is controlled by a wall switch in BR1.

The three bedrooms have ceiling-mounted incandescent lighting fixtures controlled by wall switches. All receptacle outlets in the bedrooms are installed per 210.52(A)(1). BR1 and BR3 have telephone jacks. BR1 also has a TV jack on the East wall.

A cutting plane line is shown in the East wall of BR3. This refers to Section A-A which shows details of the foundation footing, foundation wall, exterior wall, ceiling, and roof. Specific materials required are noted. The 4″ perforated drain tile ties into the Sump Pump (SP) shown on the Foundation and Unfinished Basement Plan.

Elevation Views

The **Elevation Views** show East, West, North, and South elevations. The slope symbol on the East Elevation shows that the front porch roof has a rise of 8″ for every 12″ of run. The main roof has a rise of 6″ for every 12″ of run. The slope symbol is a right triangle with the run given on the horizontal leg and the rise given on the vertical leg.

Sections and Details

Section A-A is taken in the East wall of the Floor Plan. This section shows the foundation footing through the roof shingles. A trussed roof is specified.

Details of three interior elevations are given. Detail 1 and Detail 2 show kitchen details and Detail 3 shows South LR wall details. See the Floor Plan for the reference marks for these details.

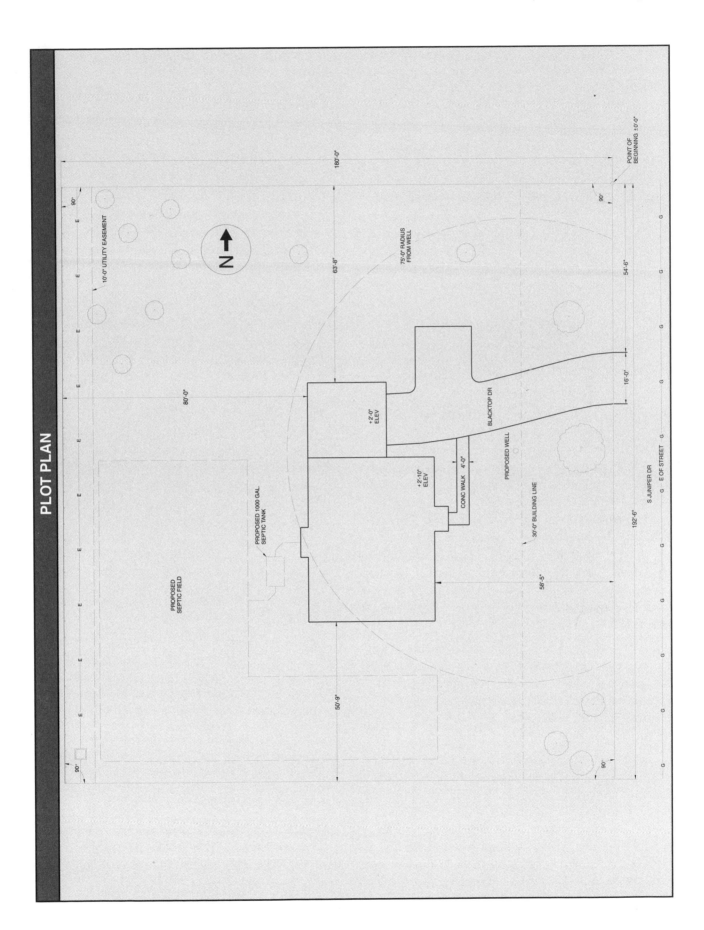

PLOT PLAN

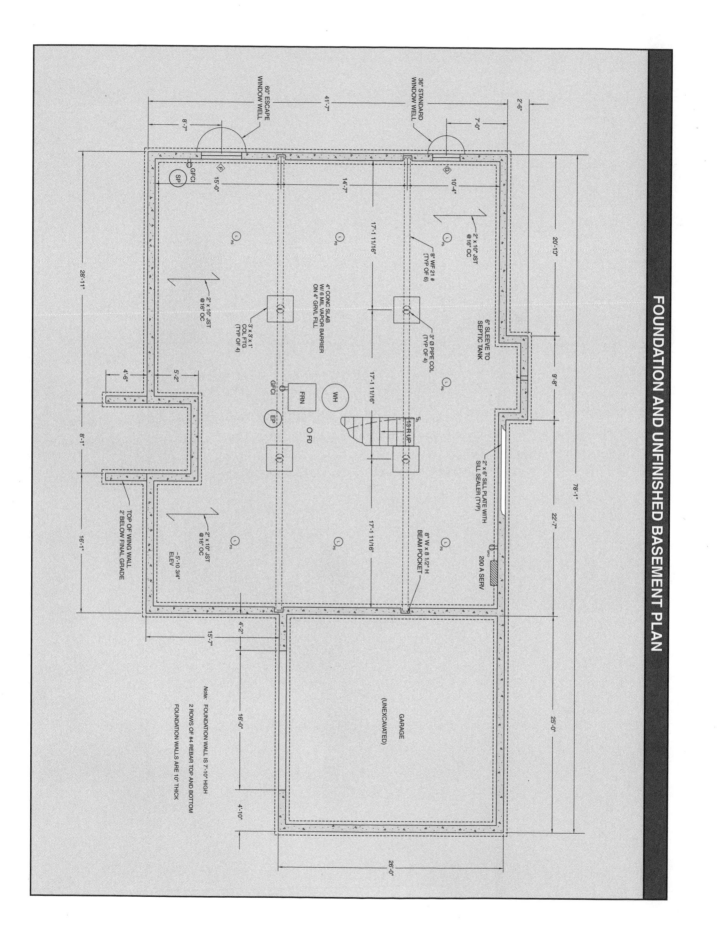

FOUNDATION AND UNFINISHED BASEMENT PLAN

FLOOR PLAN

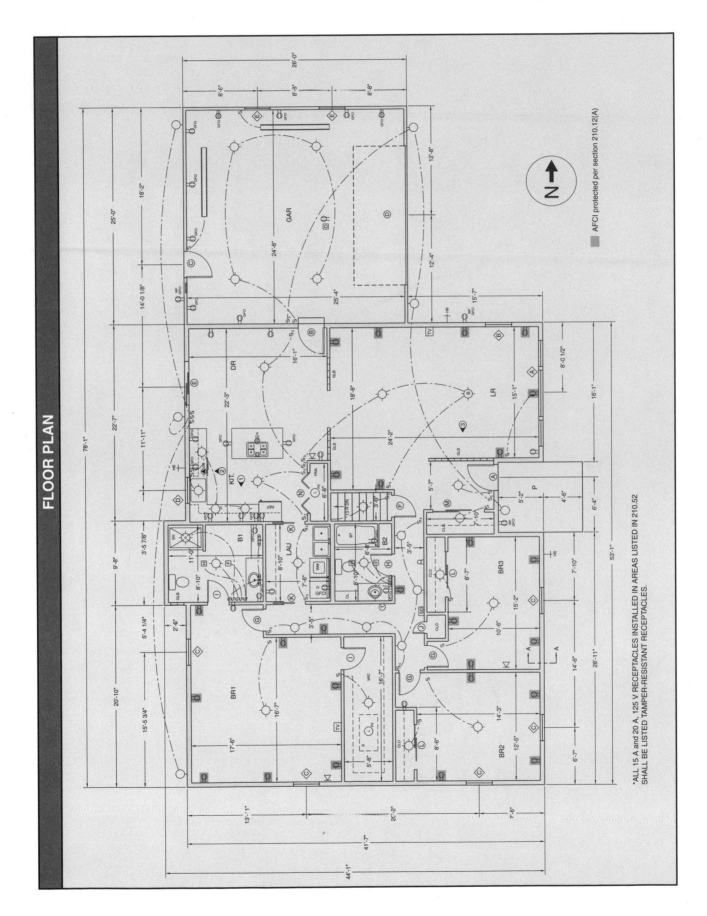

*ALL 15 A and 20 A, 125 V RECEPTACLES INSTALLED IN AREAS LISTED IN 210.52 SHALL BE LISTED TAMPER-RESISTANT RECEPTACLES.

■ AFCI protected per section 210.12(A)

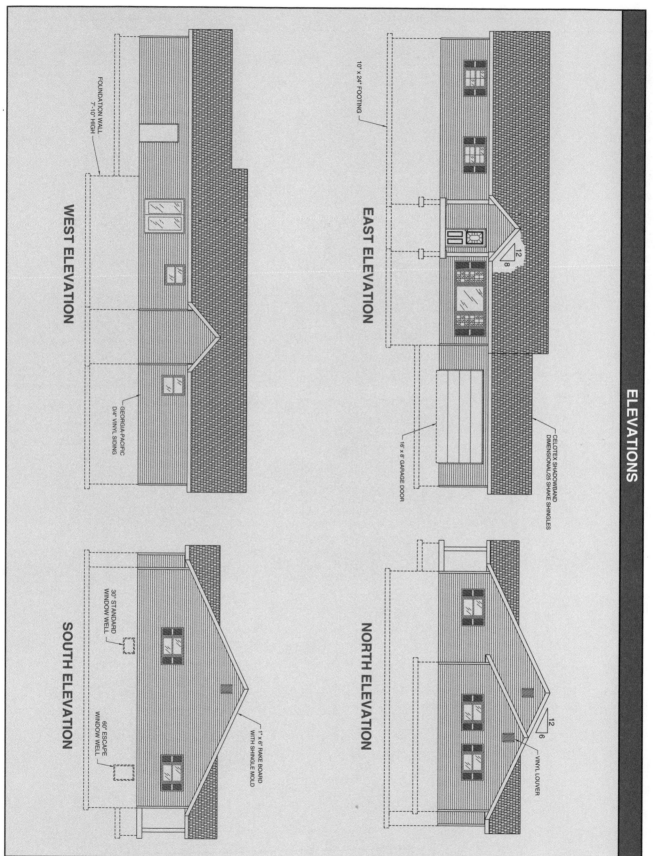

ELEVATIONS

WEST ELEVATION

FOUNDATION WALL
7'-10" HIGH

GEORGIA-PACIFIC
D/4" VINYL SIDING

EAST ELEVATION

10" X 24" FOOTING

12
8

16' X 8' GARAGE DOOR

CELOTEX SHADOWBAND
DIMENSIONAL/25 SHAKE SHINGLES

SOUTH ELEVATION

30" STANDARD
WINDOW WELL

60" ESCAPE
WINDOW WELL

1" x 6" RAKE BOARD
WITH SHINGLE MOLD

NORTH ELEVATION

12
6

VINYL LOUVER

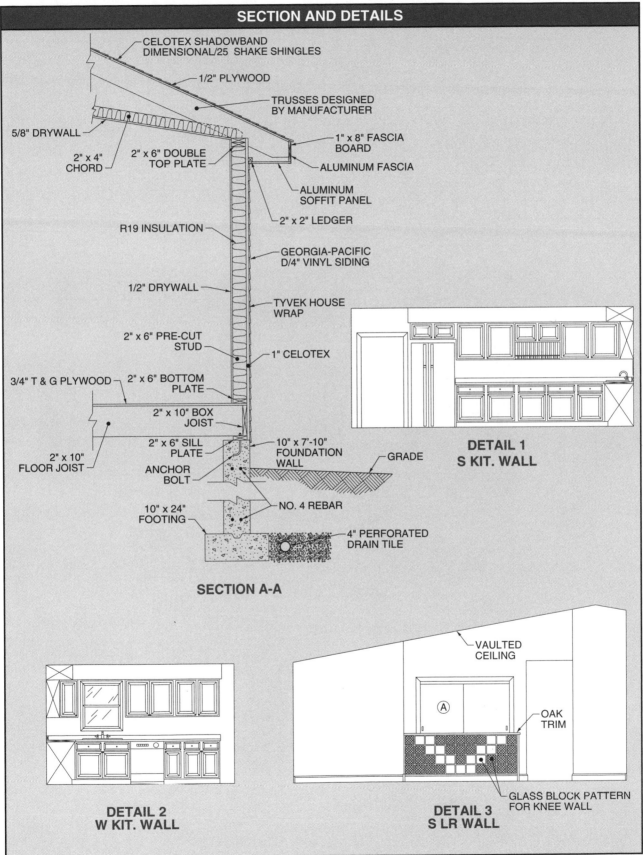

SECTION AND DETAILS

Section A-A labels:

- CELOTEX SHADOWBAND DIMENSIONAL/25 SHAKE SHINGLES
- 1/2" PLYWOOD
- TRUSSES DESIGNED BY MANUFACTURER
- 5/8" DRYWALL
- 1" x 8" FASCIA BOARD
- ALUMINUM FASCIA
- 2" x 4" CHORD
- 2" x 6" DOUBLE TOP PLATE
- ALUMINUM SOFFIT PANEL
- 2" x 2" LEDGER
- R19 INSULATION
- GEORGIA-PACIFIC D/4" VINYL SIDING
- 1/2" DRYWALL
- TYVEK HOUSE WRAP
- 2" x 6" PRE-CUT STUD
- 1" CELOTEX
- 3/4" T & G PLYWOOD
- 2" x 6" BOTTOM PLATE
- 2" x 10" BOX JOIST
- 2" x 6" SILL PLATE
- 10" x 7'-10" FOUNDATION WALL
- GRADE
- 2" x 10" FLOOR JOIST
- ANCHOR BOLT
- 10" x 24" FOOTING
- NO. 4 REBAR
- 4" PERFORATED DRAIN TILE

SECTION A-A

**DETAIL 1
S KIT. WALL**

**DETAIL 2
W KIT. WALL**

**DETAIL 3
S LR WALL**

- VAULTED CEILING
- OAK TRIM
- GLASS BLOCK PATTERN FOR KNEE WALL

SCHEDULE

DOOR SCHEDULE

CODE	QUANTITY	SIZE	SWING	REMARKS
A	1	$3^0 \times 6^8$	RH	THERMA-TRU MODEL # CC45
B	1	$3^0 \times 6^8$	LH	STEEL
C	1	$3^0 \times 6^8$	RH	STEEL
D	1	16 x 8	—	WAYNE DALTON MODEL #
E	1	$6^0 \times 6^8$	SLIDING	HURD MILLWORK MODEL # CHES-6
F	1	$2^6 \times 6^8$	LHR	HOLLOW CORE OAK LAMINATE
G	3	$2^6 \times 6^8$	RH	HOLLOW CORE OAK LAMINATE
H	1	$2^4 \times 6^8$	RH	HOLLOW CORE OAK LAMINATE
I	2	$2^4 \times 6^8$	LH	HOLLOW CORE OAK LAMINATE
J	1	$2^0 \times 6^8$	RHR	HOLLOW CORE OAK LAMINATE
K	2	$2^6 \times 6^8$	POCKET	HOLLOW CORE OAK LAMINATE
L	2	$5^0 \times 6^8$	BYPASS	STANLEY MONARCH MODEL # 1205980AGR
M	1	$6^0 \times 6^8$	BYPASS	STANLEY MONARCH MODEL # 50219961
N	1	$5^0 \times 6^8$	BI-FOLD	ARROW DOOR OAK LAMINATE MODEL # 413-1166

PANEL SCHEDULE

LEFT	RIGHT
20 A KIT. APPLIANCES	20 A DR RCPT & GAR GFCI
20 A KIT. APPLIANCES	15 A GAR LTS
15 A REF	15 A LR RCPT
15 A DW	15 A KIT. & LR LTS
15 A BR1 & HALLWAY & SD	SPARE
15 A BR1 & BR2	20 A LAU
15 A YARD SPOTS	20 A B1 & B2 GFCI
15 A BSMT S	15 A WHIRLPOOL TUB
20 A B1 HEATER	15 A FUR.
20 A B2 HEATER	15 A EP
30 A OVEN	15 A BSMT N
	SPARE
20 A WELL PUMP	30 A A/C
20 A BSMT GFCI	100 A GAR
SPARE	

WINDOW SCHEDULE

CODE	QUANTITY	SASH SIZE	ROUGH OPENING	REMARKS
A	1	2'-4"/5'-8"/2'-4" x 5'-2"	10'-9" x 5'-5"	HURD MILLWORK MODEL # 24-64-2458 24" FLANKERS/CLAD WOOD DOUBLE HUNG PICTURE COMBINATION
B	1	3'-8" x 5'-2"	3'-10" x 5'-5"	HURD MILLWORK MODEL # 40x28 CLAD WOOD DOUBLE HUNG
C	5	3'-8" x 3'-10"	3'-10" x 4'-1"	HURD MILLWORK MODEL # 40x20 CLAD WOOD DOUBLE HUNG
D	1	2'-8" x 3'-2"	2'-10" x 3'-5"	HURD MILLWORK MODEL # 28x16 CLAD WOOD DOUBLE HUNG
E	2	2'-0" x 3'-2"	2'-2" x 3'-5"	HURD MILLWORK MODEL # 20x16 CLAD WOOD DOUBLE HUNG
F	1	31" x 37"—10" GALV ESCAPE BUCK FRAME	———	KEWANEE WINDOW MODEL # ASV36G STANDARD ESCAPE INSERT
G	1	31" x 20"—10" GALV STD BUCK FRAME	———	KEWANEE WINDOW MODEL # AOV-20G STANDARD WINDOW INSERT

Name_____ Date _____

NEC®	Answer	

_____   (T)   F   **1.** The purpose of the NEC® is to protect people and property from hazards that arise from the use of electricity.

__xxxxx__   (T)   F   **2.** The key to successful NEC® proposals is proper substantiation for the proposed changes.

__xxxxx__   _D_   **3.** The NEC® is available in ___ versions.
        A. soft cover   C. electronic
        B. loose-leaf   D. all of the above

__xxxxx__   _under_   **4.** Any material in the current edition of the NEC® that has changed from the previous edition is identified by gray shading and/or a(n) ___ line in the margin.

_____   _9_   **5.** The NEC® contains an Introduction and ___ chapters.

__xxxxx__   _Print_   **6.** A(n) ___ is a detailed plan of a building which is drawn orthographically using a conventional representation of lines, and including abbreviations and symbols.

__xxxxx__   _____   **7.** A(n) ___ line is always drawn with a series of dashes of equal length.

__xxxxx__   _____   **8.** The point of ___ is a fixed point from which all measurements are made to show the building lot.

__xxxxx__   _____   **9.** An elevation is an orthographic drawing showing ___ planes of a building.

_____   T   F   **10.** The NEC® is not intended to be used as an instruction manual.

_____   _____   **11.** All applications of the NEC® are ultimately subject to approval by the ___.

__xxxxx__   _____   **12.** The most common section view of a building is taken through a(n) ___.
        A. inside partition   C. exterior door
        B. outside wall   D. window

__xxxxx__   _____   **13.** Cutting planes for section views are shown on the ___.
        A. plot plan   C. all of the above
        B. floor plan   D. none of the above

__xxxxx__   _____   **14.** The Construction Specifications Institute has developed standard procedures that are helpful in writing ___.

__xxxxx__   _____   **15.** A(n) ___ is the shortened form of a word or phrase.

__xxxxx__   T   F   **16.** The members of each CMP represent interests, such as labor, manufacturing, electrical utilities, electrical inspectors, contractor associations, and testing laboratories.

Electrical Symbols

_____ F _____   **1.** Ceiling outlet box and incandescent lighting fixture

_____ A _____   **2.** Continuous fluorescent fixture

_____ G _____   **3.** Duplex receptacle outlet

_____ C _____   **4.** Three-way switch

_____ J _____   **5.** Home run

_____ H _____   **6.** Thermostat

_____ D _____   **7.** Motor

_____ E _____   **8.** Underground direct burial cable

_____ I _____   **9.** Power panel

_____ B _____   **10.** Range outlet

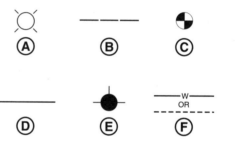

Plot Plan Symbols

_____ _____   **1.** Fire hydrant

_____ A _____   **2.** Light standard

_____ _____   **3.** Point of beginning

_____ I _____   **4.** Tree

_____ _____   **5.** Bush

_____ _____   **6.** Water line

_____ _____   **7.** Natural grade

_____ _____   **8.** Finish grade

_____ _____   **9.** Electrical service

_____ _____   **10.** Natural gas line

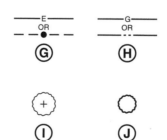

Name _____ Date _____

NEC®	**Answer**	
__XXXXX__	_____	**1.** The first step in the Code process is the receipt of ___.
__XXXXX__	_____	**2.** There are a total of ___ Code-Making Panels.
_____	T F	**3.** The NEC® is a design specification.
__XXXXX__	_____	**4.** Each print of a set of working drawings represents a part of a building drawn as a(n) ___ projection.
__XXXXX__	_____	**5.** The cutting plane for a floor plan is generally taken ___ above the floor being shown.

        A. 3'-0"   C. 5'-0"
        B. 4'-0"   D. 8'-0"

_____	_____	**6.** No point of unbroken wall space ___ or more in length measured horizontally shall be more than ___' from a receptacle in a living room of a dwelling unit.

        A. 12"; 6   C. 2'; 6
        B. 18"; 12   D. 2'; 12

__XXXXX__	_____	**7.** Division ___ of the CSI format pertains to electrical work.
__XXXXX__	_____	**8.** A(n) ___ is a graphic representation of an object.
__XXXXX__	_____	**9.** A detail is part of a(n) ___ view drawn at a larger scale.

        A. plan   C. section
        B. elevation   D. all of the above

__XXXXX__	_____	**10.** The location of all exterior walls and interior partitions is shown and dimensioned on the ___ plans.
_____	_____	**11.** The first ___ chapters of the NEC® apply generally.
__XXXXX__	T F	**12.** Other than the chairperson, membership on a Code panel is either as an alternate or principal member.
__XXXXX__	T F	**13.** The location of water, sewer, gas, and electrical utilities is given on the plot plan.
__XXXXX__	_____	**14.** A(n) ___ section is taken lengthwise through a building.
__XXXXX__	T F	**15.** All prints are drawn to ¼" = 1'-0" scale.
__XXXXX__	_____	**16.** The NEC® is ___.

        A. a legal document   C. all of the above
        B. designed to be adopted   D. none of the above
           by local and/or state
           governmental bodies

Architectural Symbols

_____   **1.** The section symbol for plywood is shown at ___.

_____   **2.** The section symbol for shingles is shown at ___.

_____   **3.** The elevation symbol for ceramic tile is shown at ___.

_____   **4.** The section symbol for structural steel is shown at ___.

_____   **5.** The section symbol for rough member is shown at ___.

_____   **6.** The section symbol for earth is shown at ___.

_____   **7.** The elevation symbol for glass block is shown at ___.

_____   **8.** The elevation symbol for concrete is shown at ___.

___ J ___   **9.** The plan symbol for loose fill insulation is shown at ___.

_____   **10.** The elevation symbol for brick is shown at ___.

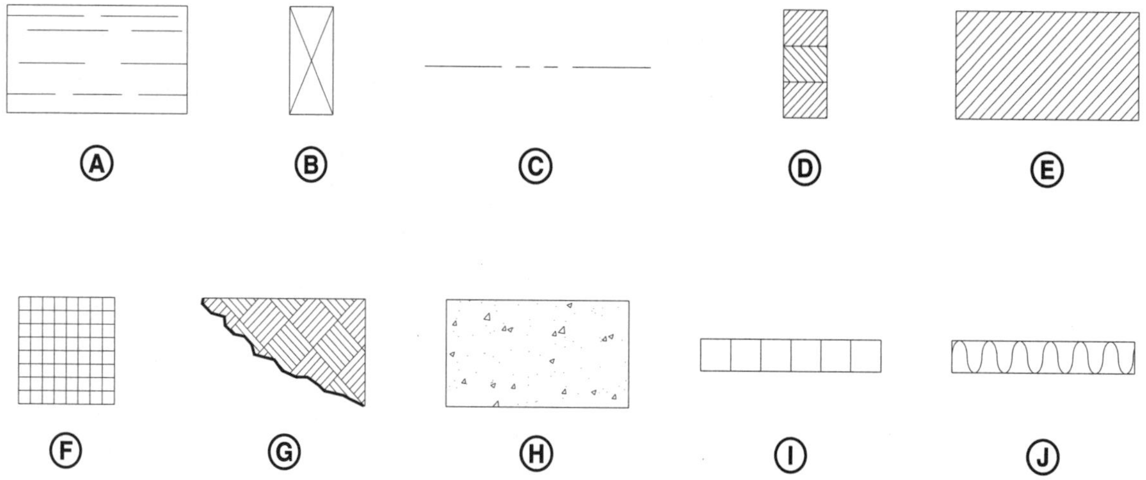

Ⓐ   Ⓑ   Ⓒ   Ⓓ   Ⓔ

Ⓕ   Ⓖ   Ⓗ   Ⓘ   Ⓙ

Abbreviations

_____   **1.** The abbreviation for volt is ___.

_____   **2.** The abbreviation for specification is ___.

_____   **3.** The abbreviation for circuit breaker is ___.

_____   **4.** The abbreviation for current is ___.

_____   **5.** The abbreviation for dryer is ___.

Name _____ Date _____

NEC®	Answer	

___xxxxx___   T   F   **1.** Anyone can submit a proposal to change the NEC® provided it contains the required information.

_____   _____   **2.** All 125 V, single-phase, 15 A and 20 A receptacles installed in dwelling bathrooms shall have ___ protection.

_____   _____   **3.** Chapter ___ contains tables and examples which are referenced throughout the entire NEC®.

_____   T   F   **4.** The AHJ must be a government employee.

_____   _____   **5.** The use of the words "___" or "shall not" indicates mandatory rules of the NEC®.

___xxxxx___   T   F   **6.** Electrical information is shown directly on the floor plans for small sets of prints.

___xxxxx___   _____   **7.** In the event of a discrepancy between specifications and prints, the ___ take precedence.

___xxxxx___   _____   **8.** Symbols are ___ in that the symbol can be understood no matter what language a person speaks.

___xxxxx___   _____   **9.** The abbreviation for overload is ___.

___xxxxx___   _____   **10.** NEC® ___ numbers are designated by a period following the article number.

___xxxxx___   _____   **11.** ___ permit alternate methods to the general rule.
   A. Subsections   C. Exceptions
   B. Informational Notes   D. none of the above

___xxxxx___   T   F   **12.** The NEC® is arranged in a simple oultine format.

___xxxxx___   T   F   **13.** The common practice today is to use a computer software program such as CAD to produce drawings.

___xxxxx___   _____   **14.** A plot plan ___.
   A. is an orthographic drawing   C. shows the contour of the land
   B. shows the location of the   D. all of the above
      structure

___xxxxx___   _____   **15.** A(n) ___ section is a section taken crosswise.

___xxxxx___   _____   **16.** On a floor plan, objects ___ the 5'-0" elevation are shown with ___ lines.
   A. above; hidden   C. all of the above
   B. below; object   D. none of the above

Abbreviations

_____   **1.** The abbreviation for benchmark is ___.

_____   **2.** The abbreviation for basement is ___.

T   F   **3.** A is the abbreviation for amps, anode, and area.

_____   **4.** The abbreviation for dining room and door is ___.

_____   **5.** The abbreviation for ground-fault circuit interrupter is ___.

_____   **6.** ___ is the abbreviation for horsepower.

_____   **7.** The abbreviation for single-pole circuit breaker is ___.

_____   **8.** ___ is the abbreviation for outside diameter.

T   F   **9.** The abbreviation for surface one side is S1S.

T   F   **10.** The abbreviation for transformer is T, TRANS, or XFMR.

_____   **11.** ___ is the abbreviation for watts or west.

_____   **12.** The abbreviation for three-wire is ___.

_____   **13.** The abbreviation for number is ___.

_____   **14.** The abbreviation for overcurrent protection device is ___.

_____   **15.** ___ is the abbreviation for point of beginning.

Alphabet of Lines

___*C*___   **1.** Object line

___*B*___   **2.** Hidden line

___*D*___   **3.** Centerline

___*G*___   **4.** Dimension line

___*E*___   **5.** Extension line

___*H*___   **6.** Leader

___*A*___   **7.** Cutting plane

___*F*___   **8.** Break line

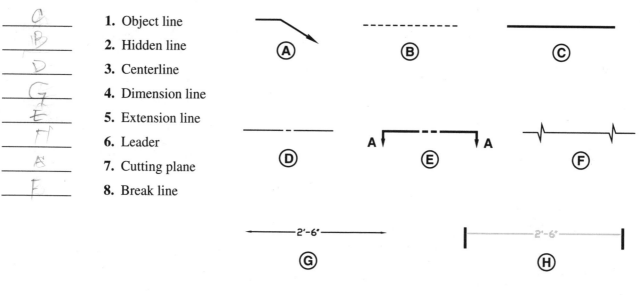

Name _____   Date _____

Refer to Floor Plan on page 19.

_____   **1.** The walk-in closet entered from BR1 measures 5'-8" × 16'.7"

____7____   **2.** B1 has a total of ___ duplex receptacle outlets with GFCI protection.

____C____   **3.** The duplex receptacle outlet on the front porch is ___.
  A. weatherproof   C. all of the above
  B. GFCI-protected   D. none of the above

___3'3"___   **4.** The stairs to the basement are ___ wide.

Ⓣ   F   **5.** The ceiling-mounted incandescent lighting fixture in the LR is controlled by a switch to the left of the front door.

Ⓣ   F   **6.** Hosebibbs are located on N, E, and W exterior walls.

_____   **7.** Cutting plane A-A indicates a ___ section.
  A. transverse   C. wall
  B. longitudinal   D. none of the above

_____   **8.** C windows on the South wall are spaced ___ OC.

_____   **9.** The front of the one-family dwelling faces ___.

_____   **10.** The duplex receptacle outlet near the telephone in the KIT. does not require GFCI protection because it does not serve a(n) ___.

_____   **11.** The duplex receptacle outlet beneath the A window in the LR is ___.
  A. GFCI-protected   C. 220 V
  B. split-wired   D. none of the above

_____   **12.** A total of ___ fluorescent fixtures are ceiling-mounted in the GAR.

T   F   **13.** All habitable rooms in the one-family dwelling contain ceiling-mounted incandescent lighting fixtures.

_____   **14.** TV jacks are located in the ___.
  A. LR and KIT.   C. LR and BR1
  B. KIT. and BR2   D. none of the above

_____   **15.** The ceiling-mounted incandescent lighting fixture in the stairway to the basement is controlled by a ___.
  A. pull switch   C. wall switch at foot of stairs
  B. wall switch at head of stairs D. both B and C

_____   **16.** The range in the island is ___.
  A. gas-operated   C. hardwired
  B. connected to a 240 V receptacle   D. connected to a 115 V receptacle

_____ **17.** Hallways in the one-family dwelling are ___ wide.

_____ **18.** The ___ is located in the hallway.
      A. smoke detector   C. doorbell
      B. thermostat   D. all of the above

_____ **19.** The garage has an area of ___ sq ft.
      A. 400   C. 600
      B. 450   D. 650

_____ **20.** A total of ___ duplex receptacle outlets are located in BR1.

T   F   **21.** All bedrooms have telephone jacks.

T   F   **22.** The PAN. doors are bi-fold doors.

T   F   **23.** The ceiling fan in BR1 is switch-controlled.

T   F   **24.** All bedroom entrance doors are the same size.

T   F   **25.** The duplex receptacle outlets in the DR are GFCI-protected.

Section A-A

Refer to Section A-A on page 21.

T   F   **1.** Double 4″ vinyl siding is used.

T   F   **2.** The foundation wall is 10″ thick.

T   F   **3.** The drain tile is perforated.

_____ **4.** The R value of insulation in the exterior walls is R ___.

_____ **5.** The 1″ × ___″ fascia board is covered with aluminum.

_____ **6.** The nominal size of the floor joists is ___.

_____ **7.** Exterior wall studs are 2″ × ___″.

_____ **8.** The sill plate is attached to the top of the foundation walls with ___.

T   F   **9.** The ceiling and walls are covered with ½″ drywall.

T   F   **10.** The soffit panel is aluminum.

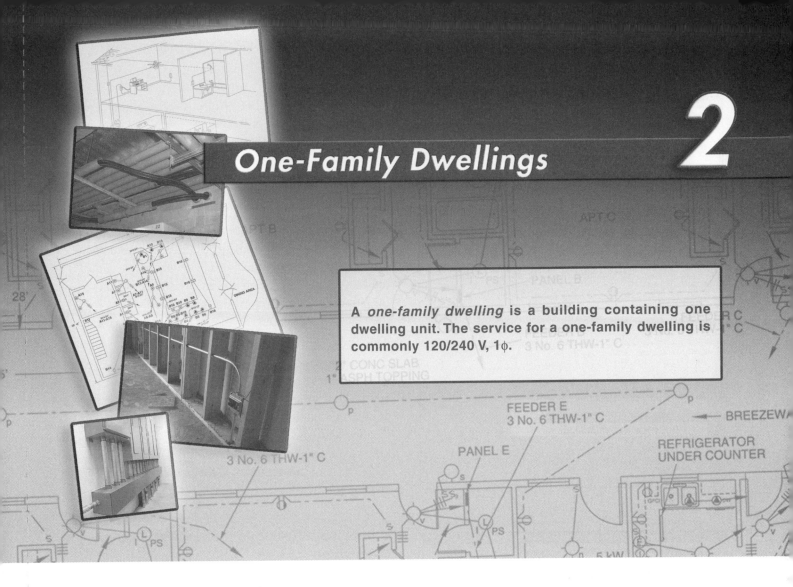

One-Family Dwellings

2

A *one-family dwelling* is a building containing one dwelling unit. The service for a one-family dwelling is commonly 120/240 V, 1φ.

ONE-FAMILY DWELLINGS

A *one-family dwelling* is a building consisting solely of one dwelling unit. A *dwelling unit* is one or more rooms used by one or more persons for housekeeping with space for eating, living, and sleeping and having permanent provisions for cooking and sanitation. To meet this definition, kitchen equipment and a bathroom are essential. Kitchen equipment includes appliances for storing food (refrigerators and freezers) and cooking food (stoves, ovens, etc.). A *bathroom* is an area with a basin and a toilet, tub, or shower.

The service for a one-family dwelling is commonly 120/240 V, 1φ. It may be either overhead or underground. The utility company installs the service to the dwelling. The *service point* is the connection point between the utility and the premises wiring. *Service equipment* is the necessary equipment to control and cut off the electrical supply. The service equipment consists of circuit breakers (CBs) or a switch and fuses normally located near the service point. Electricians install the service equipment when they wire the dwelling.

CALCULATIONS

"Calculation" is a widely used but undefined term in the NEC®. Typically, the term is used to refer to a set of procedures which are followed to arrive at a given value.

The Standard Calculation and the Optional Calculation are the two methods to calculate the load for one-family and multifamily dwellings per Article 220. Forms designed specifically for determining the Standard Calculation and the Optional Calculation standardize and simplify the operation. **See Figure 2-1.** In general, the Optional Calculation is less complex and easier to use than the Standard Calculation, and is used with larger loads. In either case, the calculation can be used to determine feeder or service loads.

STANDARD CALCULATION: ONE-FAMILY DWELLING

1. GENERAL LIGHTING: *Table 220.12*
_____ sq ft × 3 VA = _____ VA
Small appliances: *220.52(A)*
_____ VA × _____ circuits = _____ VA
Laundry: *220.52(B)*
_____ VA × 1 = _____ VA
_____ VA
Applying Demand Factors: *Table 220.42*
First 3000 VA × 100% = 3000 VA
Next _____ VA × 35% = _____ VA   **PHASES**   **NEUTRAL**
Remaining _____ VA × 25% = _____ VA
Total _____ VA   _____ VA   _____ VA

© 2011 by American Technical Publishers, Inc.
This form may be reproduced for instructional use only.
It shall not be reproduced and sold.

2. FIXED APPLIANCES: *220.53*
Dishwasher = _____ VA
Disposer = _____ VA
Compactor = _____ VA
Water heater = _____ VA
_____ = _____ VA
_____ = _____ VA
_____ = _____ VA
(120 V Loads × 75%)
Total _____ VA × 75% = _____ VA   _____ VA   _____ VA

3. DRYER: *220.54; Table 220.54*
_____ VA × _____ % = _____ VA   _____ VA × 70% = _____ VA

4. COOKING EQUIPMENT: *Table 220.55; Notes*
Col A _____ VA × _____ % = _____ VA
Col B _____ VA × _____ % = _____ VA
Col C _____ VA × _____ % = _____ VA
Total _____ VA   _____ VA × 70% = _____ VA

5. HEATING or A/C: *220.60*
Heating unit = _____ VA × 100% = _____ VA
A/C unit = _____ VA × 100% = _____ VA
Heat pump = _____ VA × 100% = _____ VA
Largest Load _____ VA   _____ VA

6. LARGEST MOTOR: *220.14(C) (220.50)*
φ _____ VA × 25% = _____ VA   _____ VA
N _____ VA × 25% = _____ VA   _____ VA

1φ service: PHASES $I = \dfrac{VA}{V} =$ _____ A
        NEUTRAL $I = \dfrac{VA}{V} =$ _____ A   _____ VA   _____ VA

220.61(B) First 200 A × 100% = 200 A
Remaining _____ A × 70% = _____ A
Total _____ A

OPTIONAL CALCULATION: ONE-FAMILY DWELLING

1. HEATING and A/C: *220.82(C)(1–6)*
Heating units (3 or less) = _____ VA × 65% = _____ VA
Heating units (4 or more) = _____ VA × 40% = _____ VA
A/C unit _____ VA × 100% = _____ VA
Heat pump _____ VA × 100% = _____ VA   **PHASES**
Largest Load _____ VA
Total _____ VA   _____ VA

© 2011 by American Technical Publishers, Inc.
This form may be reproduced for instructional use only.
It shall not be reproduced and sold.

2. GENERAL LOADS: *220.82(B)(1–4)*
General lighting: *220.82(B)(1)*
_____ sq ft × 3 VA = _____ VA
Small appliance and laundry loads: *220.82(B)(2)*
_____ VA × _____ circuits = _____ VA
Special loads: *220.82(B)(3–4)*
Dishwasher = _____ VA
Disposer = _____ VA
Compactor = _____ VA
Water heater = _____ VA
_____ = _____ VA
_____ = _____ VA
_____ = _____ VA
_____ = _____ VA
_____ = _____ VA
Total _____ VA
Applying Demand Factors: *220.82(B)*
First 10,000 VA × 100% = 10,000 VA
Remaining _____ VA × 40% = _____ VA
Total _____ VA   _____ VA

NEUTRAL (Loads from Standard Calculation)
1. General lighting = _____ VA
2. Fixed appliances = _____ VA
3. Dryer = _____ VA
4. Cooking equipment = _____ VA
5. Heating or A/C = _____ VA
6. Largest motor = _____ VA
Total _____ VA

1φ service: PHASES $I = \dfrac{VA}{V} =$ _____ A
        NEUTRAL $I = \dfrac{VA}{V} =$ _____ A   _____ VA

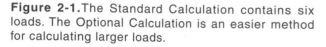

Figure 2-1. The Standard Calculation contains six loads. The Optional Calculation is an easier method for calculating larger loads.

Demand is the amount of electricity required at a given time. The concept behind the use of the term is that the total calculated load is rarely placed on the electrical system. This is because all of the electrical loads are not on at the same time. For example, a household electric range may have a nameplate rating of 14 kW. The electric range has four burners, one broiler, one heating element, and accessories such as fans, lights, and timers. Rarely, if ever, would all of these be used at the same time. Therefore, the concept of demand permits the 14 kW range to be considered as a smaller load for the purpose of determining how much load to allow for the range when calculating the total feeder or service load for the building.

Noncoincident loads are loads that are unlikely to be in use at the same time. For example, heating and air conditioning are noncoincident loads. It is permissible to drop the smaller of the two loads when completing the total load of a feeder. Heating or A/C is Load 5 on the Standard Calculation: One-Family Dwelling form and Load 1 on the Optional Calculation: One-Family Dwelling form.

One-Family Dwellings – Standard Calculation

The Standard Calculation for One-Family Dwellings contains six individual calculations that are performed before the minimum size service or feeder conductors required for the calculated load can be determined. Demand factors may be applied to four of these loads. These six loads, their NEC® references, and those to which demand factors may be applied are:

1. General Lighting – *Table 220.12 (Lighting Loads)*
2. Fixed Appliances – *220.53 (Demand Factors)*
3. Dryer – *220.54; Table 220.54 (Demand Factors)*
4. Cooking Equipment – *Table 220.55 (Demand Factors)*
5. Heating or A/C – *220.60 (Noncoincident Loads)*
6. Largest Motor – *220.50*

One-Family Dwelling – Optional Calculation

The Optional Calculation is designed for dwellings with large electric loads. It is an easier method than the

Standard Calculation for calculating feeder and service loads for dwellings. The Optional Calculation is permitted for the ungrounded conductors of 120/240 V, 3-wire or 120/208 V service-entrance or feeder conductors with an ampacity of at least 100 A.

The grounded conductor calculation for the Optional Calculation is determined per 220.61. The Optional Calculation for dwelling units consists of two groups of loads. These two groups of loads and their NEC® references are:

1. Heating and A/C – *220.82(C)(1 – 6)*

2. General Loads – *220.82(B)(1 – 4)*

The Optional Calculation can be used for large-capacity dwelling units. It permits designers of electrical systems to take advantage of the increased diversity and demand factors with these types of dwellings and utilize smaller service-entrance or feeder conductors.

Dwelling A

Dwelling A is a one-story, one-family dwelling. It has a living room (LR), kitchen (KIT.), laundry (LAU), one bathroom (B), three bedrooms (BR), and an entrance foyer leading to a hallway. These abbreviations, and others used on the plan, are standard abbreviations. **See Appendix.** Note that some architects may use abbreviations which are slightly different. Any abbreviation that forms a word is followed by a period. For example, KIT. is followed by a period.

Standard symbols are used to show building materials, fixtures, etc. For example, the symbol for a duplex receptacle is a circle with parallel lines passing through the circle. **See Appendix.**

The outside dimensions of the dwelling are 55′ × 28′. These dimensions are multiplied to find the area, which is 1540 sq ft (55′ × 28′ = 1540 sq ft). *Area* is the number of units of length multiplied by the number of units of width. Area is expressed in square units such as square inches, square feet, square yards, etc. Square feet are commonly used to express the area for dwellings.

Table 220.12 gives the General Lighting loads by occupancies. The unit load per square foot for dwellings is 3 VA. The General Lighting load is based on the area of the dwelling times the VA (1540 sq ft × 3 VA = 4620 VA), plus the loads for Small Appliances and Laundry loads per 220.14(J). Demand factors are applied to the total per Table 220.42 to determine the total General Lighting load.

By following the Standard Calculation: One-Family Dwelling form, the other five groups of loads can be calculated. The VA for these loads is taken from either the prints or specifications. For example, the loads for Fixed Appliances (220.53), Dryer (220.54; Table 220.54); Cooking Equipment (Table 220.55; Notes); Heating or AC (220.60); and the Largest Motor (220.14(C) and 220.50) are not given on the print for Dwelling A. For Dwelling A, the loads used to calculate the service are typical for this size dwelling and were taken from specifications which are not shown in this book.

The total VA is divided by the voltage to find the size of the service in amperes (A). The formula used is $I = \dfrac{VA}{V}$, where I = amps, VA = volt-amps, and V = volts. Notice that the Optional Calculation: One-Family Dwelling form can only be completed after the Standard Calculation: One-Family Dwelling form has been completed.

Dwelling B

Dwelling B is also a one-family dwelling. It has a basement in which the den (DEN), laundry room (LAU), and bathroom (B) are finished. The unfinished basement (UNFIN BSMT) area contains the water heater (WH), heating and air conditioning, and the main panelboard. The area of the basement is added to the area of the main floor to determine the area of the dwelling.

Seven window wells are shown for windows in the basement. The windows allow natural light and air to enter the basement and the window wells provide a means of egress in the event of an emergency. The window wells are shown as semicircles on the plan. Again, the loads used to calculate the service are typical for this size dwelling and may be taken from the prints or specifications, depending on where they are given. The specifications for this particular floor plan are not shown in this book.

BRANCH CIRCUIT, FEEDER, AND SERVICE CALCULATIONS

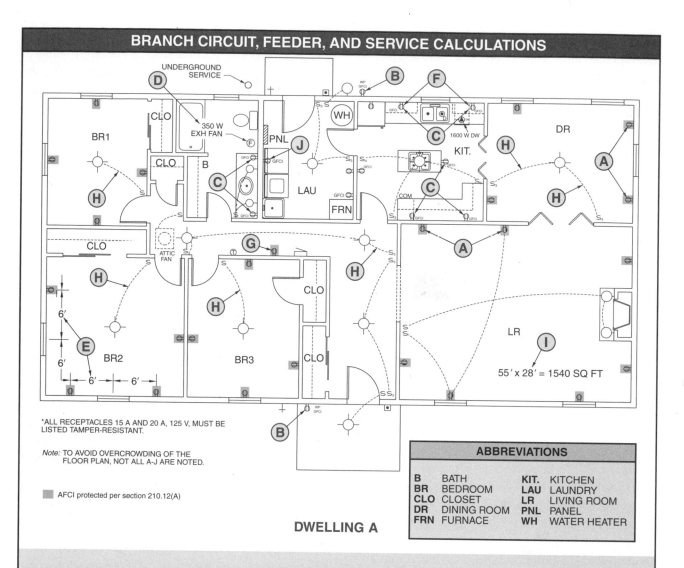

*ALL RECEPTACLES 15 A AND 20 A, 125 V, MUST BE LISTED TAMPER-RESISTANT.

Note: TO AVOID OVERCROWDING OF THE FLOOR PLAN, NOT ALL A-J ARE NOTED.

AFCI protected per section 210.12(A)

DWELLING A

ABBREVIATIONS

B	BATH	**KIT.**	KITCHEN
BR	BEDROOM	**LAU**	LAUNDRY
CLO	CLOSET	**LR**	LIVING ROOM
DR	DINING ROOM	**PNL**	PANEL
FRN	FURNACE	**WH**	WATER HEATER

A. 406.4(A) All receptacles installed on 15 A and 20 A circuits shall be of the grounding type. See Exception.

B. 210.8(A)(3) All 15 A and 20 A, 125 V receptacles installed outdoors of dwelling units shall have GFCI protection. See Exception.

C. 210.8(A)(1 – 8) Ground-fault protection for personnel in dwellings shall be provided for all 15 A and 20 A, 125 V receptacles in all (1) bathrooms, (2) garages and accessory buildings used for storage or as workshops, (3) outdoor locations, (4) crawl spaces at or below grade, (5) unfinished basements, (6) kitchen countertops, (7) within 6′ of a wet bar, laundry, utility sink, (8) boathouses. See Exceptions.

D. 210.23 Branch circuits may be installed for specific loads (exhaust fan in bath).

E. 210.52(A)(1) All general occupancy rooms in dwellings shall have receptacle outlets installed so that no point along the floor wall line is more than 6′, measured horizontally, from an outlet.

F. 210.52(B)(1) For small appliance load, including all receptacles in kitchen, pantry, breakfast room, dining room, and similar areas, two or more 20 A appliance circuits shall be provided. Such circuits shall have no other outlets. Countertop outlets to be at least two 20 A circuits. Clock outlets and receptacles for supplemental equipment on ranges, ovens, and counter-mounted cooking units are permitted on these circuits. See 210.52(B)(2), Ex. 1, 2. Refrigerator receptacle is permitted on an individual 15 A circuit. See 210.52(B)(1), Ex. 2.

G. 210.52(H) Dwelling hallways more than 10′ in length, measured from the centerline of the hall, without passing through a doorway, shall be provided with at least one receptacle outlet.

H. 210.70(A)(1) Dwelling units shall have at least one wall switch-controlled lighting outlet in every habitable room including bathrooms, hallways, stairways, and at outdoor entrances or exits. In rooms other than kitchen and bathroom, switch-controlled receptacles are permitted in lieu of lighting outlet. See Ex. 1.

I. 220.12 Unit lighting load for dwelling occupancies shall be 3 VA per sq ft. See Table 220.12. Use outside dimensions of dwelling.

Table 220.12 Receptacles other than on the two small appliance circuits are considered part of general illumination and require no allowance for additional load. See 220.14(J).

J. 210.11(C)(2) At least one 20 A circuit shall be provided for laundry receptacle outlet per 210.52(F).

210.11(B) "VA per sq ft" load shall be apportioned evenly in multioutlet branch circuits within panelboard.

220.12 For general illumination, one 15 A, 120 V branch circuit is required for every 600 sq ft of area. 3 VA per sq ft × 600 = 1800 VA ÷ 120 V = 15 A.

220.52(A) Feeder load for the two small appliance circuits of 210.11(C)(1) is to be taken at 3000 VA (1500 VA for each two-wire circuit). This load may be included with general lighting load in feeder calculations per 220.14(J).

220.52(B) Feeder load is 1500 VA for each laundry circuit as required by 210.11(C)(2). This load may be included with general lighting load in feeder calculations per 220.14(J).

230.42(B) In general, service-entrance conductors shall have an ampacity of at least 100 A. See 230.79(C).

422.11(E)(1–3) Branch circuit to single nonmotor appliance not to exceed OC rating on device. If no appliance rating, 13.3 A permitted on 20 A circuit. If over 13.3 A, the next size OCPD above 150%.

DWELLING A – 220, PARTS I, II, AND III

STANDARD CALCULATION: ONE-FAMILY DWELLING

1. GENERAL LIGHTING: *Table 220.12*

__1540__ sq ft × 3 VA = __4620__ VA

> © 2011 by American Technical Publishers, Inc.
> *This form may be reproduced for instructional use only.*
> *It shall not be reproduced and sold.*

Small appliances: *220.52(A)*

__1500__ VA × __2__ circuits = __3000__ VA

Laundry: *220.52(B)*

__1500__ VA × 1 = __1500__ VA

__9120__ VA

Applying Demand Factors: *Table 220.42*

First 3000 VA × 100% = 3000 VA

Next __6120__ VA × 35% = __2142__ VA   **PHASES**   **NEUTRAL**

Remaining __—__ VA × 25% = __—__ VA

Total __5142__ VA   __5142__ VA   __5142__ VA

2. FIXED APPLIANCES: *220.53*

Dishwasher = __1600__ VA

Disposer = __1200__ VA

Compactor = __900__ VA

Water heater = __1800__ VA

Exh Fan = __350__ VA

Attic Fan = __1200__ VA

__—__ = __—__ VA

(120 V Loads × 75%)

Total __7050__ VA × 75% = __5288__ VA   __5288__ VA   __5288__ VA

3. DRYER: *220.54; Table 220.54*

__5800__ VA × __100__ % = __5800__ VA   __5800__ VA × 70% = __4060__ VA

4. COOKING EQUIPMENT: *Table 220.55; Notes*

Col A __8000__ VA × __100__ % = __8000__ VA

Col B __—__ VA × __—__ % = __—__ VA

Col C __—__ VA × __—__ % = __—__ VA

Total __8000__ VA   __8000__ VA × 70% = __5600__ VA

5. HEATING or A/C: *220.60*

Heating unit = __—__ VA × 100% = __—__ VA

A/C unit = __—__ VA × 100% = __—__ VA

Heat pump = __16,000__ VA × 100% = __16,000__ VA

Largest Load __16,000__ VA   __16,000__ VA   __—__ VA

6. LARGEST MOTOR: *220.14(C)(220.50)*

φ __1200__ VA × 25% = __300__ VA   __300__ VA

N __—__ VA × 25% = __—__ VA   __300__ VA

1φ service: PHASES $I = \dfrac{40,530 \text{ VA}}{240 \text{ V}} = $ __168.88__ A

NEUTRAL $I = \dfrac{20,390 \text{ VA}}{240 \text{ V}} = $ __84.96__ A

     __40,530__ VA   __20,390__ VA

220.61(B); First 200 A × 100% = 200 A

Remaining __—__ A × 70% = __—__ A

Total __—__ A

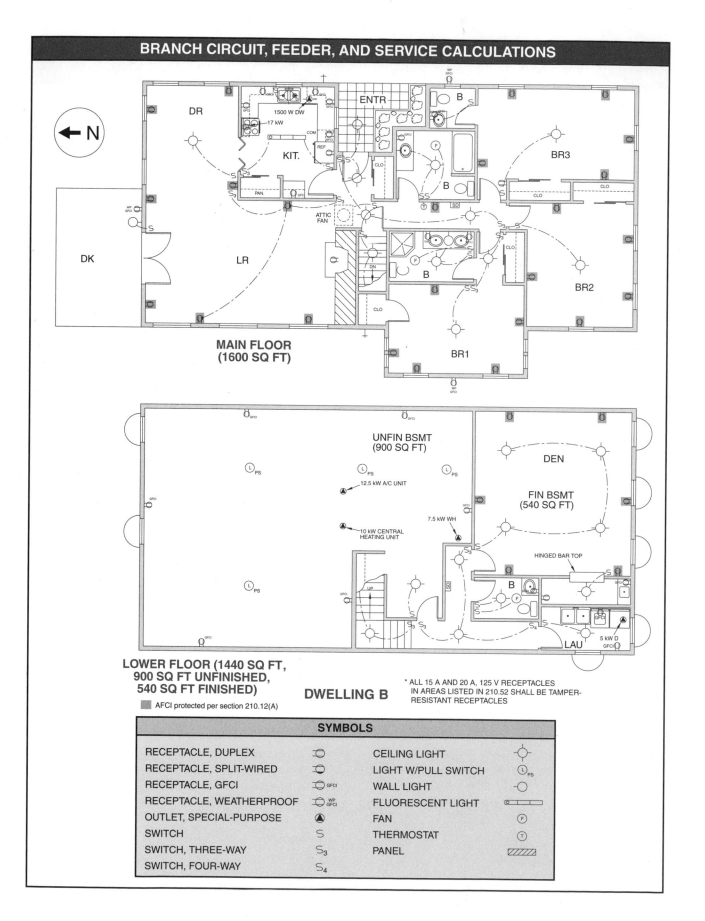

BRANCH CIRCUIT, FEEDER, AND SERVICE CALCULATIONS

MAIN FLOOR
(1600 SQ FT)

LOWER FLOOR (1440 SQ FT,
900 SQ FT UNFINISHED,
540 SQ FT FINISHED)

DWELLING B

* ALL 15 A AND 20 A, 125 V RECEPTACLES
IN AREAS LISTED IN 210.52 SHALL BE TAMPER-
RESISTANT RECEPTACLES

AFCI protected per section 210.12(A)

SYMBOLS

RECEPTACLE, DUPLEX		CEILING LIGHT	
RECEPTACLE, SPLIT-WIRED		LIGHT W/PULL SWITCH	
RECEPTACLE, GFCI	GFCI	WALL LIGHT	
RECEPTACLE, WEATHERPROOF	WP GFCI	FLUORESCENT LIGHT	
OUTLET, SPECIAL-PURPOSE		FAN	F
SWITCH	S	THERMOSTAT	T
SWITCH, THREE-WAY	S3	PANEL	
SWITCH, FOUR-WAY	S4		

DWELLING B – 220, PARTS I, II, AND III

STANDARD CALCULATION: ONE-FAMILY DWELLING

1. GENERAL LIGHTING: *Table 220.12*

__3040__ sq ft × 3 VA =   __9120__ VA

© 2011 by American Technical Publishers, Inc.
*This form may be reproduced for instructional use only.
It shall not be reproduced and sold.*

Small appliances: *220.52(A)*
__1500__ VA × ___2___ circuits =   __3000__ VA

Laundry: *220.52(B)*
__1500__ VA × 1 =   __1500__ VA
                                           __13,620__ VA

Applying Demand Factors: *Table 220.42*

First 3000 VA × 100% =   3000 VA

Next __10,620__ VA × 35% =   __3717__ VA   **PHASES**   **NEUTRAL**

Remaining ___—___ VA × 25% =   ___—___ VA
Total   __6717__ VA   __6717__ VA   __6717__ VA

2. FIXED APPLIANCES: *220.53*

Dishwasher = __1500__ VA
Disposer = __900__ VA
Compactor = __950__ VA
Water heater = __7500__ VA
2 Exh Fans = __700__ VA
Attic Fan = __1000__ VA
___ = ___—___ VA

(120 V Loads × 75%)

Total __12,550__ VA × 75% = __9413__ VA   __9413__ VA   __9413__ VA

3. DRYER: *220.54; Table 220.54*

__5000__ VA × __100__ % =   __5000__ VA   __5000__ VA × 70% = __3500__ VA

4. COOKING EQUIPMENT: *Table 220.55 ; Notes*

Col A __8000__ VA × __125__ % =   __10,000__ VA
Col B ___—___ VA × ___—___ % =   ___—___ VA
Col C ___—___ VA × ___—___ % =   ___—___ VA
Total   __10,000__ VA   __10,000__ VA × 70% = __7000__ VA

5. HEATING or A/C: *220.60*

Heating unit = __10,000__ VA × 100% =   __10,000__ VA
A/C unit = __12,500__ VA × 100% =   __12,500__ VA
Heat pump = ___—___ VA × 100% =   ___—___ VA
Largest Load   __12,500__ VA   __12,500__ VA   ___—___ VA

6. LARGEST MOTOR: *220.14(C) (220.50)*

φ __1000__ VA × 25% =   __250__ VA   __250__ VA
N ___—___ VA × 25% =   ___—___ VA   __250__ VA

1φ service: PHASES $I = \dfrac{43,880 \text{ VA}}{240 \text{ V}} = 182.83$ A

NEUTRAL $I = \dfrac{26,880 \text{ VA}}{240 \text{ V}} = 112$ A   __43,880__ VA   __26,880__ VA

220.61(B); First 200 A × 100% =   200 A
Remaining ___—___ A × 70% =   ___—___ A
Total   ___—___ A

DWELLING A – 220, PART IV

OPTIONAL CALCULATION: ONE-FAMILY DWELLING

1. HEATING and A/C: 220.82(C)(1 – 6)

Heating units (3 or less) = _____ VA × 65% = _____ VA

Heating units (4 or more) = _____ VA × 40% = _____ VA

A/C unit = _____ VA × 100% = _____ VA

Heat pump = **16,000** VA × 100% = **16,000** VA   **PHASES**

Largest Load **16,000** VA

Total **16,000** VA   **16,000** VA

2. GENERAL LOADS: 220.82(B)(1 – 4)

General lighting: 220.82(B)(1)

1540 sq ft × 3 VA   **4620** VA

Small appliance and laundry loads: 220.82(B)(2)

1500 VA × **3** circuits =   **4500** VA

Special loads: 220.82(B)(3–4)

Dishwasher = **1600** VA

Disposer = **1200** VA

Compactor = **900** VA

Water heater = **1800** VA

Exh Fan = **350** VA

Attic Fan = **1200** VA

_____ = _____ VA

_____ = _____ VA

_____ = _____ VA

7050 VA   **7050** VA

Total **16,170** VA

Applying Demand Factors: 220.82(B)

First 10,000 VA × 100% =   10,000 VA

Remaining **6170** VA × 40% =   **2468** VA

Total   **12,468** VA   **12,468** VA

NEUTRAL (Loads from Standard Calculation)

1. General lighting = **5142** VA

2. Fixed appliances = **5288** VA

3. Dryer = **4060** VA

4. Cooking equipment = **5600** VA

5. Heating or A/C = — VA

6. Largest motor = **300** VA

Total **20,390** VA

1φ service: PHASES $I = \dfrac{28,468 \text{ VA}}{240 \text{ V}} = $ **118.62** A

NEUTRAL $I = \dfrac{20,390 \text{ VA}}{240 \text{ V}} = $ **84.96** A   **28,468** VA

DWELLING B – 220, PART IV

OPTIONAL CALCULATION: ONE-FAMILY DWELLING

1. HEATING and A/C: *220.82(C)(1 – 6)*

Heating units (3 or less) = __10,000__ VA × 65% =   __6500__ VA

Heating units (4 or more) = __—__ VA × 40% =   __—__ VA

A/C unit = __12,500__ VA × 100% =   __12,500__ VA

Heat pump = __—__ VA × 100% =   __—__ VA   **PHASES**

Largest Load   __12,500__ VA

Total   __12,500__ VA   __12,500__ VA

2. GENERAL LOADS: *220.82(B)(1 – 4)*

General lighting: *220.82(B)(1)*

__3040__ sq ft × 3 VA   __9120__ VA

Small appliance and laundry loads: *220.82(B)(2)*

__1500__ VA × __3__ circuits =   __4500__ VA

Special loads: *220.82(B)(3–4)*

Dishwasher = __1500__ VA

Disposer = __900__ VA

Compactor = __950__ VA

Water heater = __7500__ VA

__2 Exh Fans__ = __700__ VA

__Attic Fan__ = __1000__ VA

__—__ = __—__ VA

__—__ = __—__ VA

__—__ = __—__ VA

__12,550__ VA   __12,550__ VA

Total __26,170__ VA

Applying Demand Factors: *220.82(B)*

First 10,000 VA × 100% =   10,000 VA

Remaining __16,170__ VA × 40% =   __6468__ VA

Total   __16,468__ VA   __16,468__ VA

NEUTRAL (Loads from Standard Calculation)

1. General lighting =   __6717__ VA
2. Fixed appliances =   __9413__ VA
3. Dryer =   __3500__ VA
4. Cooking equipment =   __7000__ VA
5. Heating or A/C =   __—__ VA
6. Largest motor =   __250__ VA

Total | __26,880__ | VA

1φ service: PHASES $I = \dfrac{28,968 \text{ VA}}{240 \text{ V}} =$ __120.7__ A

NEUTRAL $I = \dfrac{26,880 \text{ VA}}{240 \text{ V}} =$ __112__ A | __28,968__ | VA

CLEARANCES FOR SERVICE DROPS AND SERVICE LATERALS

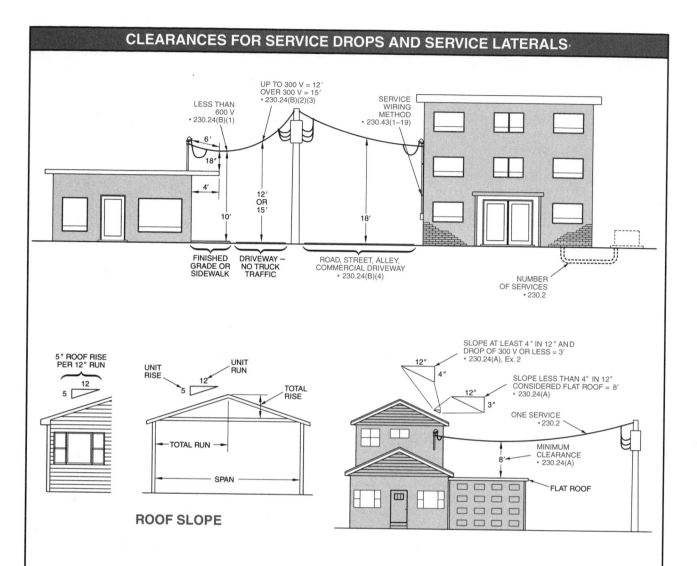

ROOF SLOPE

230.2 In general, a building or structure shall be supplied by one service. See Special Conditions (A–D). Several underground sets of No. 1/0 or larger multiple conductors, connected at the source but not at the load, are considered to be only one service. This is allowed to reduce available fault currents. See 230.2.

230.22 Overhead individual service conductors shall be insulated or covered.

230.22, Ex. The grounded conductor (neutral) of a multiconductor cable may be bare.

230.24(A) Conductors shall have 8′ minimum clearance above flat roofs and maintain 3′ minimum clearance from roof edge.

230.24(A), Ex. 2 For voltage between service drop conductors of 300 V or less and the roof slope is at least 4″ in 12″, clearances may be reduced to 3′.

230.24(A), Ex. 3 Service-drop conductors of 300 V or less may pass over a roof overhang with a clearance of 18″ or more. Up to 6′ of overhead service conductors can be run in this manner. It shall pass over no more than 4′ of overhang measured horizontally.

230.24(A), Ex. 4 Vertical clearance of 3′ is not required for the final conductor span if the service drop is connected to the side of the building.

230.24(A), Ex. 5 A reduction of clearance to 3′ shall be permitted where the voltage between conductors does not exceed 300 V and the roof is guarded or isolated.

230.24(B)(1 – 4) Overhead service conductors less than 600 V shall be a minimum of 10′ above finished grade of pedestrian areas or sidewalks. Voltage is limited to 150 V to ground with insulated conductor cabled together and supported with a grounded bare messenger.

For service drops up to 300 V to ground, the minimum clearance over residential property and driveways, and commercial areas not having truck traffic is 12′. For service drops over 300 V to ground, the minimum clearance over residential property and driveways, and commercial areas not having truck traffic, is 15′.

For service drops over public roads, streets, alleys, parking areas with truck traffic, and commercial driveways, the minimum clearance is 18′.

230.28 Service mast shall have adequate strength or be supported. All raceway fittings shall be identified for use with service masts. See 230.51 (A – C).

230.30 Underground service-lateral conductors shall be insulated for the applied voltage.

230.30, Ex., (1) An uninsulated bare copper grounded conductor is permitted in a raceway.

230.30, Ex., (2) An uninsulated bare copper grounded conductor is permitted for direct burial where suitable for soil conditions.

230.30, Ex., (3) An uninsulated bare copper grounded conductor is permitted in any soil if the cable assembly is identified for underground use.

230.30, Ex., (4) An uninsulated bare Al or Cu-clad Al conductor is permitted in a raceway or for direct burial if the conductor is part of a cable assembly identified for underground use.

SERVICE HEADS AND SERVICE-ENTRANCE CONDUCTORS

230.23(B) Service drops shall not be smaller than No. 8 Cu, No. 6 Al, or No. 6 Cu-clad Al.

230.23(B), Ex. For limited loads of a single branch circuit, service drops shall not be smaller than No. 12 hard-drawn Cu or equivalent.

230.41 Service-entrance conductors entering or on a building exterior shall be insulated.

230.41, Ex. Bare Cu conductors is permitted when used:

230.41, Ex., (1) in raceway or as part of service cable assembly.

230.41, Ex., (2) for direct burial if judged suitable for soil conditions.

230.41, Ex., (3) for direct burial in any soil if part of a cable assembly identified for underground use.

230.41, Ex., (4) Bare Al or Cu-clad Al is permitted when used as part of a cable assembly, identified for use in an underground raceway, or for direct burial.

230.41, Ex., (5) Bare conductors used in auxiliary gutter.

230.42(A) Service-entrance conductors shall have ampacity to carry loads per 220.

230.42(A)(1) Ungrounded conductors shall have an ampacity not less than the sum of the noncontinuous loads plus 125% of the continuous load. See Exception.

230.42(A)(2) Ungrounded conductors shall have an ampacity of not less than the sum of noncontinuous loads plus the continuous load if the conductors terminate in an OCPD listed for the operation at 100% of the rating.

230.42(C) The grounded conductor shall not be smaller than minimum size per 250.24(C).

230.51(A) Service entrance cables shall be supported within 12″ of service head, gooseneck, or connection to raceway or enclosures and at intervals not exceeding 30″.

230.54(A) Service raceways shall have a raintight service head at connection to service-drop or overhead service conductors.

230.54(B) Service cable shall be equipped with a service head. Type SE may be formed into a gooseneck which shall be taped and painted or taped with self-sealing, weather-resistant thermoplastic. The service head shall be listed for use in wet locations.

230.54(C) Service heads and goosenecks shall be above point of attachment of service-drop or overhead service conductors to building.

230.54(C), Ex. Service head is permitted to be located no more than 24″ from point of attachment.

230.54(D) Service entrance cables shall be securely fastened in place.

230.54(E) Service conductors of different potential shall be brought out of service head through separately bushed openings.

230.54(E), Ex. Separated bushed openings are not required for jacketed multiconductor service entrance cable without a splice.

230.54(F) Drip loops shall be formed on individual conductors.

230.54(G) Overhead service and service-entrance conductors shall be arranged to prevent water from entering service raceway or equipment.

338.2 Service-entrance cable is a single or multiconductor assembly with or without an overall covering. Type SE has a flame-retardant and moisture-resistant covering. Type USE is identified for underground use. It has a moisture-resistant covering.

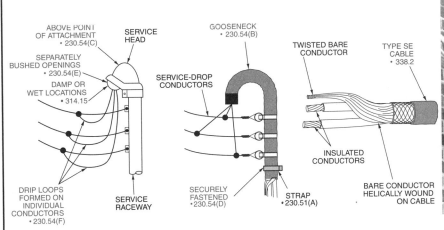

Service-drop conductors are required to be insulated or covered per 230.22.

CLEARANCES FOR SERVICE DROPS FROM BUILDING OPENINGS

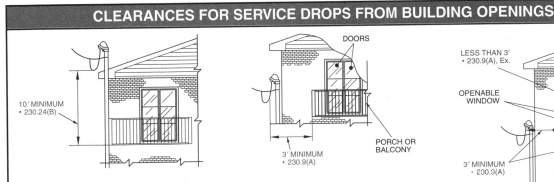

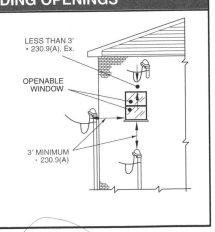

230.24(B) Overhead service conductors shall have a minimum vertical clearance of not less than 10′ from a platform provided the conductors are insulated and supported on and cabled with a grounded bare messenger and voltage limited to 150 V to ground.

230.9(A) Service conductors shall have a minimum clearance of not less than 3′ from openable windows, doors, porches, balconies, ladders, fire escapes, etc.
230.9(A), Ex. conductors run above the window are permitted to be less than 3′.

SERVICES — OVERHEAD AND UNDERGROUND

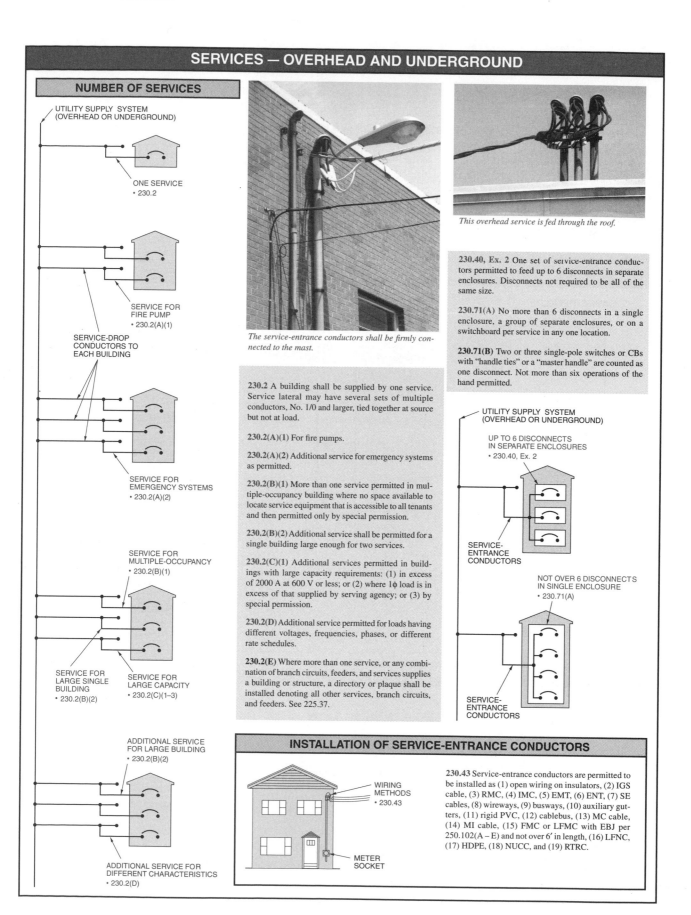

NUMBER OF SERVICES

UTILITY SUPPLY SYSTEM
(OVERHEAD OR UNDERGROUND)

ONE SERVICE
• 230.2

SERVICE FOR
FIRE PUMP
• 230.2(A)(1)

SERVICE-DROP
CONDUCTORS TO
EACH BUILDING

SERVICE FOR
EMERGENCY SYSTEMS
• 230.2(A)(2)

SERVICE FOR
MULTIPLE-OCCUPANCY
• 230.2(B)(1)

SERVICE FOR
LARGE SINGLE
BUILDING
• 230.2(B)(2)

SERVICE FOR
LARGE CAPACITY
• 230.2(C)(1–3)

ADDITIONAL SERVICE
FOR LARGE BUILDING
• 230.2(B)(2)

ADDITIONAL SERVICE FOR
DIFFERENT CHARACTERISTICS
• 230.2(D)

This overhead service is fed through the roof.

The service-entrance conductors shall be firmly connected to the mast.

230.2 A building shall be supplied by one service. Service lateral may have several sets of multiple conductors, No. 1/0 and larger, tied together at source but not at load.

230.2(A)(1) For fire pumps.

230.2(A)(2) Additional service for emergency systems as permitted.

230.2(B)(1) More than one service permitted in multiple-occupancy building where no space available to locate service equipment that is accessible to all tenants and then permitted only by special permission.

230.2(B)(2) Additional service shall be permitted for a single building large enough for two services.

230.2(C)(1) Additional services permitted in buildings with large capacity requirements: (1) in excess of 2000 A at 600 V or less; or (2) where 1φ load is in excess of that supplied by serving agency; or (3) by special permission.

230.2(D) Additional service permitted for loads having different voltages, frequencies, phases, or different rate schedules.

230.2(E) Where more than one service, or any combination of branch circuits, feeders, and services supplies a building or structure, a directory or plaque shall be installed denoting all other services, branch circuits, and feeders. See 225.37.

230.40, Ex. 2 One set of service-entrance conductors permitted to feed up to 6 disconnects in separate enclosures. Disconnects not required to be all of the same size.

230.71(A) No more than 6 disconnects in a single enclosure, a group of separate enclosures, or on a switchboard per service in any one location.

230.71(B) Two or three single-pole switches or CBs with "handle ties" or a "master handle" are counted as one disconnect. Not more than six operations of the hand permitted.

UTILITY SUPPLY SYSTEM
(OVERHEAD OR UNDERGROUND)

UP TO 6 DISCONNECTS
IN SEPARATE ENCLOSURES
• 230.40, Ex. 2

SERVICE-
ENTRANCE
CONDUCTORS

NOT OVER 6 DISCONNECTS
IN SINGLE ENCLOSURE
• 230.71(A)

SERVICE-
ENTRANCE
CONDUCTORS

INSTALLATION OF SERVICE-ENTRANCE CONDUCTORS

WIRING
METHODS
• 230.43

METER
SOCKET

230.43 Service-entrance conductors are permitted to be installed as (1) open wiring on insulators, (2) IGS cable, (3) RMC, (4) IMC, (5) EMT, (6) ENT, (7) SE cables, (8) wireways, (9) busways, (10) auxiliary gutters, (11) rigid PVC, (12) cablebus, (13) MC cable, (14) MI cable, (15) FMC or LFMC with EBJ per 250.102(A – E) and not over 6′ in length, (16) LFNC, (17) HDPE, (18) NUCC, and (19) RTRC.

DISCONNECTING MEANS

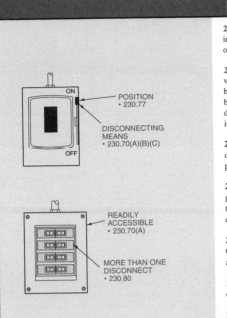

POSITION
• 230.77

DISCONNECTING MEANS
• 230.70(A)(B)(C)

READILY ACCESSIBLE
• 230.70(A)

MORE THAN ONE DISCONNECT
• 230.80

230.70(A)(1) Service disconnecting means shall be installed at a readily accessible place nearest entrance of conductors, either inside or outside of building.

230.70(A – C) Disconnecting means shall be provided for all service-entrance conductors. Each shall be suitable for use as service equipment and shall be permanently marked to identify it as a service disconnecting means. Shall be suitable for prevailing conditions.

230.74 Disconnecting means shall simultaneously disconnect all ungrounded service conductors from premises wiring system.

230.75 Means shall be provided to disconnect the grounded conductor from the premises wiring. A terminal or bus used in conjunction with pressure connectors is permissible.

230.76(1 – 2) Disconnecting means shall be a switch or CB and be manual or power operated. If power operated, it must be able to be opened by hand.

230.77 Service disconnecting means shall plainly indicate if it is in open (OFF) or closed (ON) position.

230.79 The service disconnecting means shall have a rating not less than the load as determined by Article 220 and not smaller than as given in 230.79 (A – D).

230.79(A) For limited loads of one branch circuit it may have a rating of not less than 15 A.

230.79(B) Where not more than two 2-wire branch circuits, it shall have rating of not less than 30 A.

230.79(C) For one-family dwelling the service disconnecting means shall have a rating of not less than 100 A, 3-wire.

230.79(D) All other installations require rating of not less than 60 A.

230.80 If more than one disconnect is used per 230.71, the total rating of switches or CBs shall not be less than required by 230.79.

230.205 For services exceeding 600 V, the location of the service disconnecting means shall be located in accordance with 230.70. The service disconnecting means on private property, under certain conditions, may be permitted to be in a location that is not readily accessible.

230.205(B) Each service disconnecting means shall simultaneously disconnect all ungrounded service conductors.

230.205(C) For multibuilding, industrial installations under single management, the service disconnecting means shall be permitted to be electronically operated by a readily accessible remote-control device.

SERVICE DETAILS

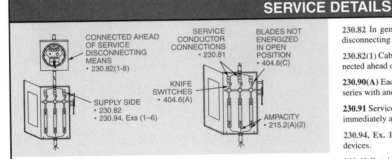

CONNECTED AHEAD OF SERVICE DISCONNECTING MEANS
• 230.82(1-8)

SERVICE CONDUCTOR CONNECTIONS
• 230.81

BLADES NOT ENERGIZED IN OPEN POSITION
• 404.6(C)

KNIFE SWITCHES
• 404.6(A)

SUPPLY SIDE
• 230.82
• 230.94, Exs (1–6)

AMPACITY
• 215.2(A)(2)

215.2(A)(3) Where a feeder carries the total service load, the ampacity of the feeder conductors shall not be less than the ampacity of the service conductors for loads of 55 A or less.

230.75 Means shall be provided for disconnecting the grounded service conductor.

230.81 Service conductors shall be attached to the disconnect by clamps, pressure connectors, or other approved means. Soldered connections are not permitted.

230.82 In general, equipment shall not be connected to supply side of service disconnecting means. See Special Conditions (1 – 9).

230.82(1) Cable limiters or other current-limiting devices are permitted to be connected ahead of service disconnecting means.

230.90(A) Each ungrounded service conductor shall have an overcurrent device in series with and not exceeding ampacity of conductor. See Ex. (1 – 5).

230.91 Service overcurrrent device shall be an integral part of disconnect or located immediately adjacent thereto.

230.94, Ex. 1 Service switch is permitted on supply side of service overcurrent devices.

240.61 Cartridge fuses and fuseholders shall be classified according to voltage and amperage ranges.

404.6(A) Single-throw knife switches shall be installed so they will not tend to close by gravity.

404.6(C) Motor-circuit switches shall be installed so blades are not energized when switch is in open position.

SERVICE GROUNDING

A. **250.52(A)(1)** Metal underground water pipe as GE.

B. **250.70** Ground clamp listed for the purpose shall be permitted to be used to connect GEC to grounding electrode.

C. **250.68(A – B)** Grounding electrode connection, where mechanical, shall be accessible. Bonding around water meters, valves, unions, etc. is required. See Exceptions for accessibility.

D. **250.64(E)** If metal enclosure is not electrically continuous, it shall be bonded at each end to grounding conductor.

E. **250.64(A – F)** GEC installation.

F. **250.64(E)** Metal GEC enclosures shall be electrically continuous from cabinet to grounding electrode.

G. Grounding electrode conductor material. (GEC). See 250.62.

H. Service-entrance equipment.

I. Neutral connection bar.

J. **250.24(A); 250.28; 250.80** GEC to also ground all metal service-entrance equipment, service raceways, and system grounded conductor (neutral).

EQUIPMENT BONDING JUMPER AND MAIN BONDING JUMPER . . .

See SERVICE GROUNDING AND BONDING.

A. **NEUTRAL SERVICE ENTRANCE CONDUCTOR** In this drawing, only the identified neutral conductor is shown. The two "ungrounded" or "hot" conductors are not shown.

B. **AC SYSTEMS TO BE GROUNDED** – 250.20 Service equipment, raceways, cable armor, cable sheaths, etc. and any service conductor that is required to be grounded by 250.20 shall be grounded per 250.

C. **METHOD OF BONDING SERVICE EQUIPMENT** – 250.92(A – B) Electrical continuity shall be assured by the method of bonding used.

D. **SERVICE EQUIPMENT ENCLOSURE** This contains the service disconnecting means.

E. **MAIN and EQUIPMENT BONDING JUMPERS ON SUPPLY SIDE** – 250.28; 250.102 The MBJ and EBJ shall be copper or other corrosion-resistant material. They shall be attached per 250.8. Per 250.102(C) and 250.28(D), the EBJ on the supply side of the service and the MBJ are sized per Table 250.66, but not less than 12½% of the largest phase conductor. Where service entrance conductors are parallel in separate raceways as permitted in 310.10(H), the size of the bonding jumper for each raceway shall be based on the size of the conductors in each raceway or cable.

F. **MAIN BONDING JUMPER** – 250.28(A – D) The MBJ is the connection between the grounded circuit conductor and the EGC at the service. See Article 100. It shall be used to connect the EGC and service equipment enclosure to the neutral within the service enclosure. The MBJ shall be a wire, bus, screw, or similar suitable conductor without a splice.

G. **EQUIPMENT GROUNDING CONDUCTOR CONNECTIONS** – 250.130 EGC connections at the service equipment shall be made for grounded and ungrounded systems. For grounded systems, the connection is made by bonding the EGC, grounded service conductor, and the GEC. For ungrounded systems, the connection is made by bonding the EGC to the GEC.

H. **METAL UNDERGROUND WATER PIPE** – 250.52(A)(1) When a metal underground water pipe is used as the grounding electrode, a supplemental electrode is required. See 250.53(D)(2).

I. **SYSTEM GROUNDING CONNECTIONS** – 250.24(A) Secondary AC systems to be grounded shall have the GEC connected at each service. The GEC is connected to the AC neutral on the supply side of the disconnecting means on the premises, preferably within the service enclosure. This connection is not to be made on the load side of the disconnecting means. See 250.24(A)(5) and 250.30(A), 250.32, 250.142.

J. **EFFECTIVE GROUNDING PATH** – 250.4(A)(5) The path to ground from circuits, equipment, and enclosures shall be permanent and electrically continuous. It shall have the capacity to safely conduct available fault currents and have low impedance so as to limit voltage to ground and to facilitate the circuit overcurrent devices.

K. **INSTALLING GROUNDING ELECTRODE CONDUCTOR** – 250.64(B) A No. 4 or larger GEC needs no protection if it is not exposed to physical damage. A No. 6 needs no protection if it follows the surface of a building and is not exposed to physical damage. A No. 8 or smaller shall be in RMC, IMC, PVC, RTRC, EMT, or cable armor.

L. **GROUNDING ELECTRODE CONDUCTOR MATERIAL** – 250.62 The GEC shall be Cu, Al, Cu-clad Al, other corrosion-resistant material, or protected against corrosion. It shall be solid or stranded, insulated, covered or bare, and shall be continuous without splices or joints. Sections of busbars are permitted to be connected together to create the GEC. Taps are permitted to the GEC where more than one service entrance enclosure is installed if such tap extends inside of such enclosure and is sized per 250.66 for largest conductors entering the enclosure. See 250.64(D).

M. **SIZE OF GROUNDING ELECTRODE CONDUCTOR** – Table 250.66 The size of the GEC is determined by the size of the largest ungrounded service entrance conductor or equivalent for parallel conductors. It need not be larger than No. 6 Cu or its equivalent in ampacity when it is the sole connection to made electrodes covered in 250.52(A)(5) and (A)(7).

N. **GROUNDING ELECTRODE CONDUCTOR** – 250.24(A) The GEC is used to connect EGCs, service equipment enclosures, and the grounded service conductor (when present) to the grounding electrode.

O. **Al or Cu-CLAD Al CONDUCTORS** – 250.64(A) These conductors are not permitted where subject to corrosive conditions or in contact with masonry or earth, and when outside shall not be terminated within 18″ of earth.

P. **EGC IN CONDUIT** – Ch 9, Note 3 to Tables When in conduit, the EGC shall be included the same as any other conductor when determining conduit fit.

Q. **CONNECTION TO ELECTRODES** – 250.70 The grounding conductor is connected to the grounding electrode by exothermic welding, listed lugs, listed pressure connectors, listed clamps, or other listed means. Solder connections are not permitted.

R. **TO GROUNDING ELECTRODE** – 250.68(A – B) The connection of the GEC to the grounding electrode shall be accessible and shall ensure a permanent and effective ground.

SIZING LOAD SIDE EBJ

What size Cu EBJ is required for a 4′ FMC with branch-circuit conductors protected by an 80 A CB?

250.102(D); Table 250.122: 80 A = No. 8 Cu
EBJ = **No. 8 Cu**

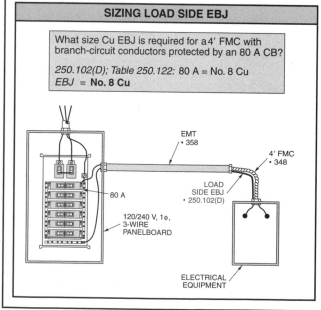

SIZING MAIN AND SUPPLY SIDE MBJs

What size Cu MBJ is required for a 120/208 V, 3ϕ, 4-wire service with three paralleled No. 500 kcmil Cu conductors?

Table 250.66: No. 500 kcmil x 3 = No. 1500 kcmil
250.28(D)(I): No. 1500 kcmil x 12½% = 187,500 cm
Ch 9, Table 8: 187,500 CM = No. 4/0 Cu
MBJ = **No. 4/0 Cu**

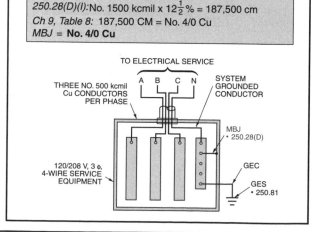

. . . EQUIPMENT BONDING JUMPER AND MAIN BONDING JUMPER

S. **NEUTRAL CONDUCTOR – 250.24(A)** The neutral (grounded conductor) shall be connected to the GEC. Per 230.75, means in the service equipment shall disconnect the neutral from premises wiring. In this case, the terminal or bus to which all grounded conductors are attached by means of pressure connectors is permitted.

T. **NEUTRAL CONDUCTOR – 250.26(2)** The neutral conductor shall be grounded for a 1ϕ, 3-wire system.

U. **BONDING OF SERVICE EQUIPMENT – 250.92(A)** Metal service equipment enclosures containing service entrance conductors, meter fittings, boxes, etc. interposed in the service raceway of armor and any conduit or armor that forms part of the grounding conductor to the service raceway shall be effectively bonded together, except as permitted by 250.80 for service raceways.

Per 250.92(B), the grounding continuity of service equipment shall be assured by (1) grounded service conductor, (2) threaded couplings or hubs, (3) threadless couplings and connections, or (4) other listed devices.

V. **SERVICE RACEWAYS AND ENCLOSURES – 250.80** Metal service raceways and enclosures shall be grounded. See Exception.

W. **GROUNDED CIRCUIT CONDUCTOR – 250.142(A)** The grounded circuit conductor can be used to ground meter enclosures, service raceways, etc. on the supply side of service disconnecting means.

X. **METER SOCKET** The utility company installs the meter into the meter socket.

Y. **THREADED CONNECTIONS – 250.92(B)(2)** Threaded couplings or hubs are acceptable means of bonding the service conduit to the meter socket.

SERVICE GROUNDING AND BONDING

DEFINITIONS

Service Equipment – All of the necessary equipment to control the supply of electrical power to a building or a structure.

Raceway – A metal or nonmetallic enclosed channel for conductors.

Grounding Conductor – The conductor that connects electrical equipment or the grounded conductor to the grounding electrode.

Grounding Electrode Conductor (GEC) – The conductor that connects the grounding electrode(s) to the grounded conductor and/or the EGC.

Equipment Grounding Conductor (EGC) – An electrical conductor that provides a low-impedance path between electrical equipment and enclosures and the system grounded conductor and GEC.

Main Bonding Jumper (MBJ) – The connection at the service equipment that ties together the EGC, the grounded conductor, and the GEC.

MASTHEAD

SERVICE PANEL

BONDING JUMPER

BUSBAR

MAIN AND EQUIPMENT BONDING JUMPERS

250.24(B) Main bonding jumper shall not be spliced.

250.28(A); 250.102(A) MBJ and EBJ shall be copper or other corrosion-resistant material.

250.28(B) If MBJ is a screw only, it shall have a green finish and be visible.

250.28(C); 250.102(B) MBJ and EBJ shall be attached per 250.8.

250.28(D); 250.102(C) EBJ on supply side of service and MBJ shall be sized per Table 250.66 but not less than 12½% of largest phase conductor. Where service-entrance conductors are paralleled in separate raceways as permitted in 310.10(4), the size of the bonding jumper for each raceway shall be based on size of conductors in each raceway. Size of the bonding jumper connected to grounding bushing for each raceway shall be based on the size of the service conductors in each raceway.

250.53(D)(2) Supplemental electrode of type specified in 250.52(A)(2) through (A)(8) when underground metal water pipe is used as grounding electrode.

250.53(G) Pipe and rod electrodes to be a minimum of 8′ long and top end of electrode to be flush or below ground level unless special protection from physical damage per 250.10. See 250.53(G).

250.68(B) To assure effective bonding of the metal piping system, bonding jumpers shall be provided around insulated joints and equipment that may need to be repaired. When used, the bonding jumpers shall be long enough to allow removal of the equipment without removing the bonding jumper.

250.8(A) Grounding electrode connection shall be by eight methods, including exothermic welding. Where used on buried electrodes, connectors shall be listed for direct burial applications. Connectors for more than one conductor shall be listed for multiple conductors.

250.70 Grounding electrode connection shall be by exothermic welding or other listed means. See (1 – 4).

Main Bonding Jumper (MBJ) – The connection between the grounded circuit conductor and the equipment grounding conductor (EGC) at the service.

Equipment Bonding Jumper (EBJ) – The connection between two or more portions of the equipment grounding conductor (EGC).

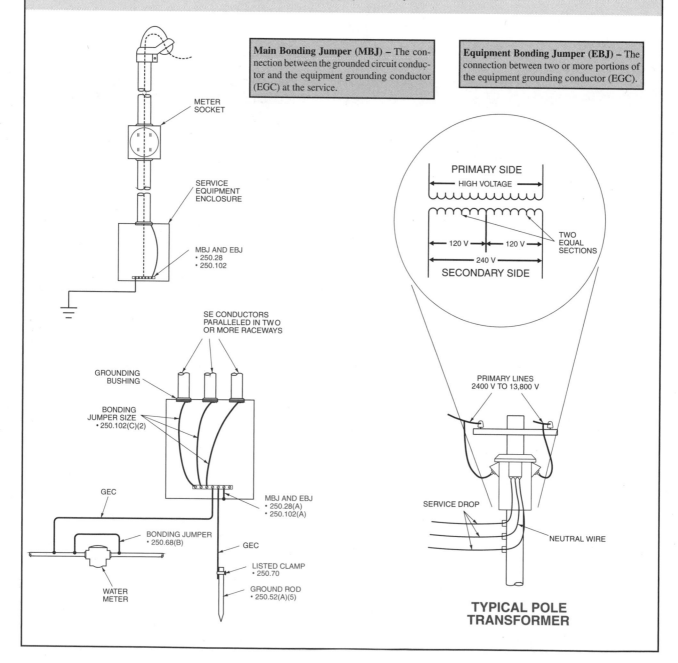

TYPICAL POLE TRANSFORMER

GROUNDED CONDUCTOR BROUGHT TO SERVICE EQUIPMENT — 250.24(C)

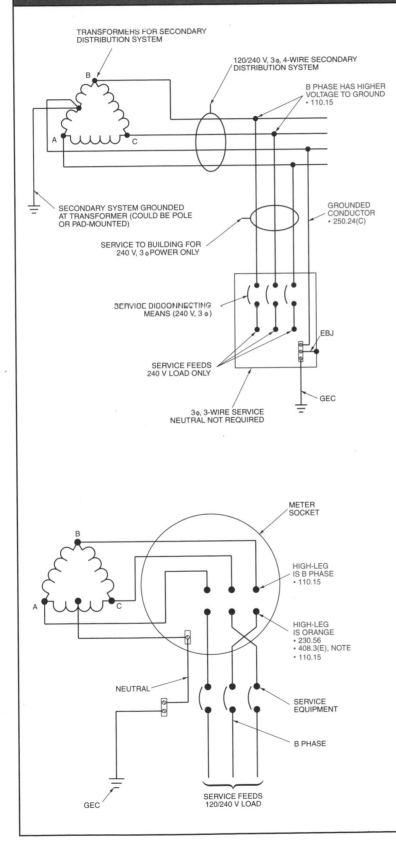

TRANSFORMERS FOR SECONDARY DISTRIBUTION SYSTEM

120/240 V, 3φ, 4-WIRE SECONDARY DISTRIBUTION SYSTEM

B PHASE HAS HIGHER VOLTAGE TO GROUND
• 110.15

SECONDARY SYSTEM GROUNDED AT TRANSFORMER (COULD BE POLE OR PAD-MOUNTED)

GROUNDED CONDUCTOR
• 250.24(C)

SERVICE TO BUILDING FOR 240 V, 3φ POWER ONLY

SERVICE DISCONNECTING MEANS (240 V, 3φ)

EBJ

SERVICE FEEDS 240 V LOAD ONLY

GEC

3φ, 3-WIRE SERVICE NEUTRAL NOT REQUIRED

METER SOCKET

HIGH-LEG IS B PHASE
• 110.15

HIGH-LEG IS ORANGE
• 230.56
• 408.3(E), NOTE
• 110.15

NEUTRAL

SERVICE EQUIPMENT

B PHASE

GEC

SERVICE FEEDS 120/240 V LOAD

110.15 High-leg or the phase conductor with the higher voltage to ground shall be identified by orange color or other effective means of each connection point where neutral is present.

250.24(C)(1 – 4) Where an AC system operating at 1000 V or less is grounded at any point, the grounded conductor shall be routed with the ungrounded conductors and connected to each service disconnecting means. This conductor shall be sized per 250.66 and, in addition, for service-phase conductors larger than 1100 kcmil Cu or 1750 kcmil Al, the grounded conductor shall not be smaller than 12½% of largest phase conductor except the grounded conductor need not be larger than the largest phase conductor.

408.3(E) High-leg of the phase conductors with the higher voltage to ground shall be the B phase.

408.3(E) High-leg to be B phase except that it may be same phase as metering equipment. On 3φ, 4-wire delta-connected systems, phase arrangement shall be such that B phase shall have the higher voltage to ground. Where additions to existing installations are made, other arrangements shall be permitted but shall be marked. See Exception.

EXAMPLE

Each service phase consists of three 600,000 cm Cu conductors in parallel. Total area of phase conductors is $3 \times 600,000$ cm, or 1,800,000 cm. The minimum size of the grounded conductor run to the service equipment is $1,800,000 \times 12\frac{1}{2}\%$, or 225,000 cm Cu, which requires a 250 kcmil conductor.

The purpose of this requirement is to ensure a low impedance path for any line to ground fault currents that could develop in the premises served. Section 250.4(A)(5) requires the path to ground shall have sufficiently low impedance to facilitate the operation of circuit protective devices in the circuit.

TRANSFORMERS

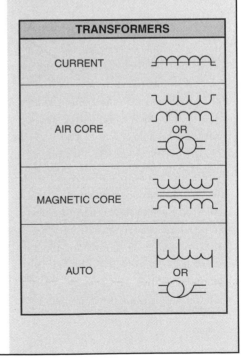

CURRENT	
AIR CORE	OR
MAGNETIC CORE	
AUTO	OR

BUILDINGS OR STRUCTURES SUPPLIED BY FEEDERS OR BRANCH CIRCUITS — 250.32

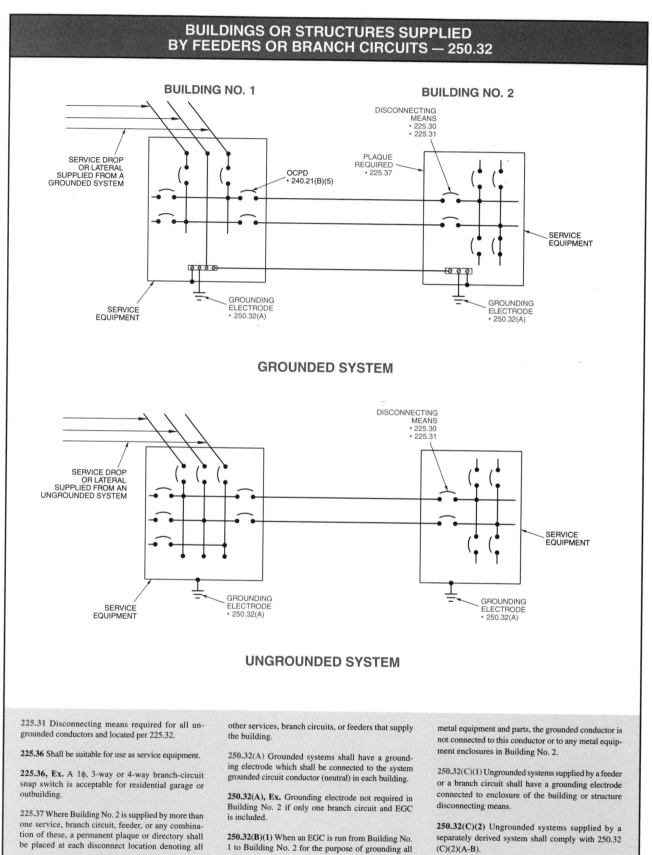

BUILDING NO. 1

BUILDING NO. 2

SERVICE DROP OR LATERAL SUPPLIED FROM A GROUNDED SYSTEM

OCPD
• 240.21(B)(5)

DISCONNECTING MEANS
• 225.30
• 225.31

PLAQUE REQUIRED
• 225.37

SERVICE EQUIPMENT

SERVICE EQUIPMENT

GROUNDING ELECTRODE
• 250.32(A)

GROUNDING ELECTRODE
• 250.32(A)

GROUNDED SYSTEM

SERVICE DROP OR LATERAL SUPPLIED FROM AN UNGROUNDED SYSTEM

DISCONNECTING MEANS
• 225.30
• 225.31

SERVICE EQUIPMENT

SERVICE EQUIPMENT

GROUNDING ELECTRODE
• 250.32(A)

GROUNDING ELECTRODE
• 250.32(A)

UNGROUNDED SYSTEM

225.31 Disconnecting means required for all un-grounded conductors and located per 225.32.

225.36 Shall be suitable for use as service equipment.

225.36, Ex. A 1φ, 3-way or 4-way branch-circuit snap switch is acceptable for residential garage or outbuilding.

225.37 Where Building No. 2 is supplied by more than one service, branch circuit, feeder, or any combina-tion of these, a permanent plaque or directory shall be placed at each disconnect location denoting all other services, branch circuits, or feeders that supply the building.

250.32(A) Grounded systems shall have a ground-ing electrode which shall be connected to the system grounded circuit conductor (neutral) in each building.

250.32(A), Ex. Grounding electrode not required in Building No. 2 if only one branch circuit and EGC is included.

250.32(B)(1) When an EGC is run from Building No. 1 to Building No. 2 for the purpose of grounding all metal equipment and parts, the grounded conductor is not connected to this conductor or to any metal equip-ment enclosures in Building No. 2.

250.32(C)(1) Ungrounded systems supplied by a feeder or a branch circuit shall have a grounding electrode connected to enclosure of the building or structure disconnecting means.

250.32(C)(2) Ungrounded systems supplied by a separately derived system shall comply with 250.32 (C)(2)(A-B).

GROUNDED CONDUCTORS FOR AC SYSTEMS — 250.26

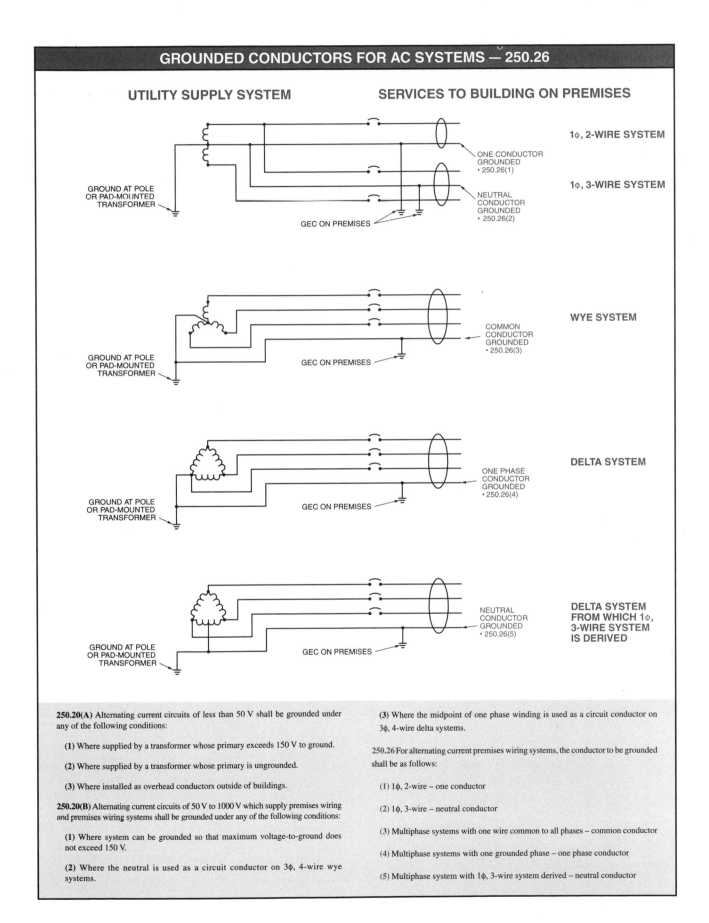

UTILITY SUPPLY SYSTEM

SERVICES TO BUILDING ON PREMISES

GROUND AT POLE OR PAD-MOUNTED TRANSFORMER

1φ, 2-WIRE SYSTEM

ONE CONDUCTOR GROUNDED
• 250.26(1)

1φ, 3-WIRE SYSTEM

NEUTRAL CONDUCTOR GROUNDED
• 250.26(2)

GEC ON PREMISES

WYE SYSTEM

COMMON CONDUCTOR GROUNDED
• 250.26(3)

GROUND AT POLE OR PAD-MOUNTED TRANSFORMER

GEC ON PREMISES

DELTA SYSTEM

ONE PHASE CONDUCTOR GROUNDED
• 250.26(4)

GROUND AT POLE OR PAD-MOUNTED TRANSFORMER

GEC ON PREMISES

NEUTRAL CONDUCTOR GROUNDED
• 250.26(5)

DELTA SYSTEM FROM WHICH 1φ, 3-WIRE SYSTEM IS DERIVED

GROUND AT POLE OR PAD-MOUNTED TRANSFORMER

GEC ON PREMISES

250.20(A) Alternating current circuits of less than 50 V shall be grounded under any of the following conditions:

(1) Where supplied by a transformer whose primary exceeds 150 V to ground.

(2) Where supplied by a transformer whose primary is ungrounded.

(3) Where installed as overhead conductors outside of buildings.

250.20(B) Alternating current circuits of 50 V to 1000 V which supply premises wiring and premises wiring systems shall be grounded under any of the following conditions:

(1) Where system can be grounded so that maximum voltage-to-ground does not exceed 150 V.

(2) Where the neutral is used as a circuit conductor on 3φ, 4-wire wye systems.

(3) Where the midpoint of one phase winding is used as a circuit conductor on 3φ, 4-wire delta systems.

250.26 For alternating current premises wiring systems, the conductor to be grounded shall be as follows:

(1) 1φ, 2-wire – one conductor

(2) 1φ, 3-wire – neutral conductor

(3) Multiphase systems with one wire common to all phases – common conductor

(4) Multiphase systems with one grounded phase – one phase conductor

(5) Multiphase system with 1φ, 3-wire system derived – neutral conductor

GROUNDING TWO TO SIX SERVICE SWITCHES FED FROM ONE SET OF SE CONDUCTORS

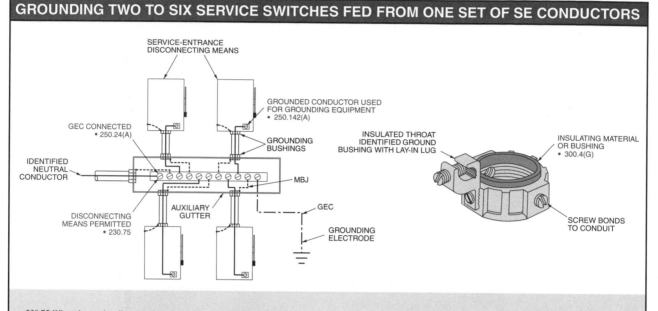

230.75 When the service disconnecting means does not disconnect the grounded conductor from premises wiring, other means shall be provided in the service equipment.

250.24(A)(1) Premises wiring supplied by an AC service that is grounded shall have a GEC connected to a grounding electrode at any accessible point from the service drop to the terminal or bus to which the grounded service conductor is connected at the service disconnecting means.

250.142(A) The grounded circuit conductor is permitted to ground noncurrent-carrying metal parts of equipment, raceways, and other enclosures at the supply side of:

(1) service disconnecting means

(2) main disconnecting means for separate buildings as per 250.32(B)

(3) disconnecting means or overcurrent devices of separately derived systems where permitted by 250.30 (A)(1).

300.4(G) Insulating material with temperature rating not less than that of installed conductors or bushing required for No. 4 or larger conductors.

SIZE OF EGCs — 250.122

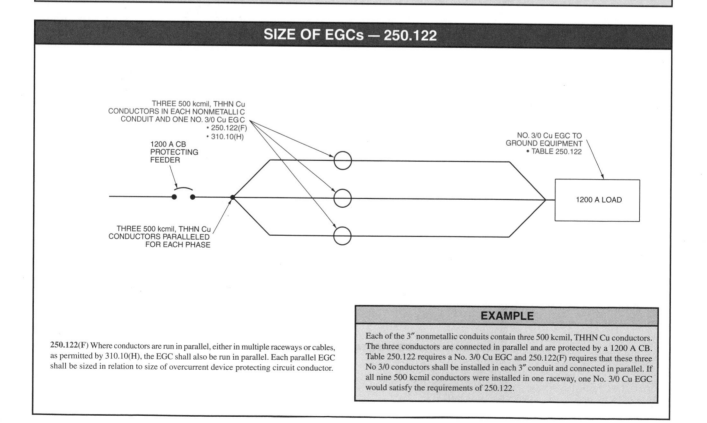

250.122(F) Where conductors are run in parallel, either in multiple raceways or cables, as permitted by 310.10(H), the EGC shall also be run in parallel. Each parallel EGC shall be sized in relation to size of overcurrent device protecting circuit conductor.

EXAMPLE

Each of the 3″ nonmetallic conduits contain three 500 kcmil, THHN Cu conductors. The three conductors are connected in parallel and are protected by a 1200 A CB. Table 250.122 requires a No. 3/0 Cu EGC and 250.122(F) requires that these three No 3/0 conductors shall be installed in each 3″ conduit and connected in parallel. If all nine 500 kcmil conductors were installed in one raceway, one No. 3/0 Cu EGC would satisfy the requirements of 250.122.

OUTLET, DEVICE, PULL AND JUNCTION BOXES, CONDUIT BODIES AND FITTINGS — 314 . . .

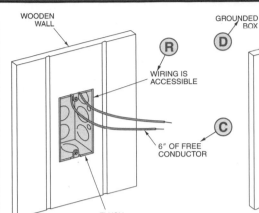

WOODEN WALL

(R) WIRING IS ACCESSIBLE

(C) 6" OF FREE CONDUCTOR

(K) FLUSH BOX

GROUNDED BOX

(D)

(L) $\frac{1}{8}''$ GAP MAXIMUM

PLASTER OR DRYWALL

FLOOR

8'

(A)

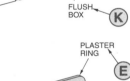

PLASTER RING (E)

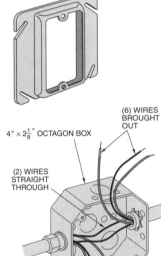

4" × 2$\frac{1}{8}$" OCTAGON BOX

(6) WIRES BROUGHT OUT

(2) WIRES STRAIGHT THROUGH

TOTAL OF (8) WIRES IN BOX

(F)

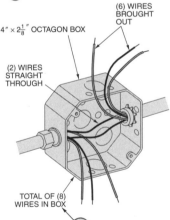

(G)

FIXTURE STUD

(9) No. 14 WIRES

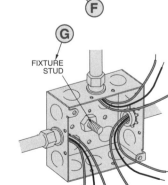

(B)

BONDING SCREW AND BOND WIRE

4" × 1$\frac{1}{8}$" SQUARE BOX

EGCs

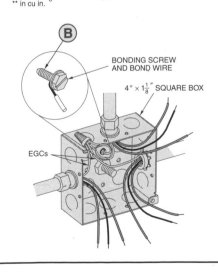

BOX FILL

Wire Size	Maximum No. of Wires* Table 314.16(A)	Volume Per Wire** Table 314.16(B)
18	14	1.5
16	12	1.75
14	10	2
12	9	2.25
10	8	2.5
8	7	3
6	4	5

* for 4" × 2$\frac{1}{8}$" round or octagonal box
** in cu in.

DEFINITIONS

100: Outlet – A point on the wiring system at which current is taken to supply utilization equipment.

100: Device – A unit of an electrical system that is intended to carry or control electric energy as its principle function.

Pull Box – A box used as a point to pull or feed electrical conductors into the raceway system.

Junction Box – A box in which splices, taps, or terminations are made.

Conduit Body – A conduit fitting that provides access to the raceway system through a removable cover at a junction or termination point.

Fitting – An electrical system accessory that performs a mechanical function.

110.13(A) Wooden plugs driven into holes in masonry, concrete, plaster, etc. shall not be used to mount electric equipment.

A. **250.110(1)** Metal boxes located within 8' vertically or 5' horizontally of ground or grounded metal objects and subject to contact by persons shall be grounded.

B. **250.148(C)** When a metal box is required to be grounded, a grounding screw or other listed grounding device (such as grounding clip) that serves no other purpose than to connect box to equipment ground wire shall be used.

C. **300.14** In general, at least 6" of free conductor shall be left at each outlet and switch point, except for conductors pulled straight through box.

314.2 Round boxes are not permitted where conduits or connectors using locknuts or bushings are to be connected.

314.3 Nonmetallic boxes shall be used only with non-metallic raceway, NMC, knob and tube wiring, ENT, or open wiring. See Exceptions.

D. **314.4** Metal boxes shall be grounded per 250 except as permitted in 250.112(I).

314.15 In damp or wet locations, boxes, conduit bodies, and fittings shall be positioned to prevent moisture from entering. Those installed in wet locations shall be listed for use in wet locations.

314.16 The volume of the box shall be calculated per 314.16(A). It shall not be less than the fill as calculated per 314.16(B). Boxes and conduit bodies containing conductors No. 4 or larger shall also comply with 314.28.

E. **314.16(A)** The volume of the original box shall be increased by cu in. marked on extension ring, plaster ring, or domed cover.

F. **314.16(A)(1)** The volume of standard boxes that are not marked with their cubic inch capacity shall be per Table 314.16(A).

314.16(A)(2) Boxes 100 cu in. or less, other than those in Table 314.16(A), and nonmetallic boxes shall be marked with their volume.

. . . OUTLET, DEVICE, PULL AND JUNCTION BOXES, CONDUIT BODIES AND FITTINGS — 314

G. 314.16(B)(1–5) Where one or more fixture studs, cable clamps, or hickeys are contained in the box, the number of conductors shall be reduced by one with an additional reduction of two wires for each strap containing one or more devices, and a further deduction of one wire for one or more ground wires for second set of EGCs passing through a box per 250.146(D).

H. 314.17(C) Nonmetallic boxes shall be suitable for the lowest temperature rated condcutor entering the box. Open wiring or concealed knob-and-tube conductors shall enter through individual holes. Conductors in flex tubing, type NM, or type UF cables shall have the outer sheath extend into the box not less than ¼″.

I. 314.17(C), Ex. Nonmetallic cables shall be secured to nonmetallic bodies so that cable cannot be pulled in or out of the box except for single *gang* boxes (not larger than 2¼″ × 4″) where cable is secured within 8″ of the box. Nonmetallic cable sheath shall extend at least ¼″ into box.

J. 314.19 Substantial support shall be provided for the device. Screws for supporting the box shall not be used for device attachment.

K. 314.20 Boxes and fittings, when installed in walls or ceilings constructed of noncombustible material, shall be permitted to be recessed no more than ¼″. Boxes in wall of combustible material must be flush with surface or project there from.

L. 314.21 Gaps in noncombustible surfaces, such as plaster, drywall, or plasterboard, shall be repaired so that they do not exceed ⅛″ at the edge of the box.

314.22 Boxes and extension rings shall be permitted for surface extensions from a box of concealed wiring system. The box or ring shall be securely fastened to the outlet box and grounded per Article 250. See Exception.

M. 314.23(A) Boxes shall be securely and rigidly fastened to the surface on which they are mounted.

N. 314.23(B) Boxes shall be rigidly supported from a structural member either directly or by using an approved metallic or wood brace.

O. 370.14(B)(2) Wood brace shall have a nominal cross section of 1″ × 2″.

P. 314.24(A – B) Outlet boxes that do not contain ventilation equipment or devices shall have a minimum internal depth of ½″.

Q. 314.25(C) Covers through which flexible cord pendants pass shall have smoothly rounded edges or bushings. Hard rubber or composition bushings are not permitted.

314.27(A) Boxes for the support of lighting fixtures shall be designed for the support and installed per 314.23.

314.28 Boxes and conduit bodies used as pull or junction boxes shall comply with 314.28(A – E).

R. 314.29 Boxes and conduit bodies shall be installed so that wiring within them can be rendered accessible without damage to the building. Listed boxes shall be permitted to be covered by gravel, soil, and light aggregate under some conditions. See Exception.

314.40(A) Metal boxes, conduit bodies, and fittings shall be corrosion resistant or similarly treated to prevent corrosion.

S. 314.40(D) A tapped hole, or equivalent means, shall be provided in each metal box to connect an EGC.

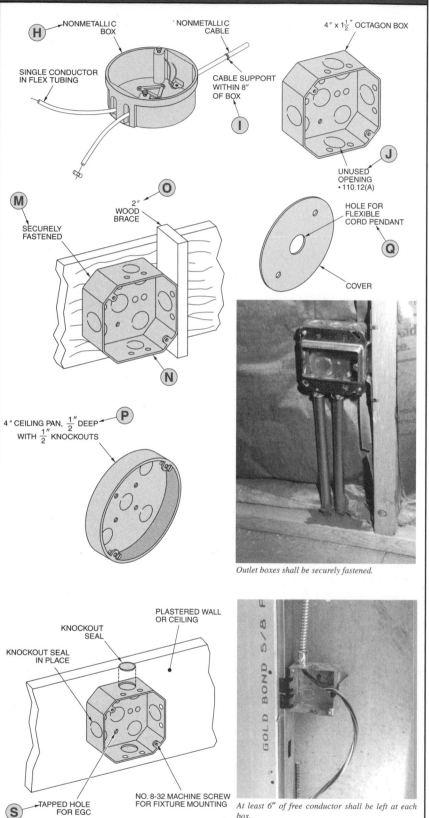

Outlet boxes shall be securely fastened.

At least 6″ of free conductor shall be left at each box.

CIRCUIT BREAKERS

FUSE AND ITCB STANDARD AMPERE RATINGS — 240.6(A)

15, 20, 25, 30, 35, 40, 45, 50, 60, 70, 80, 90, 100, 110, 125, 150, 175, 200, 225, 250, 300, 350, 400, 450, 500, 600, 700, 800, 1000, 1200, 1600, 2000, 2500, 3000, 4000, 5000, and 6000 A.

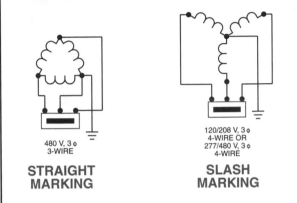

480 V, 3 φ
3-WIRE

STRAIGHT MARKING

120/208 V, 3 φ
4-WIRE OR
277/480 V, 3 φ
4-WIRE

SLASH MARKING

DEFINITIONS

Circuit Breaker (CB) – A device which opens and closes a circuit by nonautomatic means and automatically opens on a predetermined overcurrent within its rating without damaging itself.

Adjustable Trip Circuit Breaker (ATCB) – A CB which can be set to trip at various values of current, time, or both.

Instantaneous Trip Circuit Breaker (ITB) – A CB with no delay purposely introduced in the tripping action of the device.

Inverse Time Circuit Breaker (ITCB) – A CB with an intentional delay between the time when the fault or overload is sensed and the time when the CB operates.

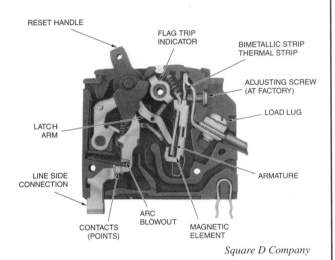

BQ1B020
I-T-E Circuit Breaker

INTERRUPTING RATING
MAX. RMS. SYM. AMPS
10,000   120/240 V.A.C.
TYPE BQ   40°C
UND. LAB. INC. LIST. ®
CIR. BRK. 1 POLE UNIT
ISSUE LM-9858
HACR TYPE
SIEMENS-ALLIS, INC.

CU/AL 60/75°C WIRE
AWG 10-8
IN. LBS. 25

RESET HANDLE
FLAG TRIP INDICATOR
BIMETALLIC STRIP THERMAL STRIP
ADJUSTING SCREW (AT FACTORY)
LOAD LUG
LATCH ARM
LINE SIDE CONNECTION
ARMATURE
CONTACTS (POINTS)
ARC BLOWOUT
MAGNETIC ELEMENT

Square D Company

110.9 Equipment intended to interrupt current at fault levels shall have an interrupting rating not less than the nominal voltage and current available at line terminals of equipment.

When checking interrupting rating of a CB, also check the voltage rating because generally the interrupting current (IC) is lower as the voltage rating is higher. For example:

at 240 V – 65,000 A IC

at 480 V – 25,000 A IC

at 600 V – 18,000 A IC

If the foregoing is not evaluated, this can result in a dangerous misapplication of the breaker insofar as interrupting rating is concerned.

110.10 CBs shall be selected and coordinated to clear a fault without extensive damage to the electrical equipment of the circuit.

240.8 CBs shall not be connected in parallel unless they are factory-assembled in parallel and listed as a unit.

A. 240.81 CBs shall clearly indicate when open (OFF) or closed (ON). When CB handles operate vertically, the up position shall be ON.

240.82 CBs shall be designed so that any alteration of the trip point or time for operation shall require dismantling or breaking the seal.

B. 240.83(A) Ampere ratings on CBs shall be durable and visible after installation. The removal of the trim or cover is permitted to see the markings.

C. 240.83(B) Ampere ratings shall be marked into handles of CBs rated at 100 A or less and 600 V or less.

D. 240.83(C) Interrupting ratings shall be marked into handles of CBs with interrupting ratings other than 5000 A. Interrupting rating marking is not required for CBs used for supplementary protection.

240.83(D) CBs used to switch 120 V or 277 V fluorescent lighting circuits shall be listed and marked SWD or HID. CBs used to switch HID lighting circuits shall be listed and marked HID.

E. 240.83(E) Voltage ratings shall be marked on all CBs.

240.85 CBs with straight marking can be used in neutral grounded, grounded wye, and ungrounded or grounded delta systems. CBs with slash markings can be used only in grounded neutral systems.

404.2(B) CBs and switches shall not disconnect the grounded conductor of a circuit.

404.2(B), Ex. CBs are permitted to disconnect the grounded conductor when all circuit conductors are disconnected at the same time. CBs are permitted to disconnect the grounded conductor when the device allows only the ungrounded conductors to be opened first.

404.3(A) CBs shall be externally operable and installed in an enclosure listed for the intended use.

404.8(A) CBs used as switches shall be located so that they may be operated for a readily accessible location. The center of the grip of the operating handle shall not be over 6′-7″ from the floor or platform. See 404.8(A), Ex. 1, 2, and 3.

GROUND-FAULT CIRCUIT INTERRUPTERS (GFCIs) — DWELLINGS

210.8(A) All 125 V, 1φ, 15 A and 20 A receptacles installed in dwelling units as specified in 210.8(A) (1 – 8) shall have GFCI protection for personnel.

210.8(A)(1) Bathrooms.

210.8(A)(2) Garages and grade-level areas of storage or work buildings.

210.8(A)(3) Outdoors. See Exception.

210.8(A)(4) Crawl spaces at or below grade.

210.8(A)(5) Unfinished basements. See Exception.

210.8(A)(6) Kitchens where the receptacles serve countertop surfaces.

210.8(A)(7) Sinks, in areas other than kitchens, where installed within 6′ of the outside edge of the sink.

210.8(A)(8) Boathouses.

Ground-Fault Circuit Interrupter (GFCI) – A device intended for the protection of personnel whose function is to de-energize a circuit within established time when current to ground exceeds a predetermined value which is less than that required to operate protective device of supply circuit.

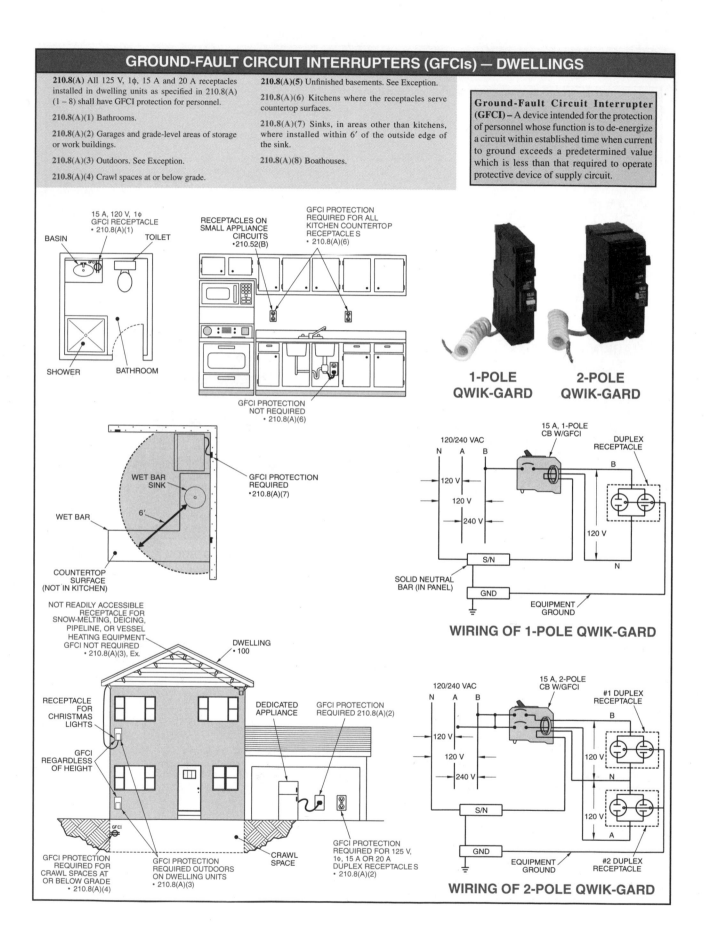

FUSETRON BOX COVER UNITS

SOU SOW SOX SOY SSU SSW SSX SSY SOY-B

SRU SRW SRX SRY SSY-I SSY-RL SCY STY SKA

Covers fit standard outlet or switch boxes. Units are UL® listed and are made with Edison-base fuseholder. Units without switch for motor are ¾ HP or smaller. Units with switch for motors are HP or smaller. Do not use on DC.

240.24(A) Overcurrent devices required to be readily accessible. See Applications (1 – 4).

240.50(D) There shall be no exposed energized parts after fuses or fuses and adapters are installed.

240.52 Edison-base fuseholders shall be installed only where made to accept Type S fuses by use of adapters.

PLUG FUSES, FUSEHOLDERS, AND ADAPTERS — 240

A. **240.30(A – B)** In general, overcurrent devices shall be enclosed in cutout boxes or cabinets and protected from physical damage.

240.32 Enclosures for overcurrent devices in damp or wet locations shall be mounted so there is a minimum of ¼″ air space between the enclosure and wall. Cabinets or cutout boxes installed in wet locations shall be weatherproof. See 312.2.

B. **240.33** Enclosures shall be mounted in vertical position except where impracticable and in compliance with 240.81.

240.50(A)(1 – 2) Plug fuses shall not be used in circuits over 125 V between conductors except in grounded circuits at not more than 150 V to ground. See 240.50(A)(2).

C. **240.50(B)** Each fuse, fuseholder, and adapter shall be marked with amp rating.

D. **240.50(C)** Fuses 15 A or less shall have hexagonal top or cap.

240.50(E) The screw shell of plug-type fuseholder shall be connected to load side of circuit.

E. **240.51(A)** Edison-base plug fuses shall not be classified over 125 V and 0 A – 30 A.

240.51(B) Plug fuses of Edison-base type to be used only as replacements where there is no evidence of overfusing or tampering.

F. **240.53(A)** Type S fuses shall be plug type and classified not over 125 V and 0 – 15, 16 – 20, and 21 – 30 A.

240.53(B) Type S fuses shall not be interchangeable with lower amp classification and designed to be used only with Type S fuseholder or Type S adapter inserted.

G. **240.54(A)** Type S adapters to fit Edison-base fuseholders.

240.54(B) Type S fuseholders and adapters are only for Type S fuses.

240.54(C) When once inserted, Type S adapters are to be nonremovable.

240.54(D) Type S fuseholders and adapters are designed to make tampering or shunting difficult.

240.54(E) Type S fuses, fuseholders, and adapters to be standardized by manufacturers to permit interchangeability regardless of the manufacturer.

H. **408.30** All panelboards shall have a rating not less than the minimum feeder capacity required for the load calculated in accordance with 220, Parts III, IV, or V as applicable.

I. **408.37** Panelboards installed in wet or damp locations must comply with the provisions of 312.2.

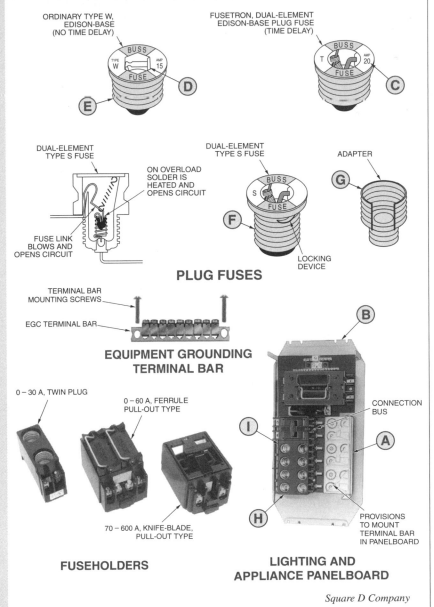

PLUG FUSES

EQUIPMENT GROUNDING TERMINAL BAR

FUSEHOLDERS

LIGHTING AND APPLIANCE PANELBOARD

Square D Company

OVERCURRENT PROTECTION — CARTRIDGE FUSES . . .

Fuse – An overcurrent protection device with a fusible link that melts and opens the circuit when an overload condition or short circuit occurs.

DUAL-ELEMENT, TIME-DELAY CARTRIDGE FUSE

The dual-element, time-delay fuse is designed to provide a time delay in the harmless low overload range and also to be fast-opening when exposed to dangerous short circuits. This fuse contains two fusible elements in series and the magnitude of a potentially damaging current flow determines which one of these elements opens. The thermal cutout element will open at currents of up to about 500% of the fuse rating (overload range), whereas the short-circuit element will open at 500% or more of the fuse rating.

DUAL-ELEMENT, TIME-DELAY CURRENT-LIMITING FUSE

Dual-element, current-limiting fuses function similar to regular dual-element, time-delay fuses except that they are faster operating in short-circuit range, thus are more current-limiting. The short-circuit element is normally made of silver, with a quartz sand arc quenching filter.

SINGLE-ELEMENT, CURRENT-LIMITING FUSE

Single-element, current-limiting fuses function similar to regular dual-element, time-delay low-overload range. This fuse has lowest let-through values in comparison with other fuses. It is used to protect circuit components having inadequate interrupting, bracing, or withstand rating, such as low IC breakers.

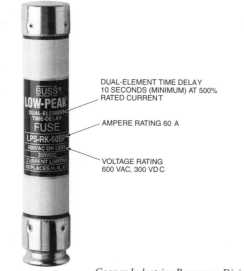

DUAL-ELEMENT TIME DELAY
10 SECONDS (MINIMUM) AT 500%
RATED CURRENT

AMPERE RATING 60 A

VOLTAGE RATING
600 VAC, 300 VDC

Cooper Industries Bussmann Division

Low-Peak® dual-element TDFs provide long time-delay for temporary motor startup.

110.9 Fuses shall have an interrupting rating not less than the voltage and the current that must be interrupted.

110.10 Fuses shall be selected and coordinated to clear fault without extensive damage to electrical equipment of the circuit.

240.1, Note Overcurrent protection opens circuit when current reaches a value that can damage either equipment or conductors. Overcurrent devices are selected based on circuit impedance, short circuit, and withstand ratings. Equipment intended to interrupt current at fault levels shall have sufficient interrupting rating. See 110.9 and 110.10.

240.3 Equipment shall be protected against overcurrent. See Table 240.3.

240.4 Conductors shall be protected at all ampacities per 310.15 except where ampacity of conductor and standard rating of fuse do not correspond, then next larger fuse may be used only up to 800 A per 240.4(B)(C).

240.6(A) Standard ratings for fuses and ITCBs shall be: 15, 20, 25, 30, 35, 40, 45, 50, 60, 70, 80, 90, 100, 110, 125, 150, 175, 200, 225, 250, 300, 350, 400, 450, 500, 600, 700, 800, 1000, 1200, 1600, 2000, 2500, 3000, 4000, 5000, and 6000 A. Additional standard ratings for fuses are: 1, 3, 6, 10, and 601 A.

240.8 Fuses shall not be connected in parallel unless they are factory assembled and listed as a unit.

240.15(A) Fuses shall be connected in series with each ungrounded conductor.

240.21 Overcurrent device connected where conductor to be protected receives supply per 240.21(A-H).

240.22 In general, overcurrent devices shall not be placed in series with any grounded conductor. See Applications for use in (2) motor overload protection and (1) simultaneous disconnection.

240.24(A) Fuses shall be readily accessible except: (1) busways – 368.17(C); (2) supplemental – 240.10; (3) OCPD – 225.40 and 230.92; (4) utilization equipment.

240.30 Fuses shall be installed in enclosures, in cabinets, or cutout boxes.

240.32 In damp or wet locations, fuses shall be in enclosures identified for such use and shall have ¼″ air space between enclosure and mounting surface. See 312.2.

240.33 Enclosures shall be mounted in vertical position except where impracticable and in compliance with 240.81.

240.40 Disconnect required on supply side of cartridge fuses where accessible to other than qualified person except: (1) service as per 230.82; (2) group operation of motors per 430.112, Ex., and fixed space heating per 424.22(C).

240.60(A) In general, cartridge fuses and fuseholders of the 300 V type are not permitted to be installed in circuits of over 300 V between conductors.

240.60(A)(1) Cartridge fuses and fuseholders are permitted in 1φ, line-to-neutral circuits supplied from a solidly grounded neutral system. However, fuses and fuseholders are permitted when the line-to-neutral voltage does not exceed 300 V.

240.60(B) Cartridge fuseholders (0 A-6000 A) shall be so designed to make it difficult to put fuse of given class into fuseholder for current lower or voltage higher than intended. Current-limiting fuseholders shall permit insertion only of current-limiting fuses.

240.60(C) Fuses shall be marked to show: (1) amp rating; (2) volt rating; (3) interrupting rating if other than 10,000 A; (4) current-limiting where applicable except where used as supplementary protection; (5) trademark of manufacturer.

240.61 Classification of cartridge fuses and fuseholders shall be on a basis of amperage and voltage. Fuses of a higher rating are permitted to be used on lower rated circuits, but fuses of a lower rating are not permitted to be used with higher voltages. 600 V rated fuses could be used on a 277 V circuit, but 250 V rated fuses could not.

240.101(A) Fuse rating in continuous amps shall not exceed 3 times amp rating of conductor. For CBs, long-time trip setting or minimum-trip setting of an electronically-activated fuse shall not exceed 6 times ampacity of conductor.

240.101(B) Conductors tapped to a feeder can be protected by the feeder OCPD where the OCPD also protects the tap conductor.

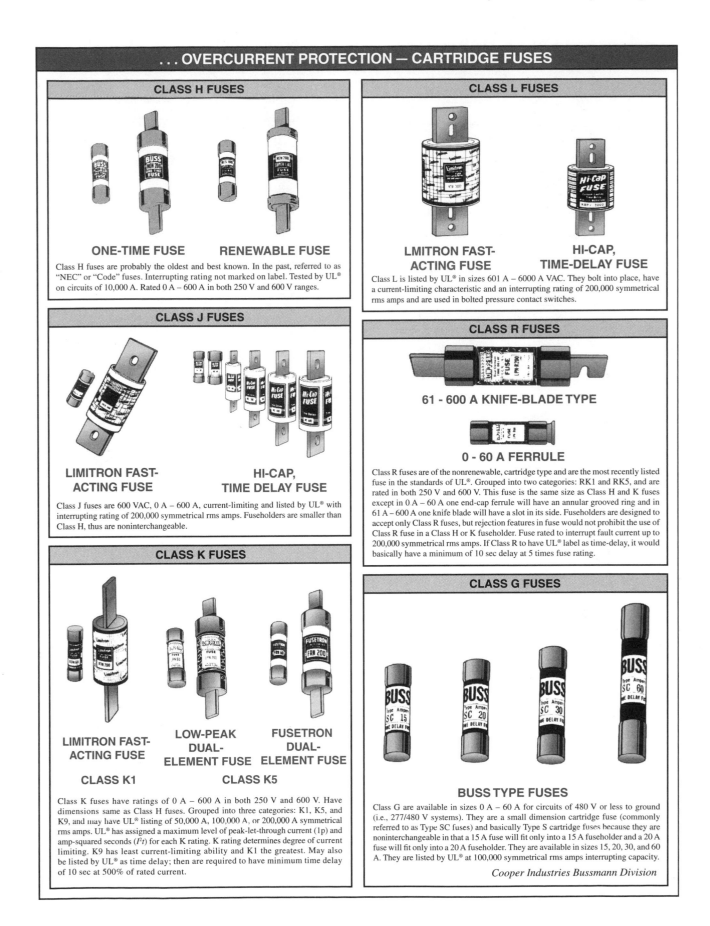

. . . OVERCURRENT PROTECTION — CARTRIDGE FUSES

CLASS H FUSES

ONE-TIME FUSE **RENEWABLE FUSE**

Class H fuses are probably the oldest and best known. In the past, referred to as "NEC" or "Code" fuses. Interrupting rating not marked on label. Tested by UL® on circuits of 10,000 A. Rated 0 A – 600 A in both 250 V and 600 V ranges.

CLASS J FUSES

LIMITRON FAST-ACTING FUSE **HI-CAP, TIME DELAY FUSE**

Class J fuses are 600 VAC, 0 A – 600 A, current-limiting and listed by UL® with interrupting rating of 200,000 symmetrical rms amps. Fuseholders are smaller than Class H, thus are noninterchangeable.

CLASS K FUSES

LIMITRON FAST-ACTING FUSE **LOW-PEAK DUAL-ELEMENT FUSE** **FUSETRON DUAL-ELEMENT FUSE**

CLASS K1 **CLASS K5**

Class K fuses have ratings of 0 A – 600 A in both 250 V and 600 V. Have dimensions same as Class H fuses. Grouped into three categories: K1, K5, and K9, and may have UL® listing of 50,000 A, 100,000 A, or 200,000 A symmetrical rms amps. UL® has assigned a maximum level of peak-let-through current (1p) and amp-squared seconds (I^2t) for each K rating. K rating determines degree of current limiting. K9 has least current-limiting ability and K1 the greatest. May also be listed by UL® as time delay; then are required to have minimum time delay of 10 sec at 500% of rated current.

CLASS L FUSES

LMITRON FAST-ACTING FUSE **HI-CAP, TIME-DELAY FUSE**

Class L is listed by UL® in sizes 601 A – 6000 A VAC. They bolt into place, have a current-limiting characteristic and an interrupting rating of 200,000 symmetrical rms amps and are used in bolted pressure contact switches.

CLASS R FUSES

61 - 600 A KNIFE-BLADE TYPE

0 - 60 A FERRULE

Class R fuses are of the nonrenewable, cartridge type and are the most recently listed fuse in the standards of UL®. Grouped into two categories: RK1 and RK5, and are rated in both 250 V and 600 V. This fuse is the same size as Class H and K fuses except in 0 A – 60 A one end-cap ferrule will have an annular grooved ring and in 61 A – 600 A one knife blade will have a slot in its side. Fuseholders are designed to accept only Class R fuses, but rejection features in fuse would not prohibit the use of Class R fuse in a Class H or K fuseholder. Fuse rated to interrupt fault current up to 200,000 symmetrical rms amps. If Class R to have UL® label as time-delay, it would basically have a minimum of 10 sec delay at 5 times fuse rating.

CLASS G FUSES

BUSS TYPE FUSES

Class G are available in sizes 0 A – 60 A for circuits of 480 V or less to ground (i.e., 277/480 V systems). They are a small dimension cartridge fuse (commonly referred to as Type SC fuses) and basically Type S cartridge fuses because they are noninterchangeable in that a 15 A fuse will fit only into a 15 A fuseholder and a 20 A fuse will fit only into a 20 A fuseholder. They are available in sizes 15, 20, 30, and 60 A. They are listed by UL® at 100,000 symmetrical rms amps interrupting capacity.

Cooper Industries Bussmann Division

SHORT CIRCUIT AND OVERLOAD PROTECTION

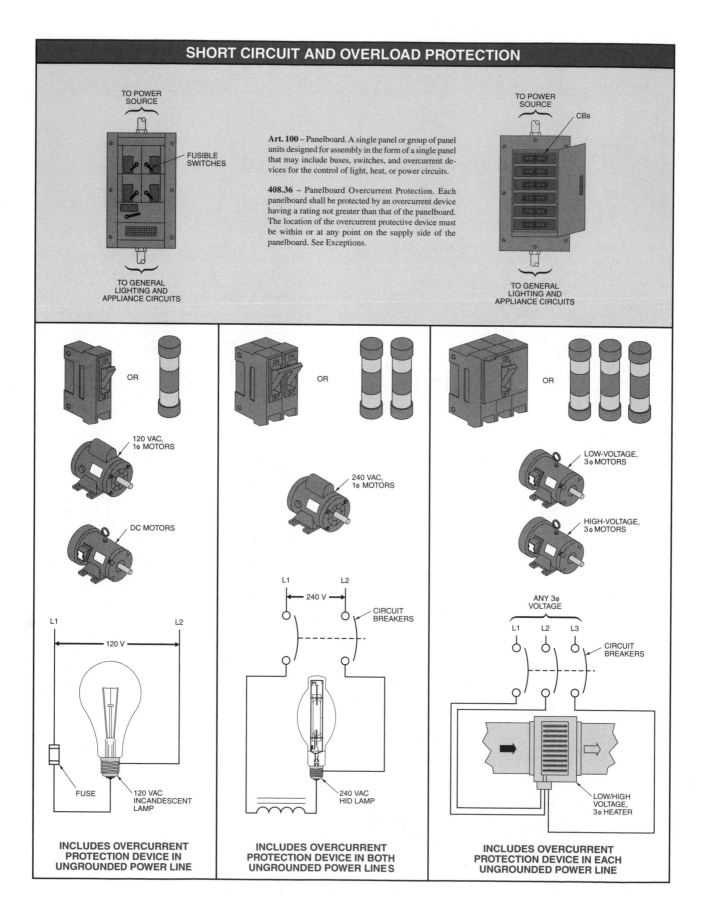

TO POWER SOURCE

FUSIBLE SWITCHES

TO GENERAL LIGHTING AND APPLIANCE CIRCUITS

Art. 100 – Panelboard. A single panel or group of panel units designed for assembly in the form of a single panel that may include buses, switches, and overcurrent devices for the control of light, heat, or power circuits.

408.36 – Panelboard Overcurrent Protection. Each panelboard shall be protected by an overcurrent device having a rating not greater than that of the panelboard. The location of the overcurrent protective device must be within or at any point on the supply side of the panelboard. See Exceptions.

TO POWER SOURCE

CBs

TO GENERAL LIGHTING AND APPLIANCE CIRCUITS

OR

120 VAC, 1ϕ MOTORS

DC MOTORS

L1   L2
120 V

FUSE   120 VAC INCANDESCENT LAMP

INCLUDES OVERCURRENT PROTECTION DEVICE IN UNGROUNDED POWER LINE

OR

240 VAC, 1ϕ MOTORS

L1   L2
240 V
CIRCUIT BREAKERS

240 VAC HID LAMP

INCLUDES OVERCURRENT PROTECTION DEVICE IN BOTH UNGROUNDED POWER LINES

OR

LOW-VOLTAGE, 3ϕ MOTORS

HIGH-VOLTAGE, 3ϕ MOTORS

ANY 3ϕ VOLTAGE
L1   L2   L3
CIRCUIT BREAKERS

LOW/HIGH VOLTAGE, 3ϕ HEATER

INCLUDES OVERCURRENT PROTECTION DEVICE IN EACH UNGROUNDED POWER LINE

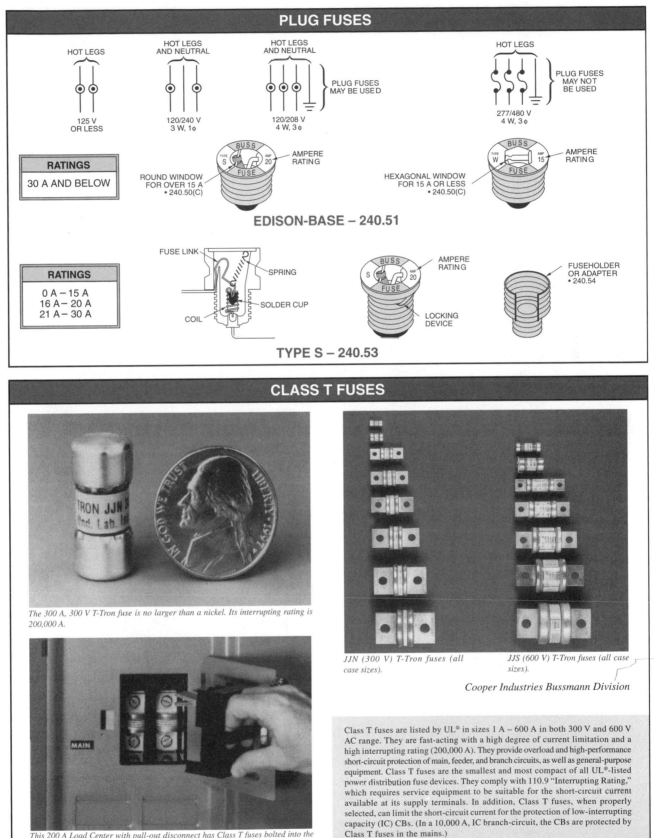

PLUG FUSES

HOT LEGS

125 V
OR LESS

HOT LEGS
AND NEUTRAL

120/240 V
3 W, 1φ

HOT LEGS
AND NEUTRAL

120/208 V
4 W, 3φ

PLUG FUSES
MAY BE USED

HOT LEGS

277/480 V
4 W, 3φ

PLUG FUSES
MAY NOT
BE USED

RATINGS

30 A AND BELOW

BUSS
TYPE S • FUSE • AMP 20

ROUND WINDOW
FOR OVER 15 A
• 240.50(C)

AMPERE
RATING

BUSS
TYPE W • FUSE • AMP 15

HEXAGONAL WINDOW
FOR 15 A OR LESS
• 240.50(C)

AMPERE
RATING

EDISON-BASE – 240.51

RATINGS

0 A – 15 A
16 A – 20 A
21 A – 30 A

FUSE LINK

SPRING

SOLDER CUP

COIL

BUSS
S • FUSE • AMP 20

AMPERE
RATING

LOCKING
DEVICE

FUSEHOLDER
OR ADAPTER
• 240.54

TYPE S – 240.53

CLASS T FUSES

The 300 A, 300 V T-Tron fuse is no larger than a nickel. Its interrupting rating is 200,000 A.

This 200 A Load Center with pull-out disconnect has Class T fuses bolted into the panelboard.

JJN (300 V) T-Tron fuses (all case sizes).

JJS (600 V) T-Tron fuses (all case sizes).

Cooper Industries Bussmann Division

Class T fuses are listed by UL® in sizes 1 A – 600 A in both 300 V and 600 V AC range. They are fast-acting with a high degree of current limitation and a high interrupting rating (200,000 A). They provide overload and high-performance short-circuit protection of main, feeder, and branch circuits, as well as general-purpose equipment. Class T fuses are the smallest and most compact of all UL®-listed power distribution fuse devices. They comply with 110.9 "Interrupting Rating," which requires service equipment to be suitable for the short-circuit current available at its supply terminals. In addition, Class T fuses, when properly selected, can limit the short-circuit current for the protection of low-interrupting capacity (IC) CBs. (In a 10,000 A, IC branch-circuit, the CBs are protected by Class T fuses in the mains.)

ARMORED CABLE (AC) — 320

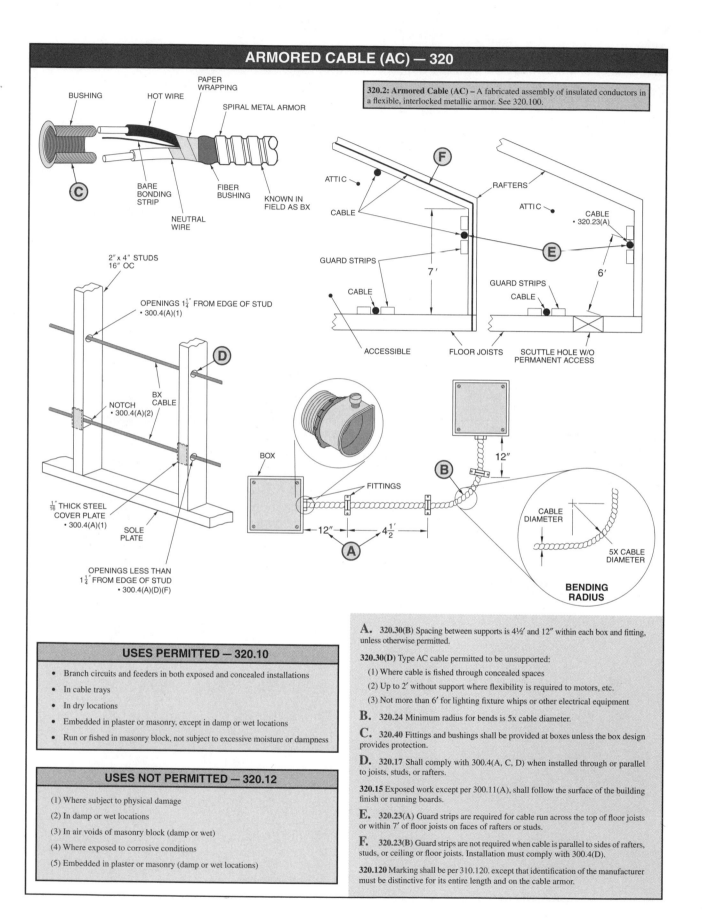

320.2: Armored Cable (AC) – A fabricated assembly of insulated conductors in a flexible, interlocked metallic armor. See 320.100.

BUSHING
HOT WIRE
PAPER WRAPPING
SPIRAL METAL ARMOR
BARE BONDING STRIP
NEUTRAL WIRE
FIBER BUSHING
KNOWN IN FIELD AS BX
C

ATTIC
CABLE
GUARD STRIPS
CABLE
ACCESSIBLE
7'
F
RAFTERS
ATTIC
CABLE • 320.23(A)
GUARD STRIPS
CABLE
FLOOR JOISTS
SCUTTLE HOLE W/O PERMANENT ACCESS
6'
E

2" x 4" STUDS 16" OC
OPENINGS 1¼" FROM EDGE OF STUD • 300.4(A)(1)
D
NOTCH • 300.4(A)(2)
BX CABLE
1/16" THICK STEEL COVER PLATE • 300.4(A)(1)
SOLE PLATE
OPENINGS LESS THAN 1¼" FROM EDGE OF STUD • 300.4(A)(D)(F)

BOX
FITTINGS
12"
4½'
A
12"
B
CABLE DIAMETER
5X CABLE DIAMETER
BENDING RADIUS

USES PERMITTED — 320.10

• Branch circuits and feeders in both exposed and concealed installations

• In cable trays

• In dry locations

• Embedded in plaster or masonry, except in damp or wet locations

• Run or fished in masonry block, not subject to excessive moisture or dampness

USES NOT PERMITTED — 320.12

(1) Where subject to physical damage

(2) In damp or wet locations

(3) In air voids of masonry block (damp or wet)

(4) Where exposed to corrosive conditions

(5) Embedded in plaster or masonry (damp or wet locations)

A. 320.30(B) Spacing between supports is 4½' and 12" within each box and fitting, unless otherwise permitted.

320.30(D) Type AC cable permitted to be unsupported:

(1) Where cable is fished through concealed spaces

(2) Up to 2' without support where flexibility is required to motors, etc.

(3) Not more than 6' for lighting fixture whips or other electrical equipment

B. 320.24 Minimum radius for bends is 5x cable diameter.

C. 320.40 Fittings and bushings shall be provided at boxes unless the box design provides protection.

D. 320.17 Shall comply with 300.4(A, C, D) when installed through or parallel to joists, studs, or rafters.

320.15 Exposed work except per 300.11(A), shall follow the surface of the building finish or running boards.

E. 320.23(A) Guard strips are required for cable run across the top of floor joists or within 7' of floor joists on faces of rafters or studs.

F. 320.23(B) Guard strips are not required when cable is parallel to sides of rafters, studs, or ceiling or floor joists. Installation must comply with 300.4(D).

320.120 Marking shall be per 310.120, except that identification of the manufacturer must be distinctive for its entire length and on the cable armor.

METAL-CLAD CABLE (MC) — 330

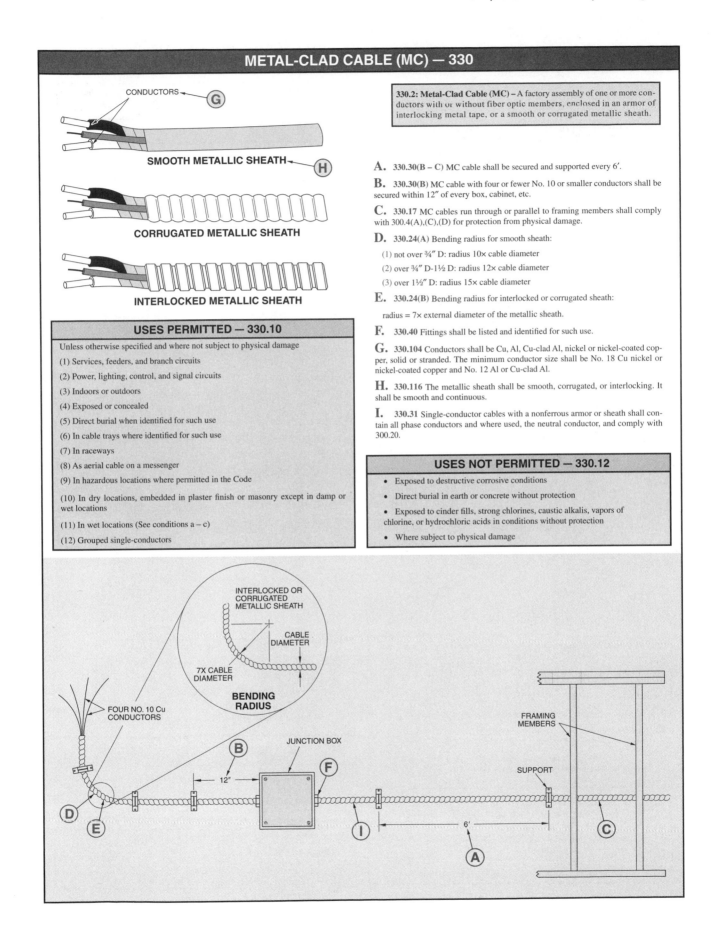

CONDUCTORS — (G)

SMOOTH METALLIC SHEATH (H)

CORRUGATED METALLIC SHEATH

INTERLOCKED METALLIC SHEATH

330.2: Metal-Clad Cable (MC) – A factory assembly of one or more conductors with or without fiber optic members, enclosed in an armor of interlocking metal tape, or a smooth or corrugated metallic sheath.

USES PERMITTED — 330.10

Unless otherwise specified and where not subject to physical damage

(1) Services, feeders, and branch circuits

(2) Power, lighting, control, and signal circuits

(3) Indoors or outdoors

(4) Exposed or concealed

(5) Direct burial when identified for such use

(6) In cable trays where identified for such use

(7) In raceways

(8) As aerial cable on a messenger

(9) In hazardous locations where permitted in the Code

(10) In dry locations, embedded in plaster finish or masonry except in damp or wet locations

(11) In wet locations (See conditions a – c)

(12) Grouped single-conductors

A. **330.30(B – C)** MC cable shall be secured and supported every 6′.

B. **330.30(B)** MC cable with four or fewer No. 10 or smaller conductors shall be secured within 12″ of every box, cabinet, etc.

C. **330.17** MC cables run through or parallel to framing members shall comply with 300.4(A),(C),(D) for protection from physical damage.

D. **330.24(A)** Bending radius for smooth sheath:

(1) not over ¾″ D: radius 10× cable diameter

(2) over ¾″ D-1½ D: radius 12× cable diameter

(3) over 1½″ D: radius 15× cable diameter

E. **330.24(B)** Bending radius for interlocked or corrugated sheath:

radius = 7× external diameter of the metallic sheath.

F. **330.40** Fittings shall be listed and identified for such use.

G. **330.104** Conductors shall be Cu, Al, Cu-clad Al, nickel or nickel-coated copper, solid or stranded. The minimum conductor size shall be No. 18 Cu nickel or nickel-coated copper and No. 12 Al or Cu-clad Al.

H. **330.116** The metallic sheath shall be smooth, corrugated, or interlocking. It shall be smooth and continuous.

I. **330.31** Single-conductor cables with a nonferrous armor or sheath shall contain all phase conductors and where used, the neutral conductor, and comply with 300.20.

USES NOT PERMITTED — 330.12

- Exposed to destructive corrosive conditions
- Direct burial in earth or concrete without protection
- Exposed to cinder fills, strong chlorines, caustic alkalis, vapors of chlorine, or hydrochloric acids in conditions without protection
- Where subject to physical damage

INTERLOCKED OR CORRUGATED METALLIC SHEATH

CABLE DIAMETER

7X CABLE DIAMETER

BENDING RADIUS

FOUR NO. 10 Cu CONDUCTORS

JUNCTION BOX

FRAMING MEMBERS

SUPPORT

12″

6′

NONMETALLIC-SHEATHED CABLE (NM) — 334

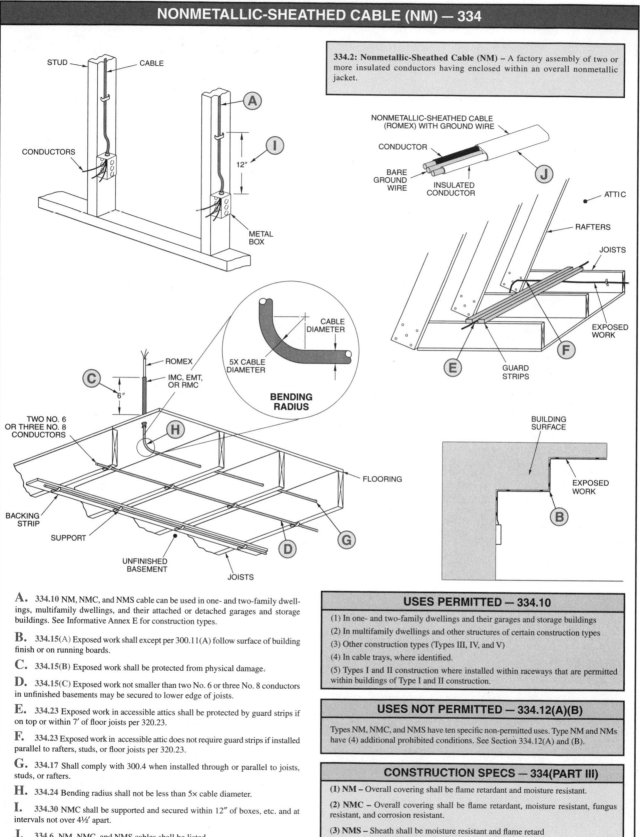

334.2: Nonmetallic-Sheathed Cable (NM) – A factory assembly of two or more insulated conductors having enclosed within an overall nonmetallic jacket.

A. 334.10 NM, NMC, and NMS cable can be used in one- and two-family dwellings, multifamily dwellings, and their attached or detached garages and storage buildings. See Informative Annex E for construction types.

B. 334.15(A) Exposed work shall except per 300.11(A) follow surface of building finish or on running boards.

C. 334.15(B) Exposed work shall be protected from physical damage.

D. 334.15(C) Exposed work not smaller than two No. 6 or three No. 8 conductors in unfinished basements may be secured to lower edge of joists.

E. 334.23 Exposed work in accessible attics shall be protected by guard strips if on top or within 7′ of floor joists per 320.23.

F. 334.23 Exposed work in accessible attic does not require guard strips if installed parallel to rafters, studs, or floor joists per 320.23.

G. 334.17 Shall comply with 300.4 when installed through or parallel to joists, studs, or rafters.

H. 334.24 Bending radius shall not be less than 5× cable diameter.

I. 334.30 NMC shall be supported and secured within 12″ of boxes, etc. and at intervals not over 4½′ apart.

J. 334.6 NM, NMC, and NMS cables shall be listed.

USES PERMITTED — 334.10

(1) In one- and two-family dwellings and their garages and storage buildings

(2) In multifamily dwellings and other structures of certain construction types

(3) Other construction types (Types III, IV, and V)

(4) In cable trays, where identified.

(5) Types I and II construction where installed within raceways that are permitted within buildings of Type I and II construction.

USES NOT PERMITTED — 334.12(A)(B)

Types NM, NMC, and NMS have ten specific non-permitted uses. Type NM and NMs have (4) additional prohibited conditions. See Section 334.12(A) and (B).

CONSTRUCTION SPECS — 334(PART III)

(1) **NM** – Overall covering shall be flame retardant and moisture resistant.

(2) **NMC** – Overall covering shall be flame retardant, moisture resistant, fungus resistant, and corrosion resistant.

(3) **NMS** – Sheath shall be moisture resistant and flame retard

SERVICE ENTRANCE CABLE (SE AND USE) — 338

A 230.24(A) Overhead service conductors shall have a vertical clearance of 8′ above roof surfaces. See Exceptions.

B. 230.24(B)(1 – 4) The minimum clearance of overhead service conductors that are not over 600 V is from 10′ – 18′ above final grade.

C. 230.51(A) Service entrance cables shall be supported within 12″ of equipment and at intervals not exceeding 30″.

D. Table 300.5 Minimum cover requirements for USE cable are per Table 300.5.

E. 338.2 USE is identified for ungrounded use.

F. 338.10(B)(4)(a) SE is permitted for use in interior wiring systems if it is installed per Part II of Article 334, excluding 334.80.

G. 338.24 Bending radius shall not be less than 5x cable diameter where the neutral is smaller than the unguarded conductors. SE cable shall be so marked. See 338.120.

H. 338.120 Marking shall be per 310.120. Where the neutral is smaller than the ungrounded conductors, SE cable shall be so marked.

338.2: Service-Entrance Cable – A single or multiconductor assembly with or without an overall covering. See 338.100.

SE – Service-entrance cable with a flame-retardant, moisture-resistant covering.

USE – Underground service-entrance cable with a moisture-resistant covering that is not flame-resistant.

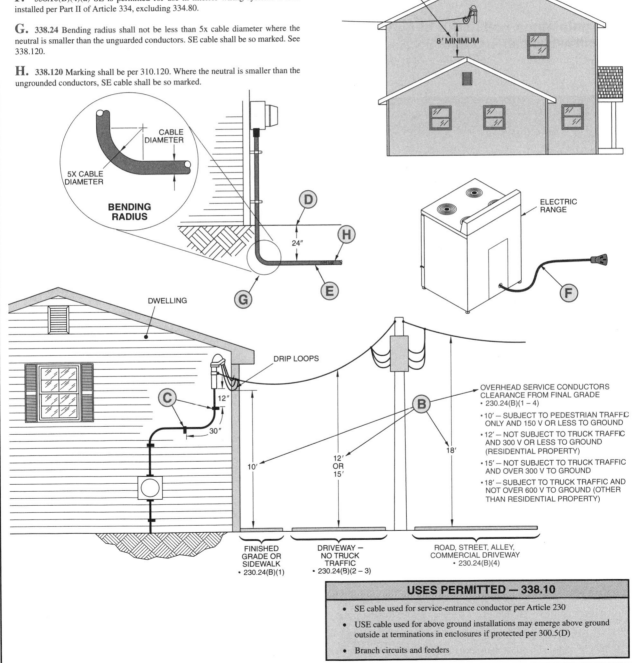

CABLE DIAMETER

5X CABLE DIAMETER

BENDING RADIUS

ELECTRIC RANGE

24″

DWELLING

DRIP LOOPS

12″

30″

10′

12′ OR 15′

18′

OVERHEAD SERVICE CONDUCTORS CLEARANCE FROM FINAL GRADE
• 230.24(B)(1 – 4)

• 10′ – SUBJECT TO PEDESTRIAN TRAFFIC ONLY AND 150 V OR LESS TO GROUND

• 12′ – NOT SUBJECT TO TRUCK TRAFFIC AND 300 V OR LESS TO GROUND (RESIDENTIAL PROPERTY)

• 15′ – NOT SUBJECT TO TRUCK TRAFFIC AND OVER 300 V TO GROUND

• 18′ – SUBJECT TO TRUCK TRAFFIC AND NOT OVER 600 V TO GROUND (OTHER THAN RESIDENTIAL PROPERTY)

FINISHED GRADE OR SIDEWALK
• 230.24(B)(1)

DRIVEWAY – NO TRUCK TRAFFIC
• 230.24(B)(2 – 3)

ROAD, STREET, ALLEY, COMMERCIAL DRIVEWAY
• 230.24(B)(4)

USES PERMITTED — 338.10

- SE cable used for service-entrance conductor per Article 230

- USE cable used for above ground installations may emerge above ground outside at terminations in enclosures if protected per 300.5(D)

- Branch circuits and feeders

RIGID METAL CONDUIT (RMC) — 344

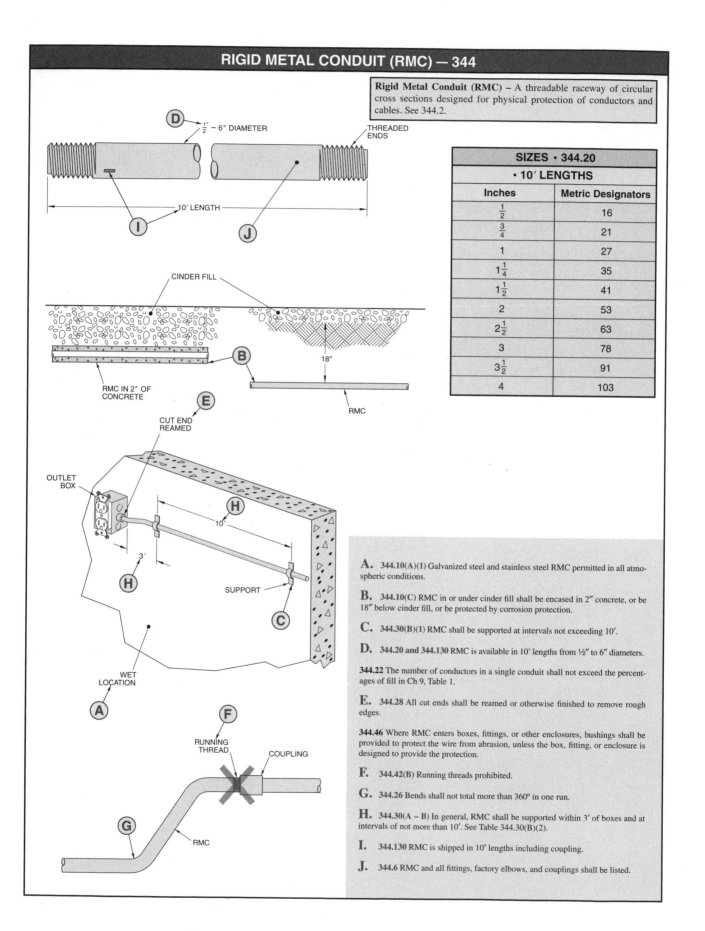

Rigid Metal Conduit (RMC) – A threadable raceway of circular cross sections designed for physical protection of conductors and cables. See 344.2.

D ½" – 6" DIAMETER

THREADED ENDS

10' LENGTH

I

J

CINDER FILL

RMC IN 2" OF CONCRETE

18"

RMC

B

CUT END REAMED

E

OUTLET BOX

H

10'

H

3'

SUPPORT

C

WET LOCATION

A

RUNNING THREAD

COUPLING

F

G

RMC

SIZES · 344.20	
· 10' LENGTHS	
Inches	**Metric Designators**
$\frac{1}{2}$	16
$\frac{3}{4}$	21
1	27
$1\frac{1}{4}$	35
$1\frac{1}{2}$	41
2	53
$2\frac{1}{2}$	63
3	78
$3\frac{1}{2}$	91
4	103

A. 344.10(A)(1) Galvanized steel and stainless steel RMC permitted in all atmospheric conditions.

B. 344.10(C) RMC in or under cinder fill shall be encased in 2" concrete, or be 18" below cinder fill, or be protected by corrosion protection.

C. 344.30(B)(1) RMC shall be supported at intervals not exceeding 10'.

D. 344.20 and 344.130 RMC is available in 10' lengths from ½" to 6" diameters.

344.22 The number of conductors in a single conduit shall not exceed the percentages of fill in Ch 9, Table 1.

E. 344.28 All cut ends shall be reamed or otherwise finished to remove rough edges.

344.46 Where RMC enters boxes, fittings, or other enclosures, bushings shall be provided to protect the wire from abrasion, unless the box, fitting, or enclosure is designed to provide the protection.

F. 344.42(B) Running threads prohibited.

G. 344.26 Bends shall not total more than 360° in one run.

H. 344.30(A – B) In general, RMC shall be supported within 3' of boxes and at intervals of not more than 10'. See Table 344.30(B)(2).

I. 344.130 RMC is shipped in 10' lengths including coupling.

J. 344.6 RMC and all fittings, factory elbows, and couplings shall be listed.

RIGID POLYVINYL CHLORIDE CONDUIT (PVC) — 352

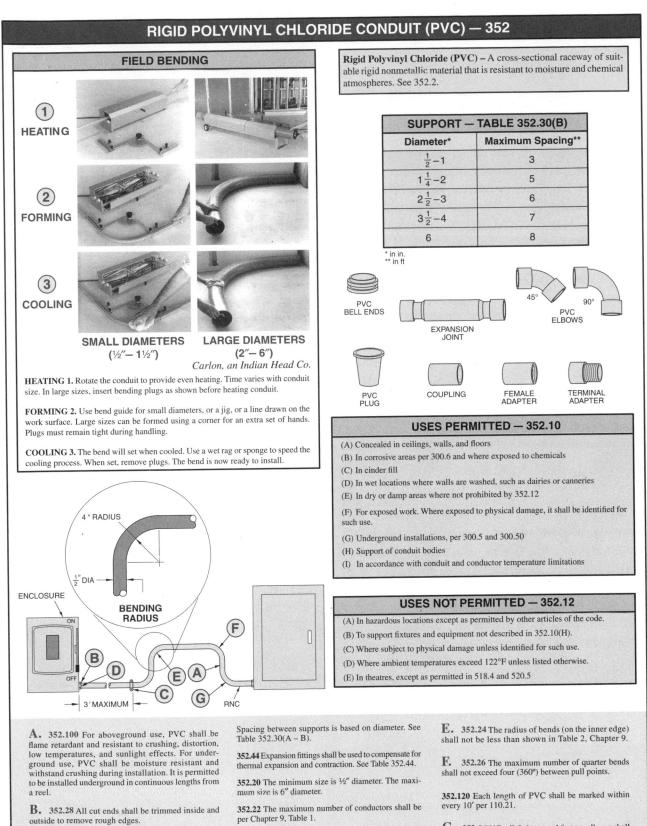

FIELD BENDING

① HEATING

② FORMING

③ COOLING

SMALL DIAMETERS
(½″ – 1½″)

LARGE DIAMETERS
(2″– 6″)

Carlon, an Indian Head Co.

HEATING 1. Rotate the conduit to provide even heating. Time varies with conduit size. In large sizes, insert bending plugs as shown before heating conduit.

FORMING 2. Use bend guide for small diameters, or a jig, or a line drawn on the work surface. Large sizes can be formed using a corner for an extra set of hands. Plugs must remain tight during handling.

COOLING 3. The bend will set when cooled. Use a wet rag or sponge to speed the cooling process. When set, remove plugs. The bend is now ready to install.

Rigid Polyvinyl Chloride (PVC) – A cross-sectional raceway of suitable rigid nonmetallic material that is resistant to moisture and chemical atmospheres. See 352.2.

SUPPORT — TABLE 352.30(B)	
Diameter*	**Maximum Spacing****
½–1	3
1¼–2	5
2½–3	6
3½–4	7
6	8

* in in.
** in ft

PVC BELL ENDS

EXPANSION JOINT

45° / 90° PVC ELBOWS

PVC PLUG

COUPLING

FEMALE ADAPTER

TERMINAL ADAPTER

USES PERMITTED — 352.10

(A) Concealed in ceilings, walls, and floors

(B) In corrosive areas per 300.6 and where exposed to chemicals

(C) In cinder fill

(D) In wet locations where walls are washed, such as dairies or canneries

(E) In dry or damp areas where not prohibited by 352.12

(F) For exposed work. Where exposed to physical damage, it shall be identified for such use.

(G) Underground installations, per 300.5 and 300.50

(H) Support of conduit bodies

(I) In accordance with conduit and conductor temperature limitations

USES NOT PERMITTED — 352.12

(A) In hazardous locations except as permitted by other articles of the code.

(B) To support fixtures and equipment not described in 352.10(H).

(C) Where subject to physical damage unless identified for such use.

(D) Where ambient temperatures exceed 122°F unless listed otherwise.

(E) In theatres, except as permitted in 518.4 and 520.5

4″ RADIUS

½″ DIA

ENCLOSURE

ON / OFF

BENDING RADIUS

3′ MAXIMUM

RNC

B D C E A G F

A. 352.100 For aboveground use, PVC shall be flame retardant and resistant to crushing, distortion, low temperatures, and sunlight effects. For underground use, PVC shall be moisture resistant and withstand crushing during installation. It is permitted to be installed underground in continuous lengths from a reel.

B. 352.28 All cut ends shall be trimmed inside and outside to remove rough edges.

C. 352.30 PVC shall be installed as a complete system per 300.18. It shall be securely fastened within 3′ of each box, conduit body, or other termination.

Spacing between supports is based on diameter. See Table 352.30(A – B).

352.44 Expansion fittings shall be used to compensate for thermal expansion and contraction. See Table 352.44.

352.20 The minimum size is ½″ diameter. The maximum size is 6″ diameter.

352.22 The maximum number of conductors shall be per Chapter 9, Table 1.

D. 352.46 Bushings are required where RNC enters a box, fitting, or other enclosure unless the box, fitting, or enclosure provides equivalent protection.

E. 352.24 The radius of bends (on the inner edge) shall be not less than shown in Table 2, Chapter 9.

F. 352.26 The maximum number of quarter bends shall not exceed four (360°) between pull points.

352.120 Each length of PVC shall be marked within every 10′ per 110.21.

G. 352.6 PVC, all fittings, and factory elbows shall be listed.

GROUNDING ELECTRODES — 250.52

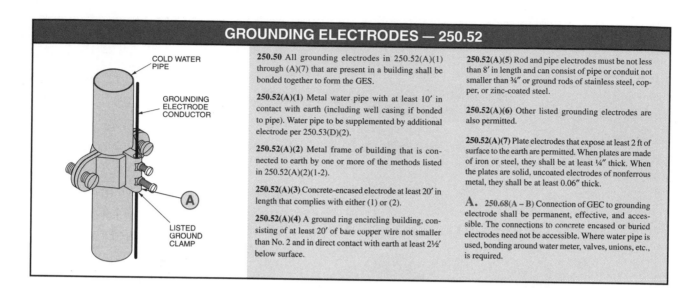

COLD WATER PIPE

GROUNDING ELECTRODE CONDUCTOR

(A)

LISTED GROUND CLAMP

250.50 All grounding electrodes in 250.52(A)(1) through (A)(7) that are present in a building shall be bonded together to form the GES.

250.52(A)(1) Metal water pipe with at least 10′ in contact with earth (including well casing if bonded to pipe). Water pipe to be supplemented by additional electrode per 250.53(D)(2).

250.52(A)(2) Metal frame of building that is connected to earth by one or more of the methods listed in 250.52(A)(2)(1-2).

250.52(A)(3) Concrete-encased electrode at least 20′ in length that complies with either (1) or (2).

250.52(A)(4) A ground ring encircling building, consisting of at least 20′ of bare copper wire not smaller than No. 2 and in direct contact with earth at least 2½′ below surface.

250.52(A)(5) Rod and pipe electrodes must be not less than 8′ in length and can consist of pipe or conduit not smaller than ¾″ or ground rods of stainless steel, copper, or zinc-coated steel.

250.52(A)(6) Other listed grounding electrodes are also permitted.

250.52(A)(7) Plate electrodes that expose at least 2 ft of surface to the earth are permitted. When plates are made of iron or steel, they shall be at least ¼″ thick. When the plates are solid, uncoated electrodes of nonferrous metal, they shall be at least 0.06″ thick.

A. 250.68(A – B) Connection of GEC to grounding electrode shall be permanent, effective, and accessible. The connections to concrete encased or buried electrodes need not be accessible. Where water pipe is used, bonding around water meter, valves, unions, etc., is required.

CONDUCTORS — GENERAL RULES

3-WIRE CIRCUIT

BLACK

WHITE

RED

200.2 Grounded conductor, on premises, shall be identified per 200.6. Grounded conductor, when insulated, shall have suitable insulation as ungrounded conductor.

200.7(A) Conductors having a white or gray covering or insulation with three continuous white stripes shall be used only for grounded conductors.

200.7(C)(1) Where necessary to use these conductors in circuits of 50 V or more, they may be permanently reidentified to indicate use and installed per (C)(1,2).

200.7(C)(1) Conductors that are part of a cable shall be reidentified and they shall be used as a supply, but not as a return, to single-pole, 3-way, or 4-way switch loops.

210.6(A) Voltage between conductors that supply terminals of lighting fixtures or cord-and-plug connected loads less than 1440 VA or ¼ HP shall not exceed 120 V.

210.19(A), Note 4 Voltage drop should not exceed 3% to the farthest outlet on branch circuit. Maximum voltage drop for feeders plus branch circuits should not exceed 5%. See 215.2(A)(3), Note 2 for feeders.

310.106(D), Ex. Conductors shall be insulated except where covered or bare conductors are specifically permitted.

- For covered service-drop conductors, See 230.22, Ex.
- For bare service-entrance grounded conductors, See 230.41, Ex.
- For bare or covered underground service grounded conductors, see 230.30, Ex.

310.110(C) Ungrounded conductors shall be clearly distinguishable from grounded and grounding conductors by colors or combination of colors plus distinguishing marking, and in either case the color shall be other than white, gray, or green. See 210.5(C) and 215.12(C) for means of identification.

210.20(A) Continuous loads shall not exceed 80% of the branch-circuit rating where the load shall continue for 3 hr. The rating of the OCPD shall not be less than the noncontinuous load plus 125% of the continuous load.

210.20(A), Ex. Circuits supplied by assembly are listed for continuous operation at 100% of their rating.

CONDUCTORS AND CIRCUITS

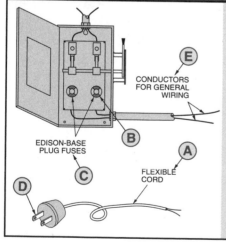

E

CONDUCTORS FOR GENERAL WIRING

B

A

EDISON-BASE PLUG FUSES

C

D

FLEXIBLE CORD

A. 240.5(A)(B)(1) Flexible cord approved for specific appliances is considered protected by branch-circuit overcurrent devices when applied within the appliance listing requirements.

B. 240.50(A)(1 – 2) Plug fuses and fuseholders shall not be used in circuits exceeding 125 V between conductors except in systems having a grounded neutral and of which no conductor operates at more than 150 V to ground.

C. 240.51(B) Edison-base plug fuses shall be used only as a replacement item in existing installations where there is no evidence of overfusing or tampering.

D. 250.114, Exceptions Some cord-and-plug connected equipment is not required to be connected to an EGC if a system of double insulation is provided.

310.10(H)(1 - 6) No. 1/0 and larger conductors may be run in parallel if of same length, same conductor material, same size, same insulation type, and terminated in the same manner. If in separate raceways or cables, these must have the same physical characteristics. See Exceptions.

E. 310.106(A) The minimum size of conductors shall be as per Table 310.106(A) unless otherwise provided in the Code.

WIRING IN ATTIC OR ROOF SPACES

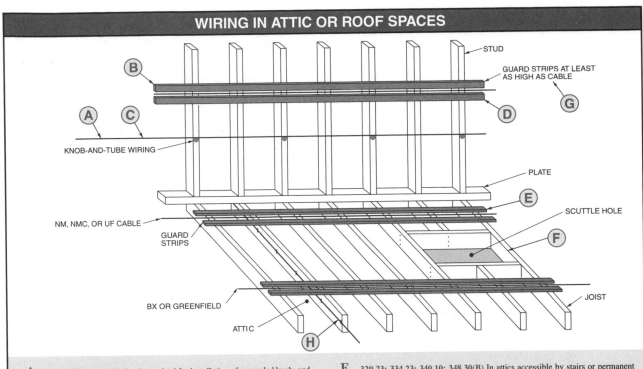

A. 394.10 Special permission is required for installation of concealed knob- and tube-wiring unless it is used for extensions of existing installations only.

B. 394.17; 320.23; 334.23; 340.10(4); 348.30(B) In general, conductors and cables shall be run through or on sides of joists, studs, and rafters and adequately supported, secured, and protected from damage.

C. 394.23(B) Ex. In unfinished attics and roof spaces not accessible by stairs or permanent ladder and head room at all points is less than 3', knob-and-tube wiring may be run on face of timber.

D. 320.23; 334.23; 340.10 NM, NMC, AC, UF, and FMC run across face of roof rafters or studding within 7' of floor or floor joists shall be protected by proper guard strips if attic is accessible by stairs or ladder.

E. 320.23; 334.23; 340.10; 348.30(B) In attics accessible by stairs or permanent ladder, BX, NM, NMC, UF, and FMC shall not be run across top of floor joists, unless protected by guard strips at least as high as the cable.

F. 320.23(A) If attic not accessible by stairs or permanent ladder, the cable need be protected only within 6' of scuttle hole.

G. 320.23(A) Guard strips shall be substantial and at least as high as the cable.

H. 320.23(B); 394.23(A); 334.23; 340.10(4) No running boards or guard strips necessary for protection on sides of joists, studs, or rafters.

WIRING RUN AT ANGLES TO JOISTS IN UNFINISHED BASEMENTS

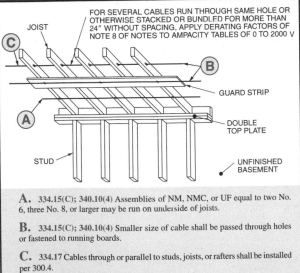

FOR SEVERAL CABLES RUN THROUGH SAME HOLE OR OTHERWISE STACKED OR BUNDLED FOR MORE THAN 24" WITHOUT SPACING, APPLY DERATING FACTORS OF NOTE 8 OF NOTES TO AMPACITY TABLES OF 0 TO 2000 V

A. 334.15(C); 340.10(4) Assemblies of NM, NMC, or UF equal to two No. 6, three No. 8, or larger may be run on underside of joists.

B. 334.15(C); 340.10(4) Smaller size of cable shall be passed through holes or fastened to running boards.

C. 334.17 Cables through or parallel to studs, joists, or rafters shall be installed per 300.4.

CABLE FISHED INTO MASONRY WALLS

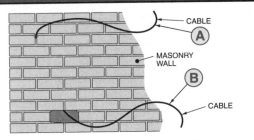

A. 334.10(C)(2); 320.10(5) Types AC, NM, NMC, or NMS may be fished in masonry walls if no moisture is present and above grade line.

334.10(B)(2); 340.10(4) If moisture is present, Types NMC, UF, or AC shall be used.

B. 334.10(C) NMS is permitted for both exposed and concealed installations in dry locations and where installed or fished in air voids of masonry walls.

334.12(B)(2) NM and NMS are not permitted to be embedded in poured cement, concrete, aggregate, or masonry.

334.12(A)(9) NM, NMC, and NMS are not permitted to be embedded in poured cement, concrete, or aggregate.

LIGHTING FIXTURES, LAMPHOLDERS, LAMPS, AND RECEPTACLES — 410 AND 406 . . .

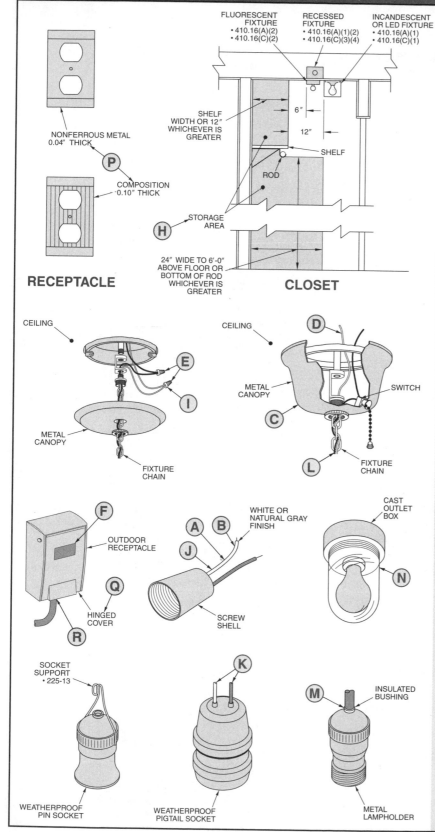

RECEPTACLE

NONFERROUS METAL 0.04" THICK

P

COMPOSITION 0.10" THICK

FLUORESCENT FIXTURE
• 410.16(A)(2)
• 410.16(C)(2)

RECESSED FIXTURE
• 410.16(A)(1)(2)
• 410.16(C)(3)(4)

INCANDESCENT OR LED FIXTURE
• 410.16(A)(1)
• 410.16(C)(1)

6"

12"

SHELF

SHELF WIDTH OR 12" WHICHEVER IS GREATER

ROD

H

STORAGE AREA

24" WIDE TO 6'-0" ABOVE FLOOR OR BOTTOM OF ROD WHICHEVER IS GREATER

CLOSET

CEILING

E

I

METAL CANOPY

FIXTURE CHAIN

CEILING

D

METAL CANOPY

C

SWITCH

L

FIXTURE CHAIN

F

OUTDOOR RECEPTACLE

HINGED COVER

Q

R

A **B**

J

SCREW SHELL

WHITE OR NATURAL GRAY FINISH

CAST OUTLET BOX

N

SOCKET SUPPORT
• 225-13

WEATHERPROOF PIN SOCKET

K

WEATHERPROOF PIGTAIL SOCKET

M

INSULATED BUSHING

METAL LAMPHOLDER

Art. 100: Luminaire – A complete lighting unit consisting of a lamp or lamps together with the parts designed to distribute the light, to position and protect the lamps and ballast (where applicable) and to connect the lamps to the power supply. *Note:* Luminaire is the international term for a lighting fixture.

A. 200.10(C) For devices with screw shells, the grounded conductor terminal shall be connected to the screw shell.

B. 200.10(D) For devices with attached leads, the conductor attached to the screw shell shall have white or gray finish.

C. 250.110(1) The metal fixture canopy or socket is required to be grounded if within 8' vertically or 5' horizontally of ground or grounded objects and subject to contact.

D. 402.8 Fixture wires used as grounded conductor to be white or gray or marked by one or more continuous white stripes or per 400.22(A – E).

E. 410.5 No live parts of fixtures, lampholders, lamps, and receptacles shall be normally exposed to contact. Exposed terminals shall be installed in metal fixture canopies or in open bases of table and floor lamps.

F. 410.10(A) Luminaries installed in wet or damp locations shall be marked.

G. 410.10(D) Cord-connected fixtures, lighting track, pendants, and ceiling fans shall not be installed in a 3' horizontal and 8' vertical zone from the bathtub rim.

H. 410.2 Closet storage space is the space above the shelf and below the rod. Storage space above the rod is shelf width or 12", whichever is greater. Storage space below the rod is 24" to the rod or 6' above the floor, whichever is greater.

410.21 Fixtures shall be constructed and installed so that temperatures in outlet boxes shall not exceed conductor rating. Wires are not permitted through outlet boxes unless identified for through-wiring.

410.23 Combustible wall or ceiling finish exposed between the edge of a fixture pan or canopy and an outlet box shall be covered with noncombustible material.

410.36(A); 314.27(A)(2) Fixtures over 50 lb shall be supported independently of the box unless listed for the weight to be supported.

I. 410.48 Fixture wiring shall be neatly arranged and not exposed to physical damage.

J. 410.50; 410.90 The grounded conductor shall be connected to the screw shell of a screw-shell lampholder.

K. 410.52 Fixture wire insulation shall be suitable for all conditions to which they will be subjected.

L. 410.56(E) Stranded conductors are required on fixture chains, movable parts, or flexible parts.

M. 410.62(A) Metal lampholders which are attached to flexible cords shall be equipped with insulating bushings.

N. 410.96 Lampholders installed in wet locations shall be listed for use in wet locations.

O. 406.3(D) Isolated ground (IG) receptacles, identified by an orange triangle on the receptacle face, are only permitted to be installed when an isolated equipment grounding conductor is provided.

...LIGHTING FIXTURES, LAMPHOLDERS, LAMPS, AND RECEPTACLES — 410 AND 406

P. **406.6(A, C)** Faceplates of nonferrous metal shall be not less than 0.04″ thick, ferrous 0.03″ thick, and composition 0.10″ thick, unless formed or reinforced to provide adequate mechanical strength. Metal faceplates shall be grounded.

406.6 Faceplates shall be installed to completely cover wall openings and seat against the surface.

Q. **406.9(A)** Receptacles in damp location shall be installed in an enclosure that is required to be weatherproof only when the plug cap is not inserted.

R. **406.9(B)(1)** All 15 A and 20 A, 125 V and 250 V receptacles in wet locations shall maintain weatherproof integrity whether or not the attachment plug cap is inserted.

S. **406.9(C)** Receptacles shall not be installed in or directly above bathroom or shower space.

T. **410.116(A)(1)** Unless identified for direct contact, recessed parts of fixture enclosures shall maintain at least ½″ spacing from combustible materials.

U. **410.116(B)** Thermal insulation not permitted within 3″ of fixture and not above fixture so as to entrap heat and prevent free air circulation except when identified as Type IC for insulation contact.

410.117(A) Conductor insulation shall be suitable for the temperature that may be encountered.

V. **410.117(B)** Branch-circuit wires permitted to terminate in the fixture if insulation is suitable for fixture temperature that may be encountered.

W. **410.117(C)** Tap wires with insulation suitable for the fixture temperature are permitted from fixture terminal to the outlet box if in a suitable raceway. The tap conductors shall be at least 18″ long but not longer than 6′.

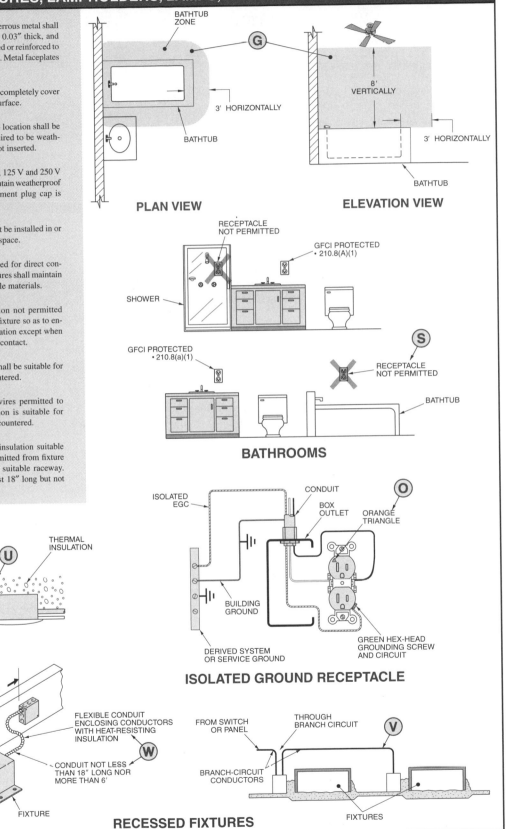

PLAN VIEW

ELEVATION VIEW

BATHROOMS

ISOLATED GROUND RECEPTACLE

RECESSED FIXTURES

WALL-MOUNTED OVEN AND COUNTERTOP RANGES

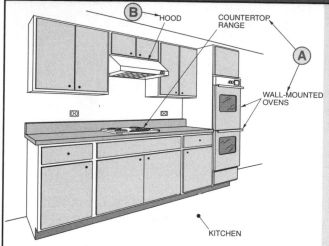

KITCHEN

A. **210.19(A)(3), Ex. 1** Where tap conductors supply a wall-mounted oven or countertop from 50 A circuit, they shall have not less than 20 A rating shall be sufficient for the load to be served and shall not be longer than necessary for servicing.

Table 220.55, Note 3 In lieu of the values given in Column C, the load for ranges over 1¾ kW but not over 8¾ kW may be taken as the sum of nameplate ratings multiplied by demand factors of Columns A and B.

Table 220.55, Note 4 The branch-circuit load for one wall-mounted oven or countertop unit shall be taken as the nameplate rating. Where countertop and not more than two wall-mounted ovens are in the same room, the nameplate ratings shall be added and the load treated as a single range.

B. **410.10(C)(1 – 4)** Luminaries are permitted within commercial cooking hoods where identified for the use.

DISCONNECTING MEANS FOR APPLIANCES AND FREE-STANDING RANGES

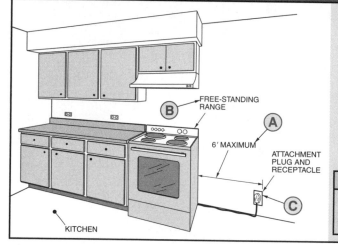

KITCHEN

A. **210.50(C)** Outlets for special appliances, such as laundry equipment, shall be within 6′ of the appliance.

B. **422.30** Each appliance shall be provided with a means for simultaneously disconnecting all underground conductors. Where supplied by more than one source, all disconnecting means shall be grouped and identified.

C. **422.33(A)** For cord-and-plug connected appliances, including free-standing household ranges and clothes dryers, a separable connector or attachment plug and receptacle may serve as the disconnecting means where accessible. See 422.31.

422.33(B) For residential ranges, plug and receptacle if accessible by removal of a drawer is acceptable.

DEMAND FACTORS

For household range from 12 kW to 27 kW, calculate per Table 220.55, Note 1.

For household ranges from 8¾ kW to 27 kW of unequal ratings, calculate per Table 220.55, Note 2.

DISCONNECTING MEANS FOR APPLIANCES

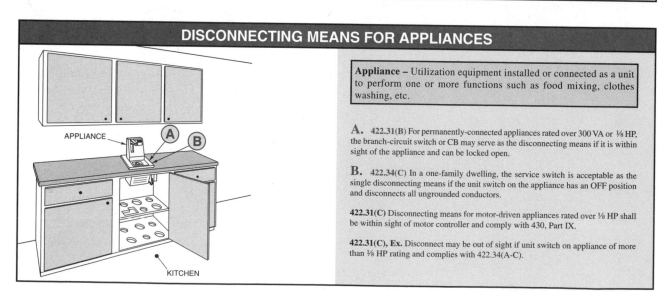

KITCHEN

Appliance – Utilization equipment installed or connected as a unit to perform one or more functions such as food mixing, clothes washing, etc.

A. **422.31(B)** For permanently-connected appliances rated over 300 VA or ⅛ HP, the branch-circuit switch or CB may serve as the disconnecting means if it is within sight of the appliance and can be locked open.

B. **422.34(C)** In a one-family dwelling, the service switch is acceptable as the single disconnecting means if the unit switch on the appliance has an OFF position and disconnects all ungrounded conductors.

422.31(C) Disconnecting means for motor-driven appliances rated over ⅛ HP shall be within sight of motor controller and comply with 430, Part IX.

422.31(C), Ex. Disconnect may be out of sight if unit switch on appliance of more than ⅛ HP rating and complies with 422.34(A-C).

Name_____ Date _____

NEC®	Answer	

_____ | (T) F | **1.** The term "service-entrance conductors overhead system" includes the term "service drops."

_____ | A | **2.** The general rule with respect to the service disconnecting means states that a grounding connection shall not be made to any grounded circuit on the ___ side.
    (A.) supply   C. either A or B
    B. load   D. both A and B

_____ | 100 | **3.** The rating of any one cord-and-plug connected utilization equipment shall not exceed ___% of the branch circuit ampere rating.

_____ | T (F) | **4.** The smallest permissible size of hard-drawn copper wire for a service drop is No. 10.

_____ | T (F) | **5.** Overhead service conductors shall have a minimum clearance from ground of less than 10'.

_____ | 6 | **6.** The GEC to a rod, pipe or plate electrode for an AC ungrounded system need not be larger than No. ___ Cu when it is the sole connection to a made electrode.

_____ | (T) F | **7.** The disconnecting means for a service shall disconnect all ungrounded conductors simultaneously.

_____ | PLASTIC | **8.** Boxes may have nonmetallic or ___ covers.

_____ | C | **9.** An EBJ is permitted outside a raceway when it is not over ___' long and is routed with the raceway or enclosure.
    A. 6   C. 10
    B. 8   D. none of the above

_____ | 125% | **10.** The branch-circuit rating shall not be less than the noncontinuous load plus ___% of the continuous load.

_____ | 6" | **11.** In general, the free length of conductor at a switch outlet shall be not less than ___".

_____ | 1 | **12.** The service disconnecting means need not be greater than 30 A where there are not more than ___ 2-wire branch circuits.

_____ | 8 | **13.** The clearance of service drops above roof surfaces shall be not less than ___'.

_____ | T (F) | **14.** The service disconnecting means shall consist of not more than eight switches or circuit breakers.

_____ | 3 | **15.** The clearance of service drops from the side or bottom of windows that are designed to open shall not be less than ___'.

_____ | 18 | **16.** The minimum clearance of service drops above a public driveway on other than residential property is ___'.
    A. 10   C. 16
    B. 12   D. 18

listed & **17.** Not more than six switches at one service location shall be used as the service disconnecting means if they are each ___ marked.

_3_ **18.** In residential occupancies, the NEC® requires ___ VA/sq ft for the general lighting load.

_100_ **19.** The service disconnect rating for a one-family dwelling shall be not less than ___ A, 1ϕ, 3-wire.
- A. 30
- B. 60
- C. 100
- D. 200

_6_ **20.** One service lateral can consist of six underground sets of multiple conductors sized at No. ___ or larger.

outside **21.** In determining VA/sq ft area, the ___ dimensions of the building shall be used.

22. A bare copper grounded service-entrance conductor is permitted to be used in a(n) ___ gutter.

_100_ **23.** Under the optional calculation for a one-family dwelling, the air conditioning load is assessed at ___%.

24. Conduit shall not be connected to the ___ of round boxes.

_600_ **25.** Unless otherwise marked, a CB with a 60°C terminal shall be loaded per the ___°C Column of Table 310.15(B)(16).

_8_ **26.** A metal fixture shall be grounded if located within ___′ vertically or 5′ horizontally of a kitchen sink.

closed **27.** Unused openings in boxes and fittings shall be effectively ___.

_1/8_ **28.** Disconnecting means for motor-driven appliances over ___ HP shall be located within sight of the motor controller.

_C_ **29.** Metal enclosures for GECs shall be ___ from the point of attachment to the GEC.
- A. rigid conduit
- B. isolated
- C. electrically continuous
- D. none of the above

Non metallic **30.** A new receptacle outlet installed on a 15 A or 20 A branch circuit shall be of the ___ type.

_1/16_ **31.** The front edge of an outlet box in a noncombustible ceiling may be ___″ back from the surface.

32. For the purpose of determining the maximum number of conductors in boxes, two conductors shall be counted for each yoke or ___.

GFCI **33.** A receptacle outlet, in a dwelling, within 6′ of a wet bar sink that is located in a living room shall be ___-protected.

_6_ **34.** Receptacles in living areas are spaced so that no point along the available floor line perimeter shall be more than ___′ from a receptacle outlet.

Rating **35.** In general, the minimum ampacity of the ungrounded service-entrance conductors shall not be less than the minimum ___ of the service disconnecting means.

_1/4_ **36.** Where necessary, a shallow outlet box not less than ___″ deep may be used, if no devices or utilization equipment is installed.

Panel **37.** In general, conductors other than service conductors shall not be installed in the same service ___ or service cable.

Name_____ Date _____

NEC®	Answer	
_____	_C_	**1.** The unit lighting load for a dwelling unit, expressed in VA/sq ft, shall be ___ VA. A. 1½ C. 3 B. 2 D. 5
_____	3.0°	**2.** The cubic inch rating of a No. 18 fixture wire is ___ cu in.
_____	T F	**3.** An insulated copper GEC used with a single service disconnecting means may be spliced by exothermic welding.
_____	AFC	**4.** Rooftop receptacle outlets for other than dwelling units used to service HACR equipment shall be ___-protected.
_____	T Ⓕ	**5.** A sheet metal strap ground clamp, constructed solely of strong sheet metal, is not permitted for connecting the GEC to the grounding electrode.
_____	_D_	**6.** A receptacle outlet is required for a hallway that has a length of ___' or more. A. 4 C. 8 B. 6 D. 10
_____	_____	**7.** Water heaters rated over ___ VA shall have a disconnecting means located within 50' and within sight.
_____	Ⓣ F	**8.** Additional plug receptacles in dwelling bedrooms, above the number required by the NEC®, shall be assessed at ½ A each.
_____	_____	**9.** Where the number of current-carrying conductors contained in the sheet metal auxiliary gutter is ___ or less, the correction factors in Table 310.15(B)(2)(a) shall not apply.
_____	_____	**10.** The minimum clearance between the lug and wall of an enclosure for a single No. 4/0 conductor being terminated where the conductor does not enter or leave through the wall opposite its terminal is ___".
_____	grounded	**11.** ___ means "connected to ground or to a conductive body that extends the ground connection."
_____	Vegtation	**12.** For other than temporary holiday wiring, ___ shall not be used for the support of overhead conductor spans.
_____	4D	**13.** When using the optional calculation for a dwelling service, all other loads above the initial 10 kVA are to be assessed at ___%. A. 5 C. 50 B. 40 D. 60
_____	disconnected	**14.** The branch-circuit switch or CB may serve as the disconnecting means for a permanently connected appliance of over 300 VA or ⅛ HP, where capable of being ___ and within sight.
_____	Ⓣ F	**15.** A No. 6 GEC that is secured to the building surface and is exposed to physical damage requires metal covering or protection.

_____ ___15___ **16.** The maximum rating of a cord-and-plug connected appliance used on a 20 A branch circuit shall not exceed ___ A.

      A. 12   C. 16

      B. 15   D. 18

_____ __120V__ **17.** In general, the voltage between conductors on a branch circuit supplying luminaires shall be not greater than ___ V.

_____ Ⓣ F **18.** In a bedroom, if a point on any available wall is 7′ from the nearest plug receptacle, an additional receptacle is needed.

_____ ___3___ **19.** For outside branch circuits and feeders, the 8′ vertical clearance above the roof level shall be maintained for ___′ in all directions from the edge of the roof.

_____ ___10___ **20.** Overhead service conductors operating at less than 300 V between conductors shall not be permitted to exceed ___′ of length with 18″ of clearance above the overhang portion of the roof.

_____ __13.6__ **21.** The maximum rating of the OCPD on a 17 A noncontinuous, single nonmotor appliance circuit is ___ A.

_____ Ⓣ F **22.** Track lighting installed in dwelling units shall be calculated at 180 VA for each 2′.

_____ ___6___ **23.** The minimum clearance between the lug and wall of an enclosure for a single No. 4/0 conductor being terminated at other than an opposite wall is ___″.

_____ ___D___ **24.** For the kitchen small appliance load in dwellings, the NEC® requires not less than ___ A circuit(s).

      A. one 15   C. two 15

      B. one 20   D. two 20

_____ ___10___ **25.** A single gang FS box (1¾″) can house five No. ___ conductors.

_____ ___5___ **26.** NMS cable shall be installed so no bend will have a radius less than ___ times the diameter of the cable.

_____ ___15___ **27.** The number of No. 12 conductors allowed in a 4″ × 1¼″ octagon box is ___.

_____ Ⓣ F **28.** A plug receptacle in the hallway outside of a kitchen shall not be connected to the small appliance circuit.

_____ __2.5__ **29.** The cubic inch rating of a No. 10 conductor is ___ cu in.

_____ ___6___ **30.** A total of ___ No. 14 conductors is permitted in a 3″ × 2″ × 2″ device box.

_____ Ⓣ F **31.** A receptacle in the middle of the dining room floor is not counted toward fulfilling minimum NEC® requirements.

_____ __120__ **32.** In dwelling occupancies, the voltage between conductors supplying screw shell lampholders shall not exceed ___ V.

_____ _____ **33.** Where raceways containing ungrounded conductors of No. ___ or larger enter a cabinet, box enclosure, or raceway, the conductors shall be protected by a substantial fitting providing a smoothly rounded insulating surface, unless the conductors are separated from the fitting or raceway by substantial insulating material securely fastened in place.

_____ _____ **34.** In general, where installed in raceways, conductors of size No. ___ and larger shall be stranded.

Name _____ Date _____

NEC®	Answer	

_____ _____ **1.** Tap conductors for recessed lighting fixtures shall be in a suitable raceway or cable at least ___″ in length but not more than 6′.

_____ T F **2.** A general-purpose branch circuit can supply both outlets for lighting and appliances.

_____ _____ **3.** If a 1½″ × 4″ octagonal box contains a fixture stud and a cable clamp, the maximum number of No. 14 conductors shall be reduced to ___.
A. 4   C. 7
B. 5   D. 8

_____ _____ **4.** A ground ring conductor used as part of the electrode system shall be buried at least ___′ below the earth's surface.

_____ T F **5.** All junction boxes shall be readily accessible.

_____ _____ **6.** Stainless steel ground rods shall be at least ___″ in diameter unless listed.

_____ _____ **7.** Stationary motors rated ___ HP or less can have their disconnecting means located in the service panel as the branch circuit OCPD.

_____ _____ **8.** Hanging lighting fixtures are not permitted within ___′ vertically from the rim of a bathtub.

_____ T F **9.** The screw shell of a plug-type fuseholder shall be connected to the load side of the circuit.

_____ _____ **10.** The demand load for a 16 kVA electric range is assessed at ___ VA.
A. 8200   C. 12,600
B. 9600   D. 14,400

_____ _____ **11.** To be considered as a GEC, a metal water pipe shall be in contact with the earth for at least ___′.

_____ _____ **12.** Where passing through a floor, NMS cable (Romex) shall be protected to a height of at least ___″ above the floor.

_____ T F **13.** Service equipment rated at 600 V or less shall be marked as suitable for use as service equipment.

_____ _____ **14.** In combustible walls or ceilings, the front edge of an outlet box or fitting may set ___″ back of the finished surface.
A. ⅛   C. ½
B. ¼   D. none of the above

_____ _____ **15.** EMT shall be securely fastened within ___′ of outlet boxes.

_____ _____ **16.** LFNC shall not be used for systems having conductors operating at over ___ V, except as permitted in 600.32(A).

_____ T   F   **17.** Wooden supports for boxes in concealed work shall not be less than nominal 1″ × 2″.

_____ _____ **18.** Cartridge fuses and fuseholders shall be classified according to ___ and amperage ranges.

_____ _____ **19.** Plug fuses are never permitted in circuits that exceed ___.
  A. 125 V between conductors   C. 150 V between conductors
  B. 125 V to ground   D. 300 V to ground

_____ _____ **20.** Receptacle outlets for special appliances such as laundry equipment shall be located within ___′ of the appliances.

_____ _____ **21.** Recessed non-Type IC lighting fixtures shall have at least ___″ clearance from combustible material.

_____ _____ **22.** The GEC can be spliced by means of the ___ welding process. The use of irreversible-type compression connectors is also permitted.

_____ _____ **23.** Securely installed boxes and fittings that are threaded into two or more conduits from two or more sides and that do not support fixtures or contain devices shall not exceed ___ cu in.
  A. 100   C. 150
  B. 125   D. 200

_____ _____ **24.** Surface-mounted fluorescent lighting fixtures shall be located at least ___″ from the storage space in a clothes closet.

_____ _____ **25.** NMS cable (Romex) shall not be used in ___ ___ studios.

_____ _____ **26.** The minimum size bare conductor permitted to be used as a concrete-encased electrode is No. ___ Cu.

_____ T   F   **27.** When two switches are mounted on the same strap, the number of conductors allowed in the box shall be reduced by one.

_____ _____ **28.** The maximum number of No. 14 conductors permitted in a 1½″ × 4″ octagonal outlet box is ___.
  A. 5   C. 8
  B. 7   D. 9

_____ _____ **29.** Outlet boxes shall have ___ depth to avoid the likelihood of damage to conductors within the box.

_____ _____ **30.** Aluminum conductors shall not be terminated within ___″ of the earth when used outside as a GEC.

_____ _____ **31.** Hanging lighting fixtures are not permitted within a zone measured ___′ horizontally from the rim of a bathtub.

_____ _____ **32.** A metal underground gas pipe shall not be used as a grounding ___.

_____ _____ **33.** Driven rods made of ___ shall not be used as electrodes.

_____ _____ **34.** The minimum size rebar permitted to be installed in concrete at the bottom of a footing is ___″ in diameter and 20′ long.

_____ _____ **35.** The smallest overhead service conductors permitted by the NEC® is No. ___ Cu or No. 6 Al.

_____ _____ **36.** For devices with ___ ___ the terminal for the grounded conductor shall be the one connected to the screw shell.

One-Family Dwellings

2

TRADE COMPETENCY TEST 4

Name ____________________ Date ____________________

NEC® **Answer**

______ (T) F **1.** All 125 V, 20 A, 1ϕ receptacles installed in bathrooms of dwelling units shall have ground-fault circuit-interrupter protection for personnel.

______ GFCI **2.** Conductor ___ protection shall not be required where the interruption of the circuit would create a hazard, such as in a fire pump circuit.

______ 60 **3.** Conductors installed inside AC cable (BX) shall have ampacity values rated at ___°C per Table 310.15(B)(16).

______ T (F) **4.** Receptacle outlets located above 5½″ from the finished floor can be counted as a required outlet.

______ ______ **5.** Where several cables are routed through the same hole at a length of more than ___″ without spacing, derating factors shall be applied.

______ (T) F **6.** Parallel conductors in a conduit run shall have an equipment ground pulled through each conduit.

______ 25 **7.** If a single electrode consisting of a rod, pipe, or plate has a resistance to ground of ___ Ω or less, a supplemental electrode is not required.

A. 25
B. 50
C. 75
D. 100

______ 6 **8.** In general, at least ___″ of free conductor shall be left at each outlet.

______ (T) F **9.** Where one or more internal cable clamps are within a box, the allowance toward the fill is based on the largest conductor entering the box.

______ Neutral **10.** The ___ conductor of all 3-wire, DC systems supplying premises wiring shall be grounded.

______ 60 **11.** Conductors installed inside nonmetallic-sheathed cable shall have a conductor insulation rating of at least ___°C.

______ A **12.** When 100′ of No. 16 fixture wire is connected to a standard 120 V circuit, the branch-circuit overcurrent protection device shall not exceed ___ A.

A. 15
B. 20
C. 30
D. 40

______ ______ **13.** Fittings such as capped elbows and service entrance elbows are classified as short radius ___.

A. conduit bodies
B. switch boxes
C. junction boxes
D. gutter boxes

______ (T) F **14.** CBs that are not marked with an interrupting rating are rated at 5000 A.

______ 12 **15.** Type NM cable is not required to be connected in a nonmetallic box where secured within ___″ of the box and where sheath extends through a cable knockout at least ¼″.

_314.16 (a)_ _____ 16. A 4″ × 1¼″ round box normally has ___ cu in. of fill space area.

_____ _____ 17. FMC installed with a total ground-fault current path over 6′ in length shall be provided with a(n) ___.

_____ _____ 18. NM cable shall be permitted for both exposed and concealed work in normally ___ locations.

_____ _____ 19. The metal raceway enclosing tap conductors for recessed fixtures shall not be less than ___′ nor more than ___′ long.
    A. 1½; 6   C. 3; 5
    B. 2; 6   D. 3; 6

_____ _5000_ 20. A fuse which does not have an IC rating marked on its surface shall be considered as having an IC rating of ___ A.

_____ _1_ 21. Conductors connected in parallel in general wiring systems shall be at least No. ___ or larger.
    A. 1/0   C. 2
    B. 1   D. 3

_____ _60_ 22. The conductors of NM cable shall be rated at 90°C, but the allowable ampacity of the conductors shall not exceed that of a ___°C rated conductor.
    A. 40   C. 75
    B. 60   D. 105

_____ T   F 23. Where not permanently accessible, nonmetallic-sheathed cable within 5′ of a scuttle hole does not require protection.

_____ _____ 24. When it is impractical to install the service drop below the weatherhead, the point of attachment can be located no farther than ___″ from the point of attachment.

_____ _____ 25. Only power service-drop conductors should be attached to a service ___.

_____ T   F 26. Lighting switches mounted in or on the wall can be located above 6′-7″ from the finished floor.

_____ _____ 27. Service drop conductors shall have a clearance of not less than ___′ from a window that opens and closes.

_____ _____ 28. The NEC® requires that recessed portions of non-type IC fixture enclosures shall be spaced at least ___″ from combustible material.
    A. ⅜   C. ⅝
    B. ½   D. ¾

_____ _____ 29. AC cable (BX) does not have to be supported where used in ___′ lengths for flexibility.

_____ _____ 30. The ampacity of UF cable shall be that of the ___°C conductors in accordance with 310.15.

_____ _____ 31. All receptacle outlets installed within ___′ of a sink in other than a kitchen and a dwelling unit shall be GFCI-protected.

_____ _____ 32. Cables or raceways installed using directional boring equipment shall be ___ for the purpose.

_____ _____ 33. Boxes installed in wet locations shall be ___ for use in wet locations.

_____ _____ 34. In general, raceways on exterior surfaces of buildings shall be ___ for use in wet locations and arranged to drain.

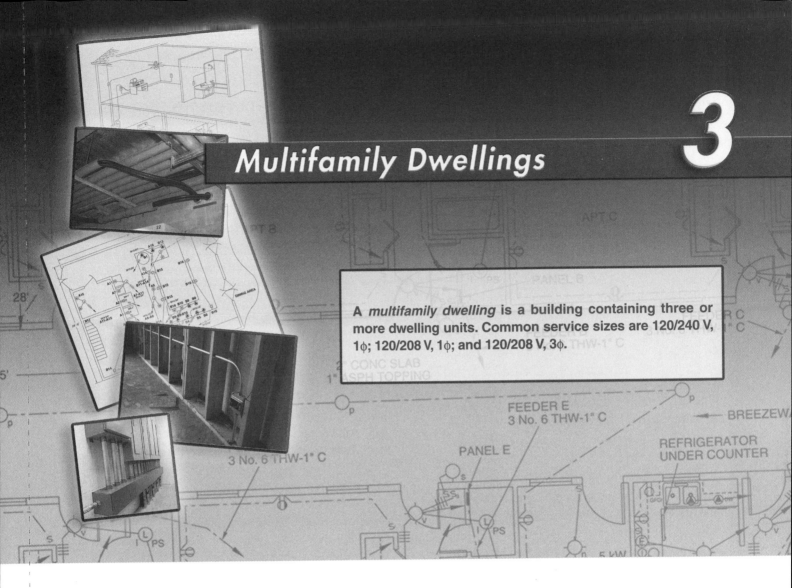

Multifamily Dwellings

3

A *multifamily dwelling* is a building containing three or more dwelling units. Common service sizes are 120/240 V, 1ϕ; 120/208 V, 1ϕ; and 120/208 V, 3ϕ.

CONDUCTORS

Table 310.15(B)(16) lists current carrying values for copper and aluminum wires when not more than three conductors are grouped in raceway, cable, or buried in the ground. Table 310.15(B)(17) lists values for copper and aluminum conductors in free air. Tables 310.15 (B)(18) through 310.15(B)(21) list ampacities of conductor insulation for specialized applications.

Table 310.15(B)(16)

The upper portion of this table deals with ampacities of conductors that are used at a nominal room temperature of 30° Celsius (C), or 86° Fahrenheit (F). Tables 310.15(B)(2)(a) and 310.15(B)(2)(b) supply correction factors for limiting ampacities of conductors at room temperatures at 30°C or 86°F and 40°C or 104°F.

Column 1 in Table 310.15(B)(16) lists AWG sizes of conductors from No. 18 to 2000 kcmil inclusive. Succeeding columns show ampacities for each size of conductor according to the type of insulation it has. The columns relating to 60°C, 75°C, and 90°C insulation in Table 310.15(B)(16) are most commonly used for inside wiring.

With size No. 8, for example, Column 2 shows that with 60°C insulation, the conductor carries 40 A without subjecting the insulation to damaging heat. If the No. 8 conductor has 7°C insulation (Column 3), it may be loaded to 50 A without harmful result. This size conductor may carry 55 A with the insulations in Column 4.

Suppose that the room temperature for this particular application is 50°C (122°F) instead of 30°C (86°F). Referring to Table 310.15(B)(2)(a), Ambient Temperature Correction Factors, a multiplying factor of 0.58 shall be used to determine the ampacity of a conductor having any of the Column 2 insulations. Use multipliers of 0.75 for Column 3 insulations and 0.82 for Column 4 insulations. The absence of a multiplying factor in some columns indicates that the insulations shall not be used for the listed ambient temperatures.

Using the indicated factors, the ampacity of No. 8 Type TW conductor used at a room temperature of 50°C (122°F) equals 40 A × 0.58, or 23 A. No. 8 Type THWN conductor is equal to 50 A × 0.75, or 37.5 A.

The table shows that Column 2 insulations are not to be used at room temperatures greater than 140°F and Column 4 insulations at temperatures greater than 194°F. Per 240.4(D), the rating of No. 14 Cu is 15 A, No. 12 Cu is 20 A, No. 10 Cu is 30 A, No. 12 Al is 15 A, and No. 10 Al is 25 A.

Table 310.15(B)(17)

This table is similar to Table 310.15(B)(16) except that it deals with the ampacities of single insulated conductors rated 0 V through 2000 V in free air. Table 310.15(B)(17) is based on an ambient air temperature of 30°C (86°F). Ampacities of conductors in free air are greater than those of conductors in raceways, cables, or earth.

Tables 310.15(B)(18) and 310.15(B)(19)

These tables are similar in construction. They deal with allowable ampacities of conductors rated 150°C to 250°C. The correction factor tables are also similar, but they also relate to higher temperatures.

Table Adjustment Factors

In addition to the tables, 310.15(B)(2) contains ampacity adjustment factors which should be carefully evaluated. Although the facts are clearly stated, notice (B)(4), which states that bare conductors have the same ampacity as the insulated conductors with which they are used, and Table 310.15(B)(3)(a), which lists derating factors where more than three current carrying conductors are installed in a raceway or cable.

Table 310.104(A) – Conductor Application and Insulations

This table lists conductor applications and the physical characteristics of the various types of insulated conductors. It should be noted that 310.10, Uses Permitted, states that the conductors listed in this table may be installed for any of the wiring methods recognized by Chapter 3 as specified in the tables.

There are several types of conductors that are given two maximum operating temperatures depending upon how they are used, such as Type THW. Type THW has a 75°C rating when used as a general purpose conductor and installed in a wet or dry location, but when it is installed within 3″ of a ballast it is given a 90°C rating as required in 410.68.

Table 310.104(A) also lists several types of conductors for specialized applications as noted, such as MI cable, which has an outer metal sheath, Types TBS and SIS, which are limited to switchboard wiring only, and Type TFE, which is limited to leads within apparatus in a dry location. Type THNN is a 90°C conductor that can be used in dry and damp locations. The ampacity is used with 60°C or 75°C based on equipment terminals.

Table 402.3 – Fixture Wire

The operating temperature of fixture wires shall not exceed the temperatures listed in Table 402.3.

Table 402.5 – Ampacity of Fixture Wires

Tables 400 5(A)(1) and (2) lists the allowable ampacities of fixture wires in sizes No. 18 through 10 AWG. The ampacities of the fixture wires listed in this table are based on an ambient temperature of 30°C (86°F).

Flexible Cords

Table 400.5(A)(1) and (2) shows ampacities of flexible cords where not more than three conductors are present. If there are more than three conductors in the group, ampacities given here are to be reduced to 80% and decreasing to 35% of listed values. It may be noted that ordinary portable cord size No. 18 will carry 7 A and size No. 16 will carry 10 A. Notes 1 through 15 of Table 400.4 should be read carefully.

Table 400.4 – Flexible Cords and Cables

This table sets forth construction features of the various cords and their conditions of use and application. Flexible cords and cables shall conform to the description provided in Table 400.4. If flexible cords and cables not included in Table 400.4 are to be used, they must undergo a special investigation as to their suitability for the intended use.

APARTMENTS – FLOOR PLAN

This print shows a six unit, one story multifamily dwelling (Apartments A–F). Five of the apartments (Apartments A–E) are identical, having living room (LR), kitchen (KIT.), bedroom (BR), and bath (B). The lighting panel, a flush type, is placed in the entrance storeroom. The kitchen has a wall mounted oven (8), a counter mounted cooktop (7), a dishwasher (10), a garbage disposal (9), and an exhaust fan (4).

The caretaker's apartment, Apartment F, has a living room (LR), a kitchen (KIT.), and a bath (B). It has no separate bedroom, a wall bed being used in the living room. Kitchen equipment is similar to that in the larger apartments.

There is a washroom (WR), a storeroom (SR), and a combination heating equipment and switchboard room (H/SR). There is a house receptacle circuit, an electric water heater circuit, and a pump motor. A house lighting circuit supplies the three workrooms and three night light standards. All these circuits originate in a special house panel, which is adjacent to service equipment in the switchboard room.

Arrows pointing in the direction of the lighting panel indicate circuit runs. The number of cross marks on the arrow shows the number of wires in the particular home run. Where no cross marks are found, it is understood that the circuit consists of two wires. Conductors are assumed to be No. 14 unless referred to in the specifications or marked adjacent to the arrow.

Alternate long and short dashed lines denote branch circuit raceways concealed in the floor. Solid black lines mark branch circuit conduit concealed in the wall or ceiling. Various types of outlets, including lights, plug receptacles, recessed units, special outlets, telephones, and fans, are shown with standard symbols. Feeders are shown by arrows originating at the various panels and pointing in the direction of the switchboard.

It should be observed that a note taken from the electrical specifications states that electrical metallic tubing is to be used in concrete floor slabs. Flexible conduit is to be used in walls and ceilings.

Load Calculations

The unit load specified for dwelling occupancies in Table 220.12 is 3 VA per square foot. The allowance for two small appliance or utility circuits per 220.52(A) is 3000 VA. The load required for two cooking units per Column B of Table 220.55 is 65% of the sum of the nameplate ratings. Based on these NEC® sections, the load for each of the five apartments, A to E inclusive, is determined as follows:

LOAD CALCULATIONS FOR APARTMENTS A–E

Lighting and Appliance		
710 sq ft × 3 VA	2130 VA	
Two appliance circuits	3000 VA	
Total	5130 VA	
Feeder Demand Factors		
1st 3000 VA at 100%	3000 VA	
Remainder		
2130 VA at 35%	745 VA	
Total	3745 VA	**3745 VA**
Oven and Cooktop		
Oven rated	4800 VA	
Cooktop rated	6700 VA	
Total	11,500 VA	
11,500 VA × 0.65	7475 VA	**7475 VA**
Dishwasher		**1500 VA**
Garbage Disposal		**600 VA**
Exhaust Fan		**350 VA**
	Total	**13,670 VA**

FEEDER SIZE (120/240 V service)
$$13{,}670 \text{ VA} \div 240 \text{ V} = \textbf{56.9 A}$$

Table 310.15(B)(16) shows that No. 6 THW Cu conductor is needed. Table C1, Annex C shows that a ¾″ EMT is required for three No. 6 THW Cu conductors.

APARTMENTS A-F

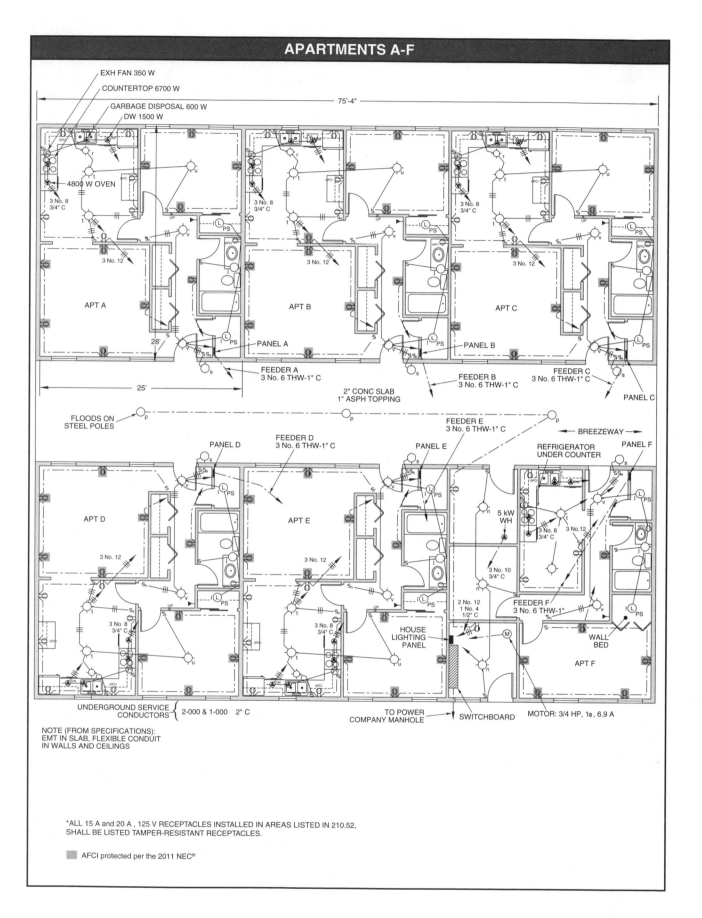

EXH FAN 350 W
COUNTERTOP 6700 W
GARBAGE DISPOSAL 600 W
DW 1500 W

75'-4"

4800 W OVEN

3 No. 8
3/4" C

3 No. 12

APT A

28'

25'

3 No. 8
3/4" C

3 No. 12

APT B

PANEL A

FEEDER A
3 No. 6 THW-1" C

2" CONC SLAB
1" ASPH TOPPING

FEEDER B
3 No. 6 THW-1" C

3 No. 8
3/4" C

3 No. 12

APT C

PANEL B

FEEDER C
3 No. 6 THW-1" C

PANEL C

FLOODS ON
STEEL POLES

PANEL D

FEEDER D
3 No. 6 THW-1" C

FEEDER E
3 No. 6 THW-1" C

BREEZEWAY

PANEL E

REFRIGERATOR
UNDER COUNTER

PANEL F

APT D

3 No. 12

3 No. 8
3/4" C

APT E

3 No. 12

3 No. 8
3/4" C

5 kW
WH

3 No. 8
3/4" C

3 No. 12

3 No. 10
3/4" C

2 No. 12
1 No. 4
1/2" C

HOUSE
LIGHTING
PANEL

FEEDER F
3 No. 6 THW-1"

WALL
BED

APT F

UNDERGROUND SERVICE
CONDUCTORS   2-000 & 1-000   2" C

TO POWER
COMPANY MANHOLE

SWITCHBOARD

MOTOR: 3/4 HP, 1ϕ, 6.9 A

NOTE (FROM SPECIFICATIONS):
EMT IN SLAB, FLEXIBLE CONDUIT
IN WALLS AND CEILINGS

*ALL 15 A and 20 A , 125 V RECEPTACLES INSTALLED IN AREAS LISTED IN 210.52,
SHALL BE LISTED TAMPER-RESISTANT RECEPTACLES.

AFCI protected per the 2011 NEC®

LOAD CALCULATIONS FOR APARTMENT F

Lighting and Appliance

448 sq ft × 3 VA	1344 VA	
Two appliance circuits	3000 VA	
Total	4344 VA	

Feeder Demand Factors

1st 3000 VA at 100%	3000 VA	
Remainder		
1344 VA at 35%	470 VA	
Total	3470 VA	**3470 VA**
Oven and Cooktop		**7475 VA**
Dishwasher		**1500 VA**
Garbage Disposal		**600 VA**
Exhaust Fan		**350 VA**
Total		**13,395 VA**

FEEDER SIZE (120/240 V service)

    13,395 VA ÷ 240 V = **55.8 A**

Table 310.15(B)(16) shows that No. 6 THW Cu conductor is needed. Table C1, Annex C shows that a ¾" EMT is required for three No. 6 THW Cu conductors.

In some cases, the neutral feeder may be smaller than the outer conductors. NEC® 220.61(B) allows a 70% demand factor on the portion of the neutral feeder load that supplies an electric range or equivalent cooking appliance. Although the specifications and prints state that three No. 6 THW Cu conductors shall be installed to each panel, it is useful to perform calculations to see if a smaller neutral conductor might have been otherwise admissible. Neutral calculations for Apartment F are determined as follows:

ADDITIONAL LOAD CALCULATIONS FOR APARTMENT F

Oven and Cooktop = 7475 VA

7475 VA × 0.7 = 5233 VA		5233 VA

Assumed Feeder Load

Lighting and Appliance		**3470 VA**
Dishwasher		**1500 VA**
Garbage Disposal		**600 VA**
Exhaust Fan		**350 VA**
Total		**11,153 VA**

NEUTRAL FEEDER SIZE

    11,153 VA ÷ 240 V = **46.5 A**

Per Table 310.15(B)(16), No. 8 THW Cu conductor is acceptable for the neutral. Maximum unbalance of 50 A per 220.61(A).

Another factor that is sometimes important when the load appears close to the current-carrying capacity of a particular size conductor is that of balancing circuits. In Apartment A, for example, a 3-wire lighting circuit, a 3-wire utility circuit, and a 3-wire cooking circuit are balanced across the ungrounded conductors and the neutral wire. But the 1500 VA, 120 V dishwasher, which is connected between one ungrounded conductor and the neutral, does not balance the 950 VA garbage disposal and exhaust fan load, which is connected between the other ungrounded conductor and the neutral.

In such a case, an imaginary (artificial) load may be assumed in parallel with the smaller load in order to gain balance, its value here being equal to 1500 VA minus 950 VA, or 550 VA. The listing is then as follows:

LOAD CALCULATION ON APARTMENT A WITH ARTIFICIAL LOAD

Lighting and Appliance	**3745 VA**
Oven and Cooktop	**7475 VA**
Dishwasher	**1500 VA**
Garbage Disposal	**600 VA**
Exhaust Fan	**350 VA**
Artificial Load	**550 VA**
Total	**14,220 VA**

ADJUSTED LOAD

    14,220 VA ÷ 240 V = **59 A**

Per Table 310.15(B)(16), No. 6 THW Cu conductors may be used.

Although the calculation has not altered the result here, the method is worth noting. Some inspection authorities insist upon balancing loads, others do not.

Service Calculations

NEC® 220.14(I) provides that a load of 180 VA per outlet shall be allowed for outlets other than for general illumination. The two plug receptacles in the work area come within such classification. NEC® 220.14(J) permits this type of load to be included with the general lighting load in applying demand factors. These plug receptacles and the three 200 VA night light standards come within scope of the section.

NEC® 220.53 states that where four or more fixed appliances other than ranges, clothes dryers, space heating units, or A/C equipment are connected to the same feeder or service in a multifamily dwelling, a demand factor of 75% may be applied to the load. The six dishwashers, six garbage disposals, six exhaust fans, and electric water heater qualify here. The service conductor sizes are governed by 230.42, which refers to Article 220, so the service conductor sizes are calculated as follows:

SERVICE CALCULATIONS

Lighting and Appliance

4220 sq ft × 3 VA =	12,660 VA	

Six Appliance Circuits

6 × 3000 VA=	18,000 VA	

Laundry Circuit 1500 VA

Three Night light Standards

3 × 200 VA =	600 VA	

Two Receptacle Outlets in Work Area

2 × 180 VA =	360 VA	
Total	33,120 VA	

1st 3000 VA at 100%	3000 VA	
Remainder		
30,120 VA at 35%	10,542 VA	
Total	13,542 VA	**13,542 VA**

Ovens and Cooktops

6 × 4800 VA =	28,800 VA	
6 × 6700 VA =	40,200 VA	
Total	69,000 VA	

Service Demand Factor

32% for 12 ranges 69,000 VA × 0.32 =	**22,080 VA**	

Other Loads

Dishwashers	6 × 1500 = 9000 VA	
Disposals	6 × 600 = 3600 VA	
Exhaust fans	6 × 350 = 2100 VA	
Water heater	1 × 5000 = 5000 VA	
Total	19,700 VA	
19,700 VA × 0.75 =	**14,775 VA**	

Motor Load

¾ HP, 1φ, 230 V× 6.9 A =	1600 VA	
1600 VA × 125% =		**2000 VA**
Total		**52,397 VA**

SERVICE ENTRANCE CONDUCTORS

52,397 VA ÷ 240 V = **218 A**

Table 310.15(B)(16) shows that No. 4/0 THW Cu conductors are acceptable.

NEUTRAL CONDUCTOR

Lighting and Appliance	**13,542 VA**	
Cooking Units		
22,080 VA × 0.7 =	**15,456 VA**	
Fixed appliances	**14,775 VA**	
Total	**43,773 VA**	

43,773 VA ÷ 240 V = **182 A**

Table 310.15(B)(16) shows that No. 3/0 THW Cu conductor is acceptable as the neutral.

SERVICE SIZE

Two No. 4/0 and one No. 3/0 THW Cu conductors, all in 2″ RMC, are acceptable. *Note:* The Optional Calculation per 220.80 may be applied.

APARTMENTS – DETAILS AND ELEVATIONS

The larger print shows one of the five similar apartments (A–E). The plan view drawn to a larger scale brings out details to better advantage than is possible on the general print, so that the position of equipment may be found with greater accuracy.

The conduit in the floor slab is indicated by a dashed line with alternating long and short dashes. The conduit in a wall or ceiling is indicated by solid black lines. All wires are No. 14 except the 3-wire No. 8 circuit to the cooking units and the 3-wire No. 12 circuit to the kitchen appliance receptacles.

Circuits may be traced to learn the reason for three or four wires, as the case may be, in various conduits. Certain runs that might otherwise have been combined are kept separate to avoid the penalty of derating where more than three wires are installed in one raceway.

One point that should not be overlooked is the arrangement for connecting bedroom plug receptacles. They are supplied by conduit laid in the floor slab. This run continues up to the switch outlet box in the wall at the side of the door and then to the ceiling outlet. A small circle and the word *UP* indicate this procedure. It should also be noted that one half of the symbol designating the duplex receptacle outlet in the living room is a solid color. This indicates that one half of this receptacle outlet is controlled by a wall switch.

Triangles *A, B,* and *C* marked in a semicircle on the kitchen floor represent one scheme for directing attention to elevations seen at the right of the drawing. From these elevations, it is possible to determine exact locations of outlets such as those for plug receptacles and for cooking units. The location of the garbage disposal and its method of connection are shown in elevation A.

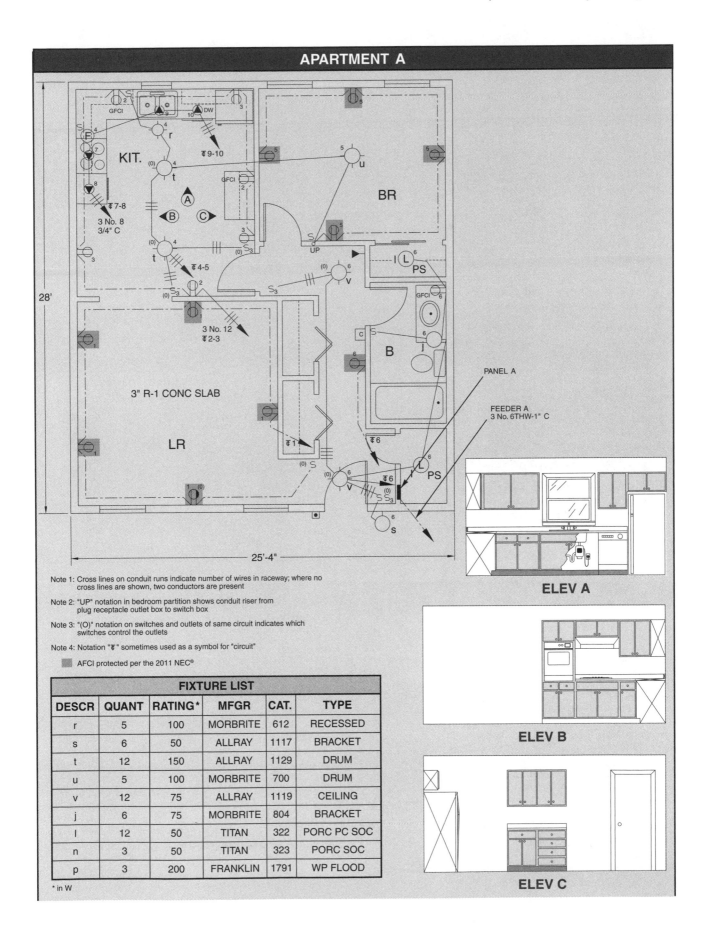

APARTMENT A

KIT.

GFCI

DW

r

₢9-10

(A)
(B) (C)

t

3 No. 8
3/4" C

₢7-8

t

₢4-5

3 No. 12
₢2-3

3" R-1 CONC SLAB

LR

₢1

GFCI

BR

u

UP

PS

v

GFCI

C

j

B

₢6

PANEL A

FEEDER A
3 No. 6THW-1" C

PS

v
₢6

S

28'

25'-4"

Note 1: Cross lines on conduit runs indicate number of wires in raceway; where no cross lines are shown, two conductors are present

Note 2: "UP" notation in bedroom partition shows conduit riser from plug receptacle outlet box to switch box

Note 3: "(O)" notation on switches and outlets of same circuit indicates which switches control the outlets

Note 4: Notation "₢" sometimes used as a symbol for "circuit"

■ AFCI protected per the 2011 NEC®

FIXTURE LIST

DESCR	QUANT	RATING*	MFGR	CAT.	TYPE
r	5	100	MORBRITE	612	RECESSED
s	6	50	ALLRAY	1117	BRACKET
t	12	150	ALLRAY	1129	DRUM
u	5	100	MORBRITE	700	DRUM
v	12	75	ALLRAY	1119	CEILING
j	6	75	MORBRITE	804	BRACKET
l	12	50	TITAN	322	PORC PC SOC
n	3	50	TITAN	323	PORC SOC
p	3	200	FRANKLIN	1791	WP FLOOD

* in W

ELEV A

ELEV B

ELEV C

SERVICE FOR MULTIPLE OCCUPANCY BUILDINGS

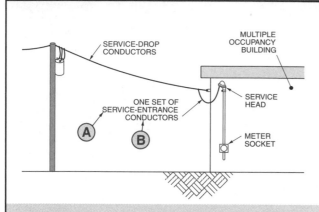

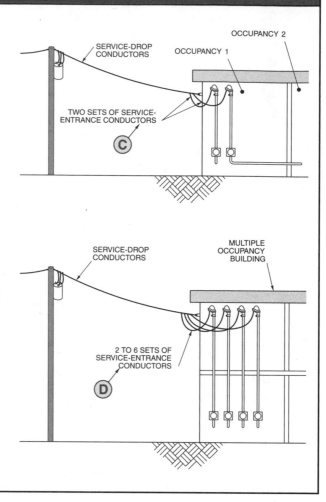

225.36 In general, when more than one building is on the same property, each building or structure shall have a separate disconnecting means. Disconnecting means shall be suitable for use as service equipment except for 3-way and 4-way snap switches for garages and out buildings on residential property. See Exception.

A. **230.2(A)** A building or other structure shall be supplied by only one service. See 230.2(A-E).

230.2(E) Where conditions permit more than one service, a permanent plaque or directory is required at each service and branch-circuit disconnect location which indicates all other services and area served by each.

B. **230.40** In general, each service drop or lateral and each set of overhead or underground service conductors shall supply only one set of service-entrance conductors.

C. **230.40, Ex. 1** A multiple occupancy building is permitted to be supplied by two separate sets of service-entrance conductors.

D. **230.40, Ex. 2** Separate sets of service conductors are permitted for 2 to 6 service disconnects.

230.40, Ex. 3 A one-family dwelling and their accessory structures can have a separate service to each.

230.40, Ex. 4 A set of service conductors can supply a common panel per 210.25.

230.40, Ex. 5 A set of service-entrance conductors can supply a fire pump or life safety per 230.82(5) or 230.82(6).

SERVICE — LIGHTING AND APPLIANCE PANELBOARD

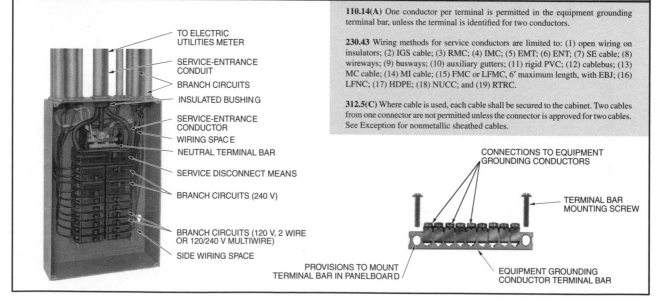

110.14(A) One conductor per terminal is permitted in the equipment grounding terminal bar, unless the terminal is identified for two conductors.

230.43 Wiring methods for service conductors are limited to: (1) open wiring on insulators; (2) IGS cable; (3) RMC; (4) IMC; (5) EMT; (6) ENT; (7) SE cable; (8) wireways; (9) busways; (10) auxiliary gutters; (11) rigid PVC; (12) cablebus; (13) MC cable; (14) MI cable; (15) FMC or LFMC, 6' maximum length, with EBJ; (16) LFNC; (17) HDPE; (18) NUCC; and (19) RTRC.

312.5(C) Where cable is used, each cable shall be secured to the cabinet. Two cables from one connector are not permitted unless the connector is approved for two cables. See Exception for nonmetallic sheathed cables.

SERVICES AND FEEDERS

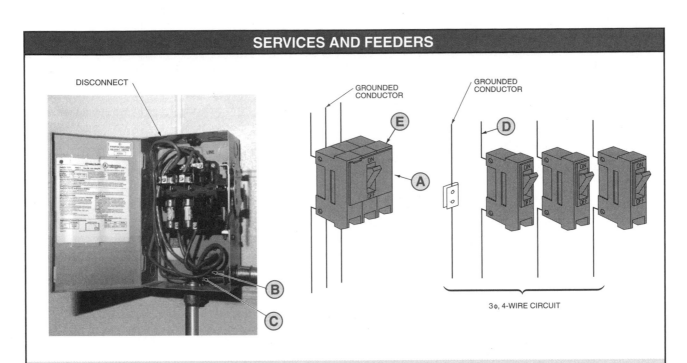

DISCONNECT

GROUNDED CONDUCTOR

GROUNDED CONDUCTOR

3φ, 4-WIRE CIRCUIT

A. **110.10** Impedance and other circuit characteristics shall be considered so that the selected overcurrent device will clear the fault before extensive damage to the equipment occurs.

B. **220.61(A)** The neutral conductor shall be large enough to carry the maximum unbalanced load.

C. **230.42(C)** The grounded (neutral) conductors shall not be smaller than required by 250.24(C).

D. **240.15** CBs are permitted as overcurrent devices and shall open all ungrounded conductors. Individual single-pole CBs may be used with handle ties in grounded systems in: (1) 1φ circuits; (2) 3-wire DC circuits; (3) 2φ, 5-wire branch circuits supplied from a system with a grounded neutral and not operating at over 150 V to ground, or 3φ, 4 wire systems; (4) 3-wire DC circuits.

250.24(C)(1 – 4) The service grounded conductor shall not be smaller than the GEC per Table 250.66 and if phase conductors are larger than 1100 kcmil Cu or 1750 kcmil Al, the grounded conductor shall not be smaller than 12½% of the area of the largest phase conductor except it need never be larger than the largest service ungrounded conductor.

E. **230.90(B)** No overcurrent device shall be inserted in a grounded service conductor except a CB which simultaneously opens all conductors.

DISTRIBUTION PANELBOARDS

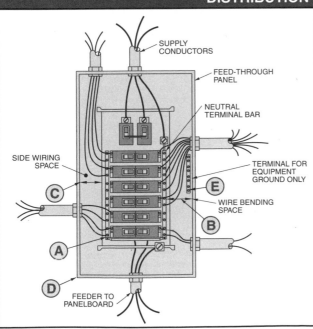

SUPPLY CONDUCTORS

FEED-THROUGH PANEL

NEUTRAL TERMINAL BAR

SIDE WIRING SPACE

TERMINAL FOR EQUIPMENT GROUND ONLY

WIRE BENDING SPACE

FEEDER TO PANELBOARD

A. **250.142(B)** A grounded circuit conductor shall not be used for grounding non-current-carrying metal parts of equipment on the load side of the service disconnecting means except for separate buildings per 250.32(B) and separately derived systems per 250.30(A)(1).

B. **312.6(B)(1 – 2)** Conductors entering and leaving the cabinet shall have wire bending space at the terminals per Table 312.6(A) and Table 312.6(B).

312.6(C) Ungrounded conductors No. 4 and larger shall be protected by a substantial insulating material, securely fastened in place. See 300.4(G).

C. **312.11(C)** Cabinets and cutout boxes which contain devices or apparatus connected to more than 8 conductors, not including the supply circuit or a continuation thereof, shall have back or side-wiring spaces.

408.36(A) Each branch-circuit panelboard shall be individually protected on the supply side by not more than two CBs with a combined rating that does not exceed that of the panelboard.

D. **408.36, Ex. 2** Individual protection is not required if the panelboard feeder has overcurrent protection not greater than the panelboard's rating.

E. **408.40** An identified terminal bar shall be provided inside cabinet and secured to cabinet for connection of all feeder and branch-circuit EGCs. The terminal bar for EGCs shall be secured to cabinet.

GROUNDING ELECTRODES — 250.52

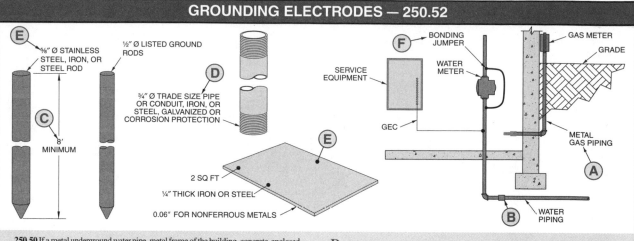

250.50 If a metal underground water pipe, metal frame of the building, concrete-enclosed electrode, or a ground ring is not available, other local metal underground systems or structures, rod and pipe electrodes, or plate electrodes shall be used. See 250.52(A)(4 – 8).

Made electrodes shall be below moisture level and shall not be painted. If more than one electrode is used, each electrode of a grounding system shall be at least 6' from any electrode of another grounding system. Two or more electrodes which are bonded together shall be considered a single system.

A. 250.52(B)(1) Underground metal gas piping shall not be used as a grounding electrode.

B. 250.52(A)(8) Other local metal underground systems or structures which are permitted include piping systems and underground tanks.

C. 250.52(A)(5) Rod and pipe electrodes shall be at least 8' in length.

D. 250.52(A)(5)(a) Conduit or pipe electrodes shall be at least ¾" in diameter and shall be corrosion-resistant.

250.52(A)(5)(b) Rod-type electrodes shall be at least ⅝" in diameter, unless listed.

250.53(G) At least 8' of the electrode shall be in contact with the soil. If rock is encountered, it shall be driven at up to a 45° angle from vertical or be buried in a trench at least 2½' deep.

E. 250.52(A)(7) Plate electrodes shall be at least 2 sq ft. Iron or steel plates shall be at least ¼" thick. Nonferrous plates shall be at least 0.06" thick.

250.53(A)(2), Ex. If resistance of single made electrodes exceeds 25 Ω, an additional electrode is required and shall be at least 6" away. See 250.53(A)(3).

F. 250.68(A – B) The mechanical elements used to terminate a GEC shall be accessible and installed in a manner that makes connection both permanent and effective. Bonding around insulated joints like water meters helps to ensure an effective ground path.

GROUNDING MULTIPHASE SYSTEMS

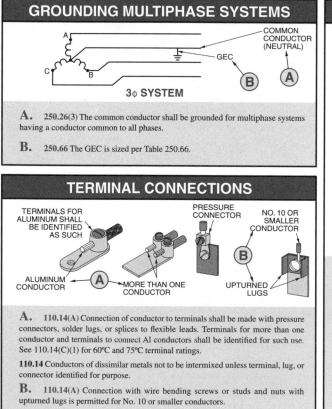

3φ SYSTEM

A. 250.26(3) The common conductor shall be grounded for multiphase systems having a conductor common to all phases.

B. 250.66 The GEC is sized per Table 250.66.

TERMINAL CONNECTIONS

TERMINALS FOR ALUMINUM SHALL BE IDENTIFIED AS SUCH

PRESSURE CONNECTOR

NO. 10 OR SMALLER CONDUCTOR

ALUMINUM CONDUCTOR

MORE THAN ONE CONDUCTOR

UPTURNED LUGS

A. 110.14(A) Connection of conductor to terminals shall be made with pressure connectors, solder lugs, or splices to flexible leads. Terminals for more than one conductor and terminals to connect Al conductors shall be identified for such use. See 110.14(C)(1) for 60ºC and 75ºC terminal ratings.

110.14 Conductors of dissimilar metals not to be intermixed unless terminal, lug, or connector identified for purpose.

B. 110.14(A) Connection with wire bending screws or studs and nuts with upturned lugs is permitted for No. 10 or smaller conductors.

SWITCH ENCLOSURES

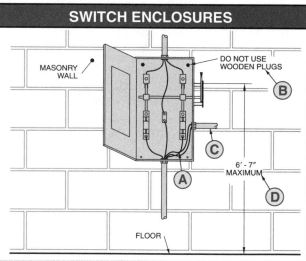

A. 110.12 Equipment shall be installed in a neat manner.

B. 110.13(A) Electrical equipment to be securely fastened to mounting surface. Wooden plugs driven into masonry or concrete are not acceptable for support.

110.3(B) Listed and labeled equipment must be installed and used in accordance with any instructions included in the listing or labeling.

C. 312.8 Switch enclosures are not to be used as auxiliary gutters, raceways, or junction boxes unless adequate space is provided. These conductors permitted if not more than 40% of cross-sectional area is used. For splices and taps, not over 75% of area may be used. See 312.8(3) for special labeling requirements.

D. 404.8(A) Center of grip of operating handle when in highest position shall not be more than 6'-7" from floor or platform.

SWITCHES

A. 404.14(A) AC general-use snap switch, suitable for use on AC only, may be used to control resistive and inductive loads not to exceed the ampere rating of the switch at the voltage involved, or tungsten-filament loads of rating of switch at 120 V, or motor loads not to exceed 80% of rating of switch of rated voltage. Snap switches rated 20 A or less shall be listed and marked CO/ALR when used with aluminum wire. See 404.14(C).

B. 404.8(A), Ex. 1, 2, 3 Switches and CBs shall be readily accessible and not over 6'-7" from floor or platform except: (1) busway switches permitted at same level as busway; (2) switches located adjacent to motors, appliances, or other equipment; (3) hookstick operated isolating switches.

404.8(B) Voltage between adjacent snap switches in same box shall not exceed 300 V. If voltage between adjacent devices exceeds 300 V, then identified securely installed barriers must be installed.

404.14(B)(1 - 3) AC/DC general-use snap switches suitable for AC or DC may be used to control resistive loads not exceeding the ampere rating of the switch or the voltage of the circuit, or inductive loads not exceeding 50% of the ampere rating of the switch at applied voltage, when "T" rated. See 430.83, 430.109, 430.110 for motors, and 600.6(B) for signs.

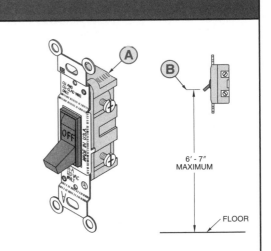

6' - 7" MAXIMUM

FLOOR

RECEPTACLE GROUNDING

A. 406.4(A) Receptacles installed on 15 A and 20 A circuits shall be of grounding type.

406.4(B – C) Receptacles and cord connectors having grounded contacts shall be grounded. Branch circuits shall include a grounding conductor unless the raceway provides grounding.

406.4(D)(1) For extension to nongrounding existing circuits, an accessible point on the GES or GEC may be used for grounding. See 250.130(C)(1 – 5).

B. 406.4(D)(3 – 6) Grounding-type receptacles to be used as replacements for existing nongrounding types except where receptacle box does not have grounding means, a GFCI or non-grounding-type receptacle is permitted.

406.8 Attachment plugs shall not be interchangeable on receptacles connected to circuits of different voltages, frequencies, or types of current.

C. 250.146(B) Contact device or yoke, if designed and listed as self-grounding, may be used to bond receptacle to flush-type boxes.

406.3(B) Receptacles and cord connectors shall be rated not less than 15 A for 125 V or 250 V and shall be of a type not suitable for use as lampholders. Also see 210.21(B).

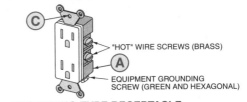

"HOT" WIRE SCREWS (BRASS)

EQUIPMENT GROUNDING SCREW (GREEN AND HEXAGONAL)

GROUNDING-TYPE RECEPTACLE

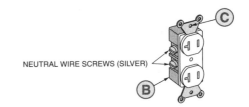

NEUTRAL WIRE SCREWS (SILVER)

NON-GROUNDING-TYPE RECEPTACLE

ELECTRIC-DISCHARGE LIGHTING

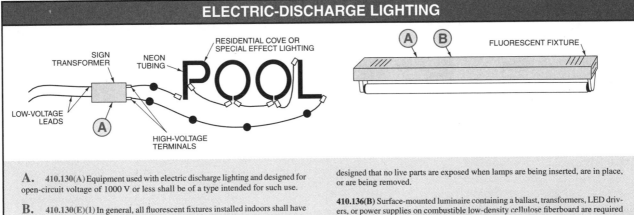

SIGN TRANSFORMER

NEON TUBING

RESIDENTIAL COVE OR SPECIAL EFFECT LIGHTING

POOL

LOW-VOLTAGE LEADS

HIGH-VOLTAGE TERMINALS

FLUORESCENT FIXTURE

A. 410.130(A) Equipment used with electric discharge lighting and designed for open-circuit voltage of 1000 V or less shall be of a type intended for such use.

B. 410.130(E)(1) In general, all fluorescent fixtures installed indoors shall have ballasts with integral thermal protection. See Special Applications for (2) simple reactance ballasts; (3) exit luminaires; and (4) egress luminaires.

410.135 Equipment having an open-circuit voltage of more than 300 V shall not be installed in dwelling occupancies. Equipment having an open-circuit voltage of more than 300 V shall not be installed in or on dwelling occupancies unless so designed that no live parts are exposed when lamps are being inserted, are in place, or are being removed.

410.136(B) Surface-mounted luminaire containing a ballast, transformers, LED drivers, or power supplies on combustible low-density cellulose fiberboard are required to have 1½" air space or listed for direct mounting.

410.136(B), Note Combustible low-density cellulose fiberboard includes sheets, panels, and tiles formed of bonded plant fiber and has a density of 20 lb or less per cubic foot. Does not include solid or laminated wood or material integrally treated with fire-retardant chemicals to limit flame spread to not over 25'.

GROUNDING — METAL ENCLOSURES

A. **250.86** Metal enclosures for other than service conductors shall be connected to the EGC. See also 250.112(I)

250.86, Ex. 1 Additions to existing installations, metal enclosures and raceways for open wiring, knob-and-tube, and NMC without an EGC need not be grounded if less than 25′ and suitably isolated or guarded against contact by persons.

B. **250.86, Ex. 2** Short sections of metal conduit used to protect cables from physical damage need not be connected to the EGC.

Table 250.122 EGC for 15 A, 20 A, and 30 A circuits shall have ampacity at least equal to overcurrent protection of circuit. For circuits over 30A, use Table 250.122.

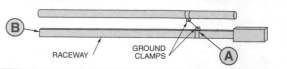

RACEWAY   GROUND CLAMPS

GROUNDING CORD-AND-PLUG-CONNECTED EQUIPMENT

A. **250.114(3)** In residential occupancies, the following cord-and-plug-connected equipment shall be connected to the EGC: clothes washers and dryers, refrigerators, ranges, freezers, air conditioners, dishwashers, electrical aquarium equipment, sump pumps, portable hand-held power tools and motor-operated appliances such as saws, sanders, drills, lawn mowers, wet scrubbers, snow blowers, portable headlamps, and hedge clippers.

B. **250.114, Ex.** Listed tools and appliances protected by a double insulation system and so marked need not be connected to the EGC.

250.138 Non-current-carrying metal parts of cord-and-plug-connected equipment which shall be grounded may be connected to the EGC: (A) by an EGC which is part of the cable or flexible cord, and which is either bare or has a green or green with one or more yellow stripes on the outer covering; or (B) by a separate flexible wire or strap. See Exception for grounding-type plug-in GFCIs.

250.122(E) Fixture wires and cords are permitted to contain a No.18 or larger Cu EGC, provided it is not smaller than circuit conductors per 240.5.

406.10(C) A grounding terminal or grounding device shall be used only for grounding purpose.

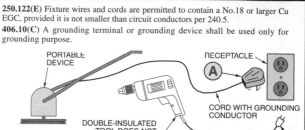

PORTABLE DEVICE

RECEPTACLE

CORD WITH GROUNDING CONDUCTOR

DOUBLE-INSULATED TOOL DOES NOT REQUIRE GROUNDING

SWITCHING CONNECTIONS

A. **404.2(A)** All 3-way and 4-way switches shall be wired so that all switching is done only in the ungrounded circuit conductor.

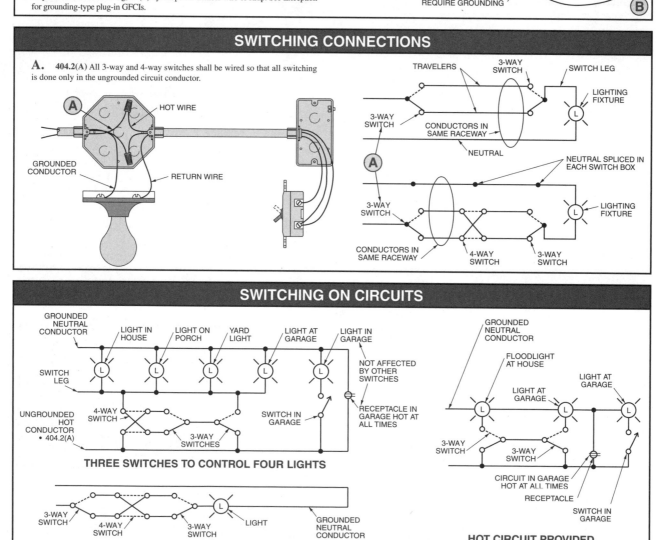

HOT WIRE

GROUNDED CONDUCTOR

RETURN WIRE

TRAVELERS   3-WAY SWITCH   SWITCH LEG

LIGHTING FIXTURE

3-WAY SWITCH   CONDUCTORS IN SAME RACEWAY

NEUTRAL

NEUTRAL SPLICED IN EACH SWITCH BOX

3-WAY SWITCH   LIGHTING FIXTURE

CONDUCTORS IN SAME RACEWAY   4-WAY SWITCH   3-WAY SWITCH

SWITCHING ON CIRCUITS

GROUNDED NEUTRAL CONDUCTOR   LIGHT IN HOUSE   LIGHT ON PORCH   YARD LIGHT   LIGHT AT GARAGE   LIGHT IN GARAGE

SWITCH LEG

NOT AFFECTED BY OTHER SWITCHES

UNGROUNDED HOT CONDUCTOR • 404.2(A)

4-WAY SWITCH

3-WAY SWITCHES

SWITCH IN GARAGE

RECEPTACLE IN GARAGE HOT AT ALL TIMES

THREE SWITCHES TO CONTROL FOUR LIGHTS

3-WAY SWITCH   4-WAY SWITCH   3-WAY SWITCH   LIGHT   GROUNDED NEUTRAL CONDUCTOR

THREE SWITCHES TO CONTROL ONE LIGHT

GROUNDED NEUTRAL CONDUCTOR

FLOODLIGHT AT HOUSE

LIGHT AT GARAGE

LIGHT AT GARAGE

3-WAY SWITCH   3-WAY SWITCH

CIRCUIT IN GARAGE HOT AT ALL TIMES

RECEPTACLE

SWITCH IN GARAGE

HOT CIRCUIT PROVIDED IN GARAGE

NONMETALLIC EXTENSIONS — 382

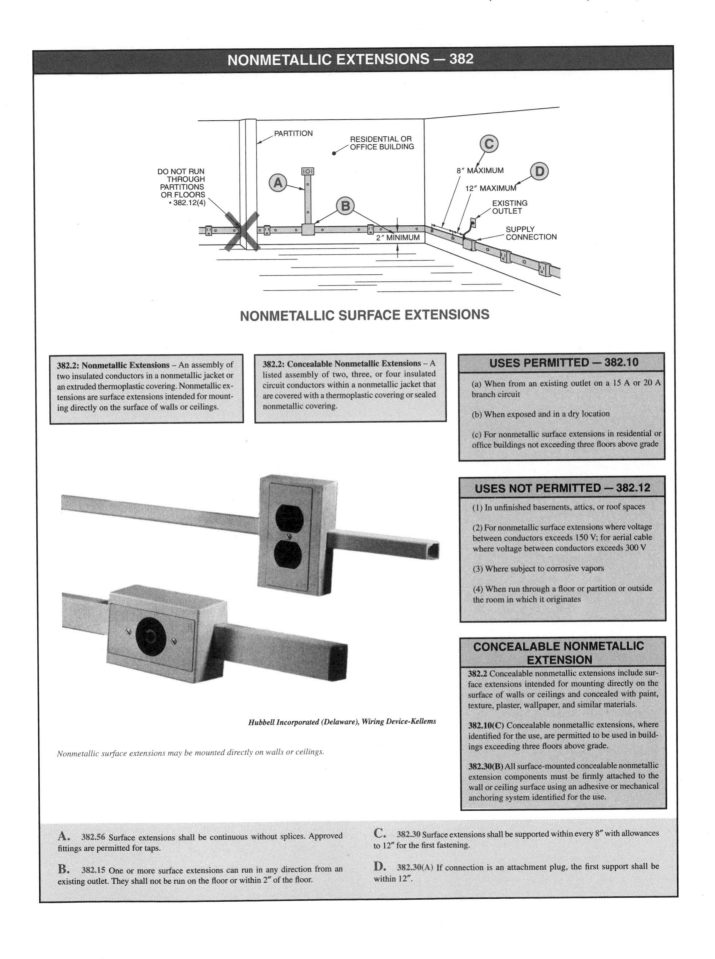

PARTITION

RESIDENTIAL OR OFFICE BUILDING

DO NOT RUN THROUGH PARTITIONS OR FLOORS
• 382.12(4)

Ⓐ

Ⓑ

Ⓒ

Ⓓ

8″ MAXIMUM

12″ MAXIMUM

EXISTING OUTLET

SUPPLY CONNECTION

2″ MINIMUM

NONMETALLIC SURFACE EXTENSIONS

382.2: Nonmetallic Extensions – An assembly of two insulated conductors in a nonmetallic jacket or an extruded thermoplastic covering. Nonmetallic extensions are surface extensions intended for mounting directly on the surface of walls or ceilings.

382.2: Concealable Nonmetallic Extensions – A listed assembly of two, three, or four insulated circuit conductors within a nonmetallic jacket that are covered with a thermoplastic covering or sealed nonmetallic covering.

USES PERMITTED — 382.10

(a) When from an existing outlet on a 15 A or 20 A branch circuit

(b) When exposed and in a dry location

(c) For nonmetallic surface extensions in residential or office buildings not exceeding three floors above grade

USES NOT PERMITTED — 382.12

(1) In unfinished basements, attics, or roof spaces

(2) For nonmetallic surface extensions where voltage between conductors exceeds 150 V; for aerial cable where voltage between conductors exceeds 300 V

(3) Where subject to corrosive vapors

(4) When run through a floor or partition or outside the room in which it originates

CONCEALABLE NONMETALLIC EXTENSION

382.2 Concealable nonmetallic extensions include surface extensions intended for mounting directly on the surface of walls or ceilings and concealed with paint, texture, plaster, wallpaper, and similar materials.

382.10(C) Concealable nonmetallic extensions, where identified for the use, are permitted to be used in buildings exceeding three floors above grade.

382.30(B) All surface-mounted concealable nonmetallic extension components must be firmly attached to the wall or ceiling surface using an adhesive or mechanical anchoring system identified for the use.

Hubbell Incorporated (Delaware), Wiring Device-Kellems

Nonmetallic surface extensions may be mounted directly on walls or ceilings.

A. 382.56 Surface extensions shall be continuous without splices. Approved fittings are permitted for taps.

B. 382.15 One or more surface extensions can run in any direction from an existing outlet. They shall not be run on the floor or within 2″ of the floor.

C. 382.30 Surface extensions shall be supported within every 8″ with allowances to 12″ for the first fastening.

D. 382.30(A) If connection is an attachment plug, the first support shall be within 12″.

FIXED ELECTRIC SPACE-HEATING EQUIPMENT — 424 . . .

A. **424.3(A)** Fixed electric space-heating equipment may be supplied by any size individual branch circuit. For two or more outlets for fixed electric space-heating equipment, branch circuits shall be 15 A, 20 A, 25A or 30 A.

424.3(A) In other than a dwelling unit, fixed infrared heating equipment may be supplied from branch circuits not exceeding 50 A.

424.3(B) Branch-circuit conductor's ampacity and the rating of OCPDs for electric space-heating equipment and motors shall not be less than 125% of the total heating load (continuous load).

424.9 Factory-installed receptacles or outlets which are part of a separate listed assembly are permitted to be installed in place of those in 210.50(B).

B. **424.11** Fixed electric space heaters with supply conductors of over 60°C shall be clearly and permanently marked.

C. **424.12(A)** Fixed electric space heaters shall be adequately protected in an approved manner where exposed to physical damage.

D. **424.12(B)** Fixed electric space heaters installed in damp or wet locations shall be listed for such use.

424.13 Unless listed for direct contact with combustible material, fixed electric space heaters shall have required spacing.

E. **424.20(A)** Thermostatic devices are permitted as both controllers and switching devices provided they:

(1) have a marked OFF position

(2) directly open all ungrounded conductors when in the OFF position

(3) are so designed that they will not energize automatically after being placed in the OFF position

(4) are located per 424.19

424.20(B) Thermostats that do not open all ungrounded conductors and thermostats that operate remote-control circuits are not required to meet 424.20(A) and shall not be used as disconnecting means.

F. **424.34** Heating cables shall have factory-assembled nonheating leads at least 7′ in length.

G. **424.35** Nonheating leads of heating cable shall be marked with the identifying name or symbol, catalog number, and ratings in volts and watts or volts and amps. The following color identification shall be used to mark nominal voltage on each nonheating lead within 3″ of the terminal end:

120 V– yellow

208 V– blue

240 V– red

277 V– brown

480 V– orange

H. **424.38(A)** Heating cables shall not extend beyond the room or area where they originate.

424.38(B) Heating cables shall not be installed in closets, over walls or partitions that extend to the ceiling, or over cabinets without proper clearance from the ceiling.

424.38(C) Heating cables may be used to control relative humidity in closets when shelves or other permanent fixtures do not obstruct the space.

I. **424.39** Heating cables shall be separated at least 8″ from outlet and junction boxes, surface fixtures, and at least 2″ from any part of recessed fixtures, ventilating, or other room openings. Surface fixtures shall not be mounted under heating cables. No heating cable shall be covered by any surface-mounted equipment.

424.40 Embedded heating cables may be spliced by approved means where necessary. The length of the cable shall not be altered.

424.41(A) In general, heating cables are not permitted in walls.

424.41(A) Single runs of heating cable are permitted to run down a vertical surface to reach a dropped ceiling.

J. **424.41(B)** Adjacent runs of heating cable not exceeding 2¾ W per foot shall be installed at least 1½″ OC.

K. **424.41(C)** Heating cables shall be applied only to gypsum board, plaster lath, or other nonconducting surfaces. A layer of plaster shall be applied to metal lath or conducting surfaces before applying heating cables.

424.41(E) The entire ceiling surface shall have a finish of thermally noninsulating material at least ½″ in nominal thickness or noninsulating material suitable for the use and installed per specifications.

L. **424.41(F)** Heating cables shall be secured at intervals not exceeding 16″ by means of staples, tape, plaster, or other approved means. Metal fasteners shall not be used with metal lath or other conducting surfaces.

424.41(H) Cables shall not contact metal or other electrically conducting surface.

M. **424.43(A)** Free nonheating leads from a junction box to a location within the ceiling shall be installed with approved wiring methods. Conductors in raceways, Type UF, NMC, MI cable, or other approved conductors may be used.

N. **424.43(B)** At least 6″ of free nonheating lead shall be within the junction box. Markings of the lead shall be visible.

424.43(C) Excess leads shall not be cut. They shall be embedded in plaster or other approved material

O. **424.44(A)** Constant wattage heating cables in concrete or poured masonry floors shall not exceed 16½ W per linear foot.

P. **424.44(B)** The minimum spacing between adjacent runs of heating cable in concrete or poured masonry floors is 1″ OC.

Q. **424.44(C)** Heating cables in concrete shall be held in place by nonmetallic frames or other approved means while concrete is poured.

424.44(D) Spacing shall be maintained between heating cable and metal in concrete floors unless the cable is grounded MC cable.

R. **424.44(E)** Leads leaving the concrete floor shall be protected by RMC, IMC, RNC, EMT, or other approved means.

S. **424.44(F)** Bushings or approved fittings shall be used in the floor slab where the leads emerge.

424.44(G) GFCI protection must be provided for cables installed in electrically heated floors of bathrooms, kitchens, and hydromassage bathtub locations.

424.45 Cable installations shall be inspected and approved before cables are covered or concealed.

424.58 Heaters installed in air ducts shall be identified as suitable for such use.

424.65 Controller equipment for duct heaters shall be accessible with the disconnect at or within sight of the controller or as permitted by 424.19(A).

424.91: Heating Panel – A complete assembly with a junction box or length of flexible conduit for connection.

424.91: Heating Panel Set – A rigid or nonrigid assembly with nonheating leads or a terminal junction identified as being suitable for connection.

T. **424.92(A)** Panels to be marked in a location that is visible before finish is applied.

424.92(B) Heating panels shall be identified as being suitable for the installation.

U. **424.92(C)** Each heating panel shall be marked with identifying name or symbol, catalog number, volts and watts, or volts and amps.

424.93(A)(2) Heating panels and heating panel sets shall not be:

(1) installed in or behind surfaces where subject to physical damage

(2) run through or above walls and ceilings

(3) run in or through thermal insulation

V. **424.93(A)(3)** Panels shall be separated at least 8″ from outlet and junction boxes for surface fixtures and at least 2″ from any part of recessed fixtures, ventilating, or other room openings. Surface fixtures shall not be mounted under heating panels.

424.93(B)(4) Heating panel sets shall be installed complete unless identified for field cutting.

W. **424.94** Wiring installed above a heated ceiling shall be considered as operating at an ambient temperature of 50°C and shall be derated accordingly. Wiring shall be spaced at least 2″ above the heated ceiling.

X. **424.94** Wiring located above thermal insulation which is at least 2″ thick need not be derated.

Y. **424.95(A)** Wiring methods are per Article 300 and 310.15(A)(3).

...FIXED ELECTRIC SPACE-HEATING EQUIPMENT — 424

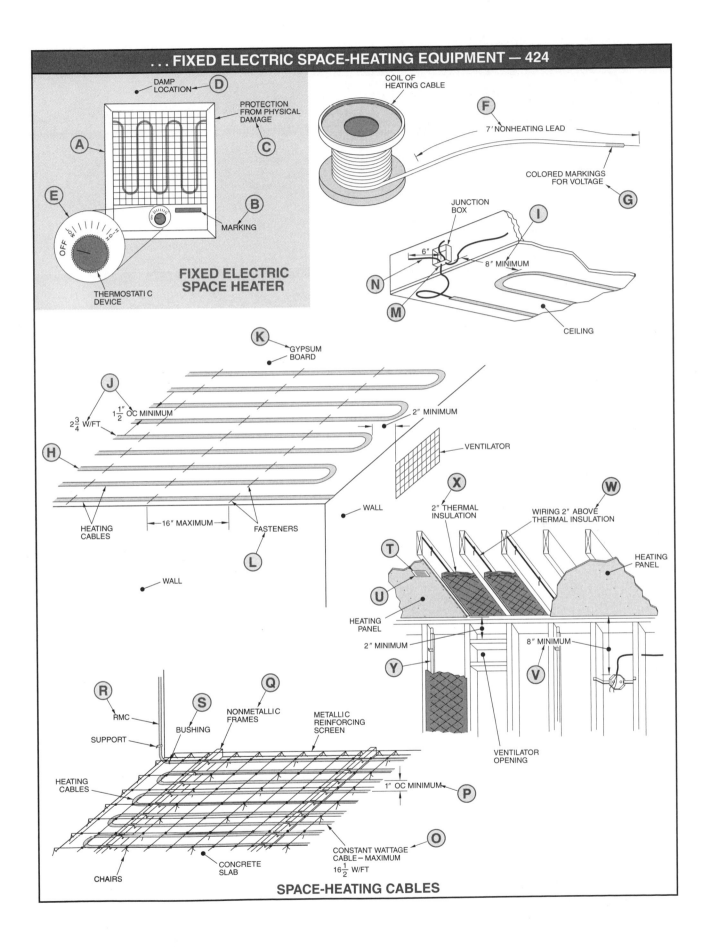

DAMP LOCATION — D

PROTECTION FROM PHYSICAL DAMAGE — C

A

E

B — **MARKING**

THERMOSTATIC DEVICE

FIXED ELECTRIC SPACE HEATER

OFF

COIL OF HEATING CABLE

F — 7' NONHEATING LEAD

COLORED MARKINGS FOR VOLTAGE — G

JUNCTION BOX

I

6"

8" MINIMUM

N

M

CEILING

K — **GYPSUM BOARD**

J — $1\frac{1}{2}$" OC MINIMUM

$2\frac{3}{4}$ W/FT

2" MINIMUM

H

VENTILATOR

X — 2" THERMAL INSULATION

W — WIRING 2" ABOVE THERMAL INSULATION

HEATING PANEL

HEATING CABLES

16" MAXIMUM

FASTENERS

L

WALL

T

U

HEATING PANEL

2" MINIMUM

Y

8" MINIMUM

V

WALL

R — RMC

SUPPORT

S — BUSHING

Q — NONMETALLIC FRAMES

METALLIC REINFORCING SCREEN

VENTILATOR OPENING

1" OC MINIMUM — P

HEATING CABLES

CHAIRS

CONCRETE SLAB

CONSTANT WATTAGE CABLE — MAXIMUM $16\frac{1}{2}$ W/FT — O

SPACE-HEATING CABLES

FIXED OUTDOOR ELECTRIC DEICING AND SNOW-MELTING EQUIPMENT — 426

Heating System–A complete system of components such as heating elements, fasteners, nonheating circuit wiring, leads, temperature controllers, safety signs, junction boxes, raceways, and fittings.

Resistance Heating Element – A specific element to generate heat that is embedded in or fastened to the surface.

Impedance Heating System – A system in which heat is generated by pipe(s) and rod(s) by a nonvoltage electrical source from an isolating transformer.

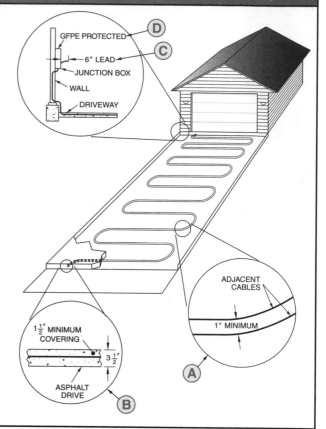

426.4 Branch circuit conductors for fixed electric deicing and snow-melting equipment shall have an ampacity of at least 125% of total heater load (continuous load).

426.10(1) Equipment for outdoor electric deicing and snow melting shall be suitable for the chemical, thermal, and physical environment.

426.10(2) Equipment for outdoor electric deicing and snow melting shall be installed per the manufacturer's plans and instructions.

426.20(A) Panels shall not exceed 120 W/sq ft of heated area.

A. **426.20(B)** Spacing between adjacent cables shall be at least 1′ OC.

B. 426.20(C) Units, panels, or cables shall be installed:

(1) On 2″ of asphalt or masonry base with at least 1½″ of covering

(2) On other approved bases at least 3½″ thick and 1½″ minimum top covering

(3) Equipment listed for other forms of installation shall be installed in the manner for which it has been identified.

426.20(E) Cables and panels shall not bridge expansion joints without provisions for expansion and contraction.

426.22(A) Nonheating leads with a grounding sheath or braid may be embedded in asphalt or masonry without other protection.

426.22(B) All but 1″ to 6″ of nonheating leads not having a grounded conductor shall be enclosed in RMC, EMT, IMC, or other raceways within asphalt or masonry.

C. 426.23(A) Not less than 6″ of nonheating leads shall be in junction box.

D. 426.28 GFPEs shall be provided for fixed outdoor electric deicing and snow-melting equipment.

OVERCURRENT PROTECTION OF FIXED ELECTRIC SPACE-HEATING EQUIPMENT — 424.22

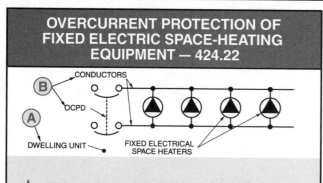

A. 220.51 Fixed electric space-heating loads are computed at 100% of the total connected load. The feeder load current rating shall not be smaller than the rating of the largest branch circuit supplied.

220.51, Ex. The AHJ may grant permission for feeder conductors to have an ampacity less than 100% when units operate on duty cycle, intermittently, or not at one time.

B. 424.3(B) The branch-circuit conductors and overcurrent protective devices which supply fixed electric space-heating equipment and motors shall be rated not less than 125% of the total load of the motors and heaters (continuous load).

424.22(A) Fixed electric space-heating equipment, except motor-operated equipment per 430 and 440, is permitted to be protected against overcurrent when supplied by a branch circuit per 210.

424.22(B) When resistance-type heating elements of over 48 A are used in space-heating equipment, they shall be subdivided into smaller circuits of not over 48 A and protected at not more than 60 A. The rating of the overcurrent device shall be per 424.3(B) where the subdivided load is less than 48 A.

INFRARED LAMP INDUSTRIAL HEATING APPLIANCES — 422.48

A. 422.48(A) Infrared heating lamps rated at 300 W or less may be used with lampholders of the medium-base unswitched type or other types identified for the purpose.

422.48(B) Infrared lamps over 300 W shall not use screw-shell lampholders unless the lampholders are identified for use at over 300 W.

LOADS FOR ADDITION TO EXISTING BUILDINGS — 20.16

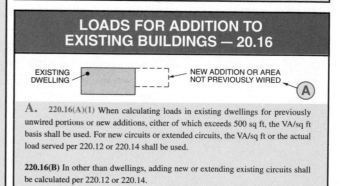

A. 220.16(A)(1) When calculating loads in existing dwellings for previously unwired portions or new additions, either of which exceeds 500 sq ft, the VA/sq ft basis shall be used. For new circuits or extended circuits, the VA/sq ft or the actual load served per 220.12 or 220.14 shall be used.

220.16(B) In other than dwellings, adding new or extending existing circuits shall be calculated per 220.12 or 220.14.

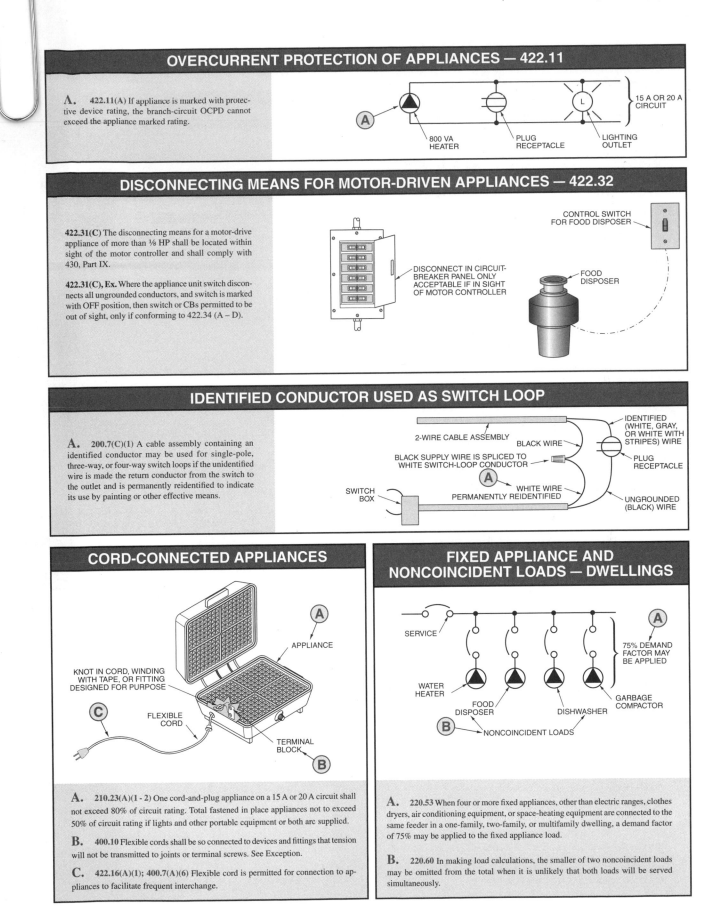

OVERCURRENT PROTECTION OF APPLIANCES — 422.11

A. **422.11(A)** If appliance is marked with protective device rating, the branch-circuit OCPD cannot exceed the appliance marked rating.

(A) → 800 VA HEATER — PLUG RECEPTACLE — LIGHTING OUTLET — 15 A OR 20 A CIRCUIT

DISCONNECTING MEANS FOR MOTOR-DRIVEN APPLIANCES — 422.32

422.31(C) The disconnecting means for a motor-drive appliance of more than ⅛ HP shall be located within sight of the motor controller and shall comply with 430, Part IX.

422.31(C), Ex. Where the appliance unit switch disconnects all ungrounded conductors, and switch is marked with OFF position, then switch or CBs permitted to be out of sight, only if conforming to 422.34 (A – D).

CONTROL SWITCH FOR FOOD DISPOSER
DISCONNECT IN CIRCUIT-BREAKER PANEL ONLY ACCEPTABLE IF IN SIGHT OF MOTOR CONTROLLER
FOOD DISPOSER

IDENTIFIED CONDUCTOR USED AS SWITCH LOOP

A. **200.7(C)(1)** A cable assembly containing an identified conductor may be used for single-pole, three-way, or four-way switch loops if the unidentified wire is made the return conductor from the switch to the outlet and is permanently reidentified to indicate its use by painting or other effective means.

2-WIRE CABLE ASSEMBLY
BLACK WIRE
IDENTIFIED (WHITE, GRAY, OR WHITE WITH STRIPES) WIRE
BLACK SUPPLY WIRE IS SPLICED TO WHITE SWITCH-LOOP CONDUCTOR
PLUG RECEPTACLE
(A)
SWITCH BOX
WHITE WIRE PERMANENTLY REIDENTIFIED
UNGROUNDED (BLACK) WIRE

CORD-CONNECTED APPLIANCES

(A) APPLIANCE
KNOT IN CORD, WINDING WITH TAPE, OR FITTING DESIGNED FOR PURPOSE
(C) FLEXIBLE CORD
TERMINAL BLOCK (B)

A. **210.23(A)(1 - 2)** One cord-and-plug appliance on a 15 A or 20 A circuit shall not exceed 80% of circuit rating. Total fastened in place appliances not to exceed 50% of circuit rating if lights and other portable equipment or both are supplied.

B. **400.10** Flexible cords shall be so connected to devices and fittings that tension will not be transmitted to joints or terminal screws. See Exception.

C. **422.16(A)(1); 400.7(A)(6)** Flexible cord is permitted for connection to appliances to facilitate frequent interchange.

FIXED APPLIANCE AND NONCOINCIDENT LOADS — DWELLINGS

SERVICE
(A) 75% DEMAND FACTOR MAY BE APPLIED
WATER HEATER
FOOD DISPOSER
DISHWASHER
GARBAGE COMPACTOR
(B) NONCOINCIDENT LOADS

A. **220.53** When four or more fixed appliances, other than electric ranges, clothes dryers, air conditioning equipment, or space-heating equipment are connected to the same feeder in a one-family, two-family, or multifamily dwelling, a demand factor of 75% may be applied to the fixed appliance load.

B. **220.60** In making load calculations, the smaller of two noncoincident loads may be omitted from the total when it is unlikely that both loads will be served simultaneously.

ELECTRIC RANGES

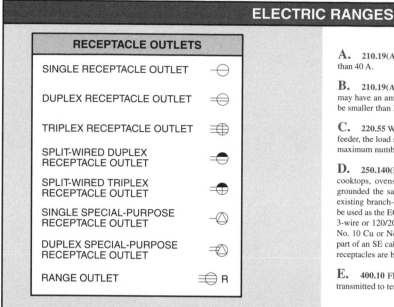

RECEPTACLE OUTLETS

SINGLE RECEPTACLE OUTLET	
DUPLEX RECEPTACLE OUTLET	
TRIPLEX RECEPTACLE OUTLET	
SPLIT-WIRED DUPLEX RECEPTACLE OUTLET	
SPLIT-WIRED TRIPLEX RECEPTACLE OUTLET	
SINGLE SPECIAL-PURPOSE RECEPTACLE OUTLET	
DUPLEX SPECIAL-PURPOSE RECEPTACLE OUTLET	
RANGE OUTLET	R

A. 210.19(A)(3) Conductors for ranges 8¾ kW and larger shall not be rated less than 40 A.

B. 210.19(A)(3), Ex. 2 The neutral conductor for a household electric range may have an ampacity not less than 70% of the branch-circuit rating and shall not be smaller than No. 10.

C. 220.55 Where two or more single-phase ranges are supplied by a 3ϕ, 4-wire feeder, the load shall be computed per Table 220.55 on a demand basis of twice the maximum number between any two phases.

D. 250.140(Ex.) For new branch-circuit installations, frames of electric ranges, cooktops, ovens, dryers, and their associated outlet or junction boxes shall be grounded the same as other electrical equipment. See 250.134 and 250.138. For existing branch-circuit installations, the grounded circuit conductor (neutral) may be used as the EGC, if all following conditions are met: (1) circuit is 120/240 V, 1ϕ, 3-wire or 120/208 V derived from a 3ϕ, 4-wire system; (2) grounded conductor is No. 10 Cu or No. 8 Al; (3) grounded conductor is insulated or is uninsulated and part of an SE cable and is fed from service equipment; or (4) grounded contacts of receptacles are bonded to equipment.

E. 400.10 Flexible cords and cables shall be connected so that tension is not transmitted to terminals.

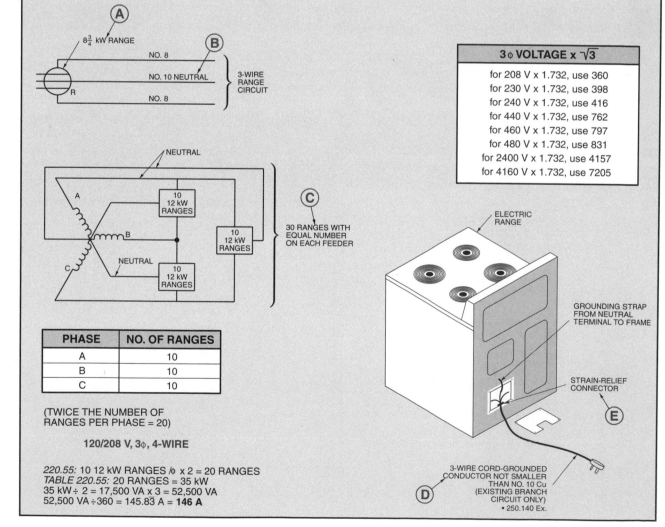

3ϕ VOLTAGE x √3

for 208 V x 1.732, use 360
for 230 V x 1.732, use 398
for 240 V x 1.732, use 416
for 440 V x 1.732, use 762
for 460 V x 1.732, use 797
for 480 V x 1.732, use 831
for 2400 V x 1.732, use 4157
for 4160 V x 1.732, use 7205

PHASE	NO. OF RANGES
A	10
B	10
C	10

(TWICE THE NUMBER OF RANGES PER PHASE = 20)

120/208 V, 3ϕ, 4-WIRE

220.55: 10 12 kW RANGES /ϕ x 2 = 20 RANGES
TABLE 220.55: 20 RANGES = 35 kW
35 kW ÷ 2 = 17,500 VA x 3 = 52,500 VA
52,500 VA ÷ 360 = 145.83 A = **146 A**

AIR CONDITIONING EQUIPMENT — 440

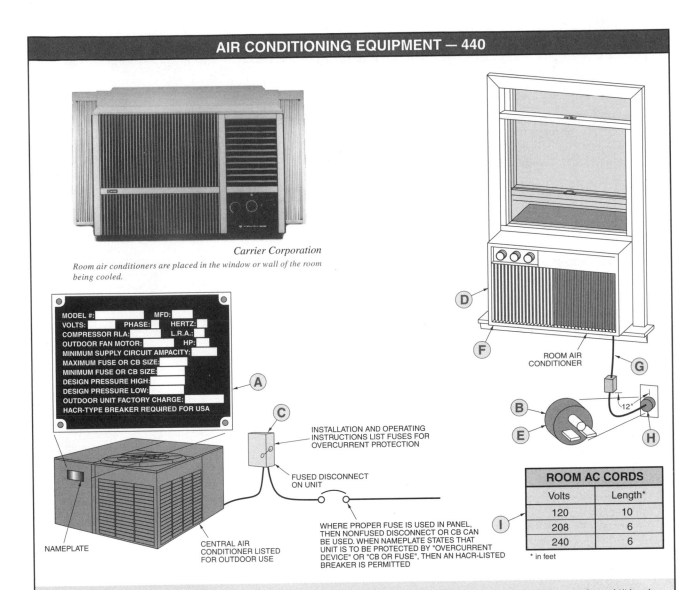

Carrier Corporation

Room air conditioners are placed in the window or wall of the room being cooled.

MODEL #: _____ MFD: _____
VOLTS: ____ PHASE: __ HERTZ: ____
COMPRESSOR RLA: ____ L.R.A.: ____
OUTDOOR FAN MOTOR: ____ HP: ____
MINIMUM SUPPLY CIRCUIT AMPACITY: ____
MAXIMUM FUSE OR CB SIZE: ____
MINIMUM FUSE OR CB SIZE: ____
DESIGN PRESSURE HIGH: ____
DESIGN PRESSURE LOW: ____
OUTDOOR UNIT FACTORY CHARGE: ____
HACR-TYPE BREAKER REQUIRED FOR USA

INSTALLATION AND OPERATING INSTRUCTIONS LIST FUSES FOR OVERCURRENT PROTECTION

FUSED DISCONNECT ON UNIT

WHERE PROPER FUSE IS USED IN PANEL, THEN NONFUSED DISCONNECT OR CB CAN BE USED. WHEN NAMEPLATE STATES THAT UNIT IS TO BE PROTECTED BY "OVERCURRENT DEVICE" OR "CB OR FUSE", THEN AN HACR-LISTED BREAKER IS PERMITTED

NAMEPLATE

CENTRAL AIR CONDITIONER LISTED FOR OUTDOOR USE

ROOM AIR CONDITIONER

ROOM AC CORDS	
Volts	Length*
120	10
208	6
240	6

* in feet

12"

A. 110.3(B) Equipment that is listed or labeled shall be installed or used per the instructions included in the listing or labeling.

250.110 Exposed non-current-carrying metal parts of fixed equipment shall be connected to the EGC if: (1) within 8' vertically or 5' horizontally of ground or grounded objects and possible contact by persons; (2) in damp or wet location and not isolated; (3) in electrical contact with metal; (4) in hazardous location; (5) if fed by metal enclosed wiring (except for 25' of isolated raceway per 250.86, Ex 2.; or (6) if equipment operates at over 150 V to ground. See Ex. 1 – 3.

B. 250.114 Exposed metal non-current-carrying parts of cord-and-plug-connected equipment shall be grounded if: (1) in hazardous location; (2) if equipment operates at over 150 V to ground; (3) in residential occupancies: refrigerators, freezers, air conditioners, ranges, clothes washers, clothes dryers, dishwashers, sump pumps, and portable tools unless double insulated. See 250.114 (4) for grounding of cord-and-plug equipment in other than residential occupancies. See Exceptions.

440.12(A) Disconnect selected on basis of nameplate rated-load current or branch-circuit selection current (whichever is greater), and locked-rotor current:

(A)(1) ampere rating at least 115% of FLA where non-horsepower disconnect (such as CB) is used; (A)(2) to determine size of HP rated disconnect, as required by 430.109, the HP corresponding to the FLA in Tables 430.248, 430.249, 430.250, the HP corresponding to the LRA in Tables 430.251(A) or (B), where different HP ratings are obtained, the largest HP shall be used.

C. 440.14 Disconnect within sight and readily accessible for air conditioning or refrigerating equipment except for appliances connected by cord and plug. Permitted on or within equiment. See Ex. 2 for cord-and-plug-connected appliances.

D. 440.60 Generally, a room air conditioner must be in the window or wall or room being cooled, incorporate a hermetic-refrigerant motor-compressor, and operate at not over 250 V, 1ϕ. A 3ϕ unit rated over 250 V shall be directly wired with a recognized wiring method from Chapter 3 of the NEC®

E. 440.61 Grounding of room air conditioning units shall be in accordance with 250.110, 250.112, and 250.114.

F. 440.62(A) For determining branch-circuit requirements, a room air conditioner shall be considered a single motor when it is (1) cord-and-plug-connected; (2) rating is less than 40 A at 250 V, 1ϕ (3) total rated-

load current is provided on nameplate; and (4) branch-circuit, short-circuit, and ground-fault protective device is less than ampacity of branch-circuit conductors or receptacle rating.

G. 440.62(B) The rating of a cord-and-plug-connected room air conditioner shall not exceed 80% of its individual branch-circuit rating where no other loads are supplied.

440.62(C) The total load of cord-and-plug-connected room air conditioning equipment shall not exceed 50% of the rating of a branch circuit which also supplies lighting units or appliances.

H. 440.63 A plug and receptacle or cord connector may serve as disconnecting means for a 1ϕ air conditioner rated 250 V or less when the manual controls are readily accessible and within 6' of floor, or an approved manually operable switch is readily accessible and within sight of unit.

I. 440.64 Supply cords for room air conditioners shall not exceed 10' at 120 V or 6' at 208 V or 240 V.

J. 440.65 LCDI or AFCI protection required for all 1ϕ cord-and-plug-connected room A/C units.

VOLTAGE DROP ...

VOLTAGE DROP

Voltage drop is the voltage that is lost due to the resistance of conductors. The longer the conductor for a given size, the larger a conductor is selected, or the voltage may be increased (if available). Per 215.(A), Note 2, the maximum recommended voltage drop for conductors for feeders is 3%. The maximum recommended voltage drop for conductors feeders and branch circuits is 5%. The voltage drop is figured to the last outlet.

Chapter 9, Table 8, Conductor Properties, lists wire sizes from 18 AWG through 2000 kcmil. The circular mil (CM) area is given for 18 AWG through 4/0 AWG sizes. For example, 4 AWG has a cross-sectional area of 41,740 CM. The CMs for No. 250 kcmil through No. 2000 kcmil can be determined by adding three zeroes to the kcmil size. For example, No. 500 kcmil conductor has a cross-sectional area of 500,000 CM.

Resistance of conductors is given in the last three columns of Ch 9, Table 8. These resistances are used to find exact K and in solving voltage drop problems.

VOLTAGE DROP VARIABLES

V = Voltage (in V)

VD = Voltage drop (in V)

$\%VD$ = Percent voltage drop (in %)

V_l = Voltage loss (in V)

V_s = Supply voltage (in V)

K = Resistivity of conductor (in Ω)

R = Resistance of conductor (in Ω/kft)

CM = Circular mils (in area)

I = Current (in A)

L = Length of conductor (in ft)

1000 = 1000′ or less of conductor

$0.866 = \dfrac{\sqrt{3}}{2}$

Note: x2 for 1ϕ
   x2 × 0.866 for 3ϕ

VOLTAGE DROP FORMULAS

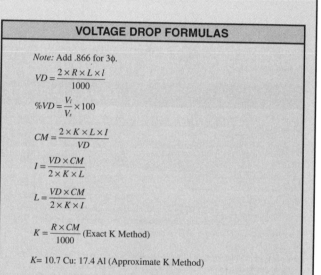

Note: Add .866 for 3ϕ.

$$VD = \frac{2 \times R \times L \times I}{1000}$$

$$\%VD = \frac{V_l}{V_s} \times 100$$

$$CM = \frac{2 \times K \times L \times I}{VD}$$

$$I = \frac{VD \times CM}{2 \times K \times L}$$

$$L = \frac{VD \times CM}{2 \times K \times I}$$

$$K = \frac{R \times CM}{1000} \text{ (Exact K Method)}$$

K= 10.7 Cu: 17.4 Al (Approximate K Method)

The *K factor (constant)* is the resistance of a circular-mil foot of wire at a set temperature. A circular-mil-foot of wire is a wire 1′ long with a cross-sectional area of kcmil. Two methods are used to find K: one for approximate K and one for exact K. Either K may be used in the voltage drop formula. The approximate K is most commonly used.

The approximate K is 10.7 Ω for Cu and 17.4 Ω for Al for all size conductors. The exact K is found by selecting the resistance for the conductor from Ch 9, Table 8 and moving the decimal point three places o the right and applying the formula:

Ch 9, Table 8: No. 10 = 10,380 CM

$$K = \frac{R \times CM}{1000}$$

For example, the exact K for a No. 10 solid, uncoated Cu conductor is 12.56 Ω.

$$K = \frac{R \times CM}{1000}$$

$$K = \frac{1.21 \times 10,380}{1000}$$

$$K = \mathbf{12.56}\ \Omega$$

AC VOLTAGE MEASUREMENT

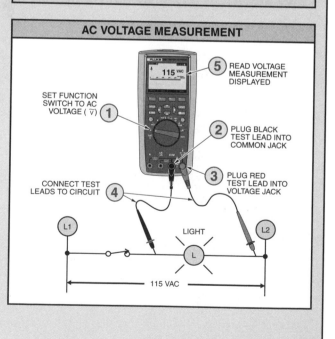

115 VAC

① SET FUNCTION SWITCH TO AC VOLTAGE ($\overline{v}$)

② PLUG BLACK TEST LEAD INTO COMMON JACK

③ PLUG RED TEST LEAD INTO VOLTAGE JACK

④ CONNECT TEST LEADS TO CIRCUIT

⑤ READ VOLTAGE MEASUREMENT DISPLAYED

L1   LIGHT   L2

L

115 VAC

. . . VOLTAGE DROP . . .

CIRCUIT A

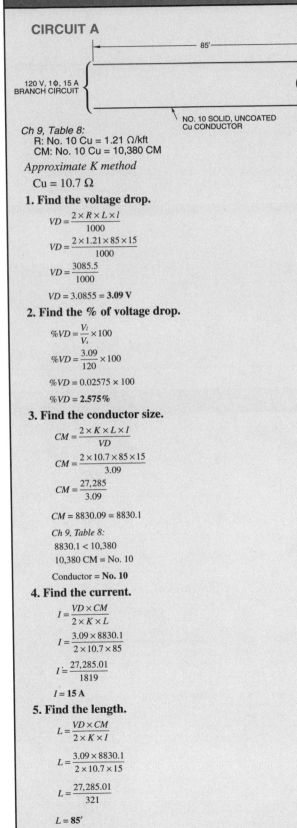

85'

120 V, 1Φ, 15 A
BRANCH CIRCUIT

L

NO. 10 SOLID, UNCOATED
Cu CONDUCTOR

Ch 9, Table 8:
R: No. 10 Cu = 1.21 Ω/kft
CM: No. 10 Cu = 10,380 CM

Approximate K method

Cu = 10.7 Ω

1. Find the voltage drop.

$$VD = \frac{2 \times R \times L \times l}{1000}$$

$$VD = \frac{2 \times 1.21 \times 85 \times 15}{1000}$$

$$VD = \frac{3085.5}{1000}$$

$$VD = 3.0855 = \textbf{3.09 V}$$

2. Find the % of voltage drop.

$$\%VD = \frac{V_l}{V_s} \times 100$$

$$\%VD = \frac{3.09}{120} \times 100$$

$$\%VD = 0.02575 \times 100$$

$$\%VD = \textbf{2.575\%}$$

3. Find the conductor size.

$$CM = \frac{2 \times K \times L \times I}{VD}$$

$$CM = \frac{2 \times 10.7 \times 85 \times 15}{3.09}$$

$$CM = \frac{27,285}{3.09}$$

$$CM = 8830.09 = 8830.1$$

Ch 9, Table 8:
8830.1 < 10,380
10,380 CM = No. 10

Conductor = **No. 10**

4. Find the current.

$$I = \frac{VD \times CM}{2 \times K \times L}$$

$$I = \frac{3.09 \times 8830.1}{2 \times 10.7 \times 85}$$

$$I = \frac{27,285.01}{1819}$$

$$I = \textbf{15 A}$$

5. Find the length.

$$L = \frac{VD \times CM}{2 \times K \times I}$$

$$L = \frac{3.09 \times 8830.1}{2 \times 10.7 \times 15}$$

$$L = \frac{27,285.01}{321}$$

$$L = \textbf{85'}$$

Exact K Method

Ch 9, Table 8:
R: No. 10 Cu = 1.21 Ω/kft
CM: No. 10 Cu = 10,380 CM

$$K = \frac{R \times CM}{1000}$$

$$K = \frac{1.21 \times 10,380}{1000}$$

$$K = \frac{12,559.8}{1000}$$

$$K = 12.559 = \textbf{12.56 Ω}$$

1. Find the voltage drop.

$$VD = \frac{2 \times R \times L \times l}{1000}$$

$$VD = \frac{2 \times 1.21 \times 85 \times 15}{1000}$$

$$VD = \frac{3085.5}{1000}$$

$$VD = 3.0855 = \textbf{3.09 V}$$

2. Find the % of voltage drop.

$$\%VD = \frac{V_l}{V_s} \times 100$$

$$\%VD = \frac{3.09}{120} \times 100$$

$$\%VD = 0.02575 \times 100$$

$$\%VD = \textbf{2.575\%}$$

3. Find the conductor size.

$$CM = \frac{2 \times K \times L \times I}{VD}$$

$$CM = \frac{2 \times 12.56 \times 85 \times 15}{3.09}$$

$$CM = \frac{32,028}{3.09}$$

$$CM = 10,365$$

Ch 9, Table 8:
10,365 < 10,380
10,380 CM = No. 10

Conductor = **No. 10**

4. Find the current.

$$I = \frac{VD \times CM}{2 \times K \times L}$$

$$I = \frac{3.09 \times 10,365}{2 \times 12.56 \times 85}$$

$$I = \frac{32,027.85}{2135.2}$$

$$I = 14.99 = \textbf{15 A}$$

5. Find the length.

$$L = \frac{VD \times CM}{2 \times K \times I}$$

$$L = \frac{3.09 \times 10,365}{2 \times 12.56 \times 15}$$

$$L = \frac{32,027.85}{376.8}$$

$$L = 84.99 = \textbf{85'}$$

. . . VOLTAGE DROP

CIRCUIT B

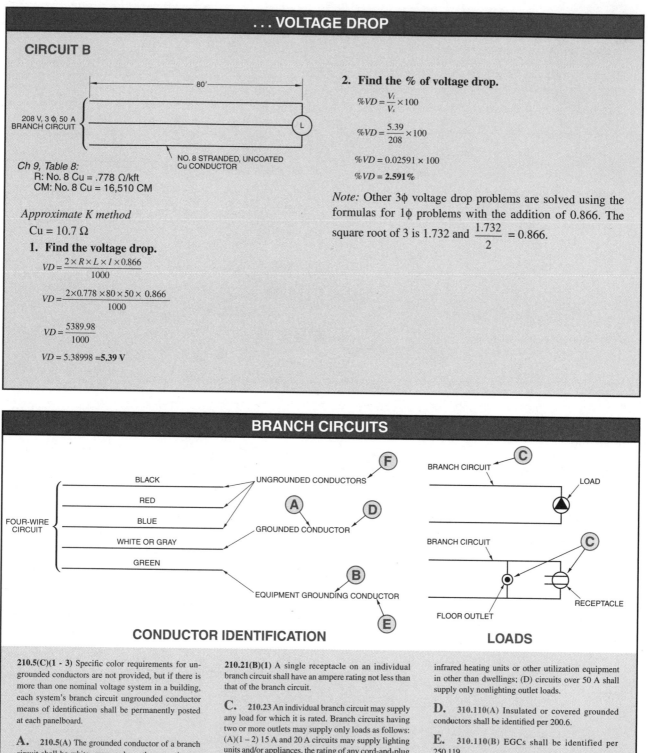

208 V, 3 Ø, 50 A
BRANCH CIRCUIT

80′

NO. 8 STRANDED, UNCOATED
Cu CONDUCTOR

Ch 9, Table 8:
R: No. 8 Cu = .778 Ω/kft
CM: No. 8 Cu = 16,510 CM

Approximate K method
Cu = 10.7 Ω

1. Find the voltage drop.

$$VD = \frac{2 \times R \times L \times I \times 0.866}{1000}$$

$$VD = \frac{2 \times 0.778 \times 80 \times 50 \times 0.866}{1000}$$

$$VD = \frac{5389.98}{1000}$$

$$VD = 5.38998 = \textbf{5.39 V}$$

2. Find the % of voltage drop.

$$\%VD = \frac{V_t}{V_s} \times 100$$

$$\%VD = \frac{5.39}{208} \times 100$$

$$\%VD = 0.02591 \times 100$$

$$\%VD = \textbf{2.591\%}$$

Note: Other 3ϕ voltage drop problems are solved using the formulas for 1ϕ problems with the addition of 0.866. The square root of 3 is 1.732 and $\frac{1.732}{2} = 0.866$.

BRANCH CIRCUITS

FOUR-WIRE CIRCUIT

BLACK
RED
BLUE
WHITE OR GRAY
GREEN

Ⓕ UNGROUNDED CONDUCTORS
Ⓐ Ⓓ
GROUNDED CONDUCTOR
Ⓑ
EQUIPMENT GROUNDING CONDUCTOR
Ⓔ

Ⓒ BRANCH CIRCUIT
LOAD

BRANCH CIRCUIT Ⓒ

FLOOR OUTLET
RECEPTACLE

CONDUCTOR IDENTIFICATION

LOADS

210.5(C)(1 - 3) Specific color requirements for ungrounded conductors are not provided, but if there is more than one nominal voltage system in a building, each system's branch circuit ungrounded conductor means of identification shall be permanently posted at each panelboard.

A. **210.5(A)** The grounded conductor of a branch circuit shall be white, gray, or have three continuous white stripes its entire length. Each other system in the same raceway or enclosure shall have the grounded conductor (if required) identified by outer covering of white with colored stripe (not green). See 200.6.

B. **250.119** The EGC of a branch circuit shall be green, green with one or more yellow stripes, or bare.

210.21(B)(1) A single receptacle on an individual branch circuit shall have an ampere rating not less than that of the branch circuit.

C. **210.23** An individual branch circuit may supply any load for which it is rated. Branch circuits having two or more outlets may supply only loads as follows: (A)(1 – 2) 15 A and 20 A circuits may supply lighting units and/or appliances, the rating of any cord-and-plug connected appliance not exceeding 80% of circuit, and total rating of fixed appliances not over 50%; (B) 30 A circuits may supply heavy-duty lampholders in other than dwellings, or appliances anywhere the rating of a cord-and-plug-connected appliance does not exceed 80% of circuit; (C) 40 A or 50 A branch circuits may supply fixed cooking appliances in any occupancy or fixed lighting units with heavy-duty lampholders or

infrared heating units or other utilization equipment in other than dwellings; (D) circuits over 50 A shall supply only nonlighting outlet loads.

D. **310.110(A)** Insulated or covered grounded conductors shall be identified per 200.6.

E. **310.110(B)** EGCs shall be identified per 250.119.

F. **310.110(C)** Ungrounded conductors shall be identified by colors or a combination of colors plus a distinguishing marking, other than white, gray, or green. These colors shall clearly distinguish the ungrounded conductor from the grounded or grounding conductor and shall not conflict with markings required by 310.120(B)(1). See 215.12 for feeder identification.

Multifamily Dwellings

TRADE COMPETENCY TEST 1

3

Name _____ Date _____

NEC®	Answer	

_____   _B_   1. The minimum size of a copper EGC required for equipment connected to a 40 A circuit is No. _10_.

     A. 8   C. 12

     B. 10   D. 14

_____   _electrically_   2. Splices that will be soldered shall first be made mechanically and ___ secure.

_____   (T)   F   3. If the resistance of a single driven ground rod exceeds 25 Ω to ground, a supplemental electrode shall be added.

_____   _____   4. Rod-type grounding electrodes of zinc shall be at least ___″ in diameter, unless listed.

_____   _____   5. For a run of rigid metal raceway interrupted by an expansion joint, the sections of rigid conduit shall be bonded by a(n) ___ jumper or other means.

_____   _40_   6. In general, the minimum branch circuit rating to a single electric range of over 8¾ kVA is _40_ A.

_____   (T)   F   7. Stranded wires No. 10 and smaller may be connected to equipment by means of screws used with terminal plates having upturned lugs.

_____   _300_   8. Double locknuts are permitted for bonding where rigid metal raceways in other than service runs enter a box or cabinet, if the raceway contains any wire of more than ___ V to ground.

_____   T   (F)   9. The grounding conductor in a cable assembly for cord-and-plug-connected equipment need not be insulated.

_____   _C_   10. The three wires of a 1ϕ,120/240 V service are designated as *A*, *N*, and *B*. A 100 A, 240 V load is connected across *A* and *B*. A 50 A, 120 V load is connected across *A* and *N*. The neutral conductor shall be No. ___ TW Cu.

     A. 4   C. 8

     B. 6   D. 10

_____   T   F   11. It is recommended that metal raceways of an electrical system should be bonded or properly spaced from lightning protection conductors.

_____   _current_   12. The neutral feeder conductor shall be capable of carrying the maximum ___ load.

_____   _____   13. The minimum area of exposed surface offered by a plate electrode shall be ___ sq ft.

     A. 1   C. 3

     B. 2   D. 4

_____   _50_   14. Wiring run above heated ceilings and within thermal insulation shall be derated on the basis of ___°C ambient temperature.

_____   _____   **15.** Receptacles connected to a 30 A branch circuit supplying two or more outlets shall be rated not less than ___ A.

_____   _____   **16.** Non-current-carrying metal parts of cord-and-plug-connected equipment may be grounded by a(n) ___ run with the power supply conductors.

_____   T   F   **17.** Iron wire shall not be used as an equipment bonding jumper.

_____   _____   **18.** A 15 A or 20 A receptacle can be installed on a ___ A general purpose circuit.
    A. 20   C. 30
    B. 25   D. none of the above

_____   _____   **19.** The maximum standard ampere rating of a fuse or inverse time circuit breaker shall be considered to be ___ A.

_____   T   F   **20.** In general, switch enclosures shall not be used for junction boxes.

_____   _____   **21.** If a change occurs in the size of the ___ conductor, a similar change may be made in the size of the grounded conductor.

_____   _____   **22.** Leads of 240 V heating cables are marked with ___.

_____   _____   **23.** The maximum permissible open-circuit voltage of electric discharge lighting equipment used in a dwelling is ___ V.
    A. 500   C. 1500
    B. 1000   D. 2000

_____   _____   **24.** ___ are permitted around service conduits entering concentric or eccentric knockouts or other impaired connections.

_____   T   F   **25.** A single 15 A receptacle may be installed on an individual 20 A branch circuit.

_____   _____   **26.** A metal elbow for a service raceway installed at least ___″ underground does not have to be grounded.

_____   _____   **27.** The ___ shall be grounded in a 1φ, 3-wire, AC system.

_____   _____   **28.** Ungrounded conductors shall be clearly distinguishable from ___ and grounding conductors by color or other marking.

_____   _____   **29.** Ground rods are permitted to be buried if the trench is at least ___′ deep where rock bottom is encountered.

_____   _____   **30.** Grounding electrodes of pipe or conduit shall not be smaller than ___″ trade size.
    A. ⅜   C. ¾
    B. ½   D. 1

_____   _____   **31.** Flexible cord shall not be used as a substitute for ___ wiring of a structure.

_____   _____   **32.** Unless special precautions are taken, exposed live parts operating at ___ V or more shall be guarded against accidental contact.

_____   _____   **33.** Separately mounted ballasts that are ___ for direct connection to a wiring system shall not be required to be additionally enclosed.

_____   _____   **34.** A single grounding electrode is permitted without the use of a supplemental electrode, when the resistance to ground does not exceed ___ Ω.
    A. 5   C. 15
    B. 10   D. 25

_____   _____   **35.** Heating cables on ceilings shall be kept free from contact with metal or other ___ surfaces.

Name _____ Date _____

NEC® **Answer**

_____ T F **1.** The metal well casing enclosing a submersible pump shall be bonded to the EGC of the pump circuits.

_____ _____ **2.** Fittings and connectors shall be designed and ___ for the specific wiring method with which they are used.

_____ _____ **3.** Heating panels shall be separated from outlet boxes to be used for mounting surface lighting fixtures by not less than ___".
  A. 6   C. 10
  B. 8   D. 12

_____ T F **4.** Not less than 10' of No. 4 Cu bare conductor encased under the concrete footing in contact with the earth is an approved made electrode.

_____ _____ **5.** The service disconnecting means for a one-family dwelling shall have a rating of not less than ___ A.

_____ T F **6.** NMS cable shall be permitted to be used in storage battery rooms.

_____ _____ **7.** The earth shall not be considered to be a(n) ___ ground-fault current path.

_____ _____ **8.** The MBJ in a panelboard may be a screw, a bus, a(n) ___, or a similar suitable conductor.

_____ _____ **9.** It is recommended that metal raceways, enclosures, frames, and other non-current-carrying metal parts of electric equipment be properly spaced from lightning rod conductors, unless they are ___ to such conductors.

_____ _____ **10.** Wiring located above electric heating panels shall be spaced not less than ___" above the heated ceiling.
  A. 2   C. 4
  B. 3   D. 5

_____ T F **11.** Time switches shall be of the enclosed type or mounted in suitable enclosures.

_____ _____ **12.** The minimum size grounded conductor required for a service with 600 kcmil per phase is No. ___ Cu.

_____ _____ **13.** A 15 A or 20 A branch circuit shall be permitted to supply ___ loads.
  A. lighting   C. all of the above
  B. utilization equipment   D. none of the above

_____ _____ **14.** Equipment with open-circuit voltage exceeding ___ V shall not be installed in dwelling units.

_____ _____ **15.** In general, flexible cords other than tinsel cords shall not be smaller than No. ___.

_____ _____ **16.** A 30 A receptacle outlet supplying power to a cord-and-plug-connected load and one additional outlet shall not be loaded to more than ___ A.

_____ _____ **17.** If more than one nominal voltage system exists in a building, each ungrounded conductor of a branch circuit in each phase and system shall be ___.
A. black   C. the same
B. orange   D. identified

_____ _____ **18.** Wiring that supplies equipment, located behind panels designed to allow ___, shall be adequately supported and kept off the ceiling tiles.

_____ _____ **19.** The general rule requires a(n) ___ means to be installed at an auxiliary building on the same property.

_____ _____ **20.** The minimum burial depth of RMC installed under a street or parking lot is ___″. (Circuit conductors operate at 120 V.)

_____ _____ **21.** A single receptacle installed on an individual branch circuit shall have a(n) ___ rating not less than that of the branch circuit.

_____ _____ **22.** In general, copper conductors of size No. ___ and larger comprising each phase, neutral, or grounded circuit conductor shall be permitted to be connected in parallel.
A. 1/0   C. 3/0
B. 2/0   D. 4/0

_____ _____ **23.** Ground-fault circuit protection for personnel is required for all 1ɸ, 125 V, 15 A and 20 A receptacles installed in the ___ of a dwelling unit.
A. attic   C. laundry
B. bathroom   D. utility room

_____ _____ **24.** ___ boxes shall not be used where conduits or connectors requiring the use of locknuts or bushings are to be connected to the side of the box.

_____ _____ **25.** The minimum burial depth for listed PVC conduit routed beneath a building is ___″. (Circuit conductors operate at 120 V.)

_____ _____ **26.** An insulated grounded conductor size No. 6 or smaller shall be identified by a white or ___ outer finish or white with stripes.

_____ _____ **27.** ___ conductors shall be used for wiring on luminaire chains.

_____ _____ **28.** Disconnecting means for circuits to feeders and branch circuits shall be legibly ___ to indicate the purpose.

_____ T   F **29.** Connections depending solely on solder are not permissible as grounding equipment.

_____ _____ **30.** Heavy-duty lampholders of the admedium type shall have a rating of not less than ___ W.

_____ _____ **31.** Side- or back-wiring spaces are required in cabinets when, in addition to the supply circuit or a continuation thereof, there are devices connected to more than ___ conductors.
A. 6   C. 10
B. 8   D. 12

_____ _____ **32.** Screw-shell lampholders shall not be used with infrared lamps of over ___ W except when identified as suitable for such use.

_____ _____ **33.** Direct buried cables operating at 480 V that are routed up a pole shall be protected at least ___′ above grade.

_____ _____ **34.** The smaller of two noncoincident loads may be omitted in calculating the load on a feeder if it is unlikely that both loads will be connected ___.

_____ T   F **35.** Permanently installed 125 V, 15 A and 20 A receptacles that are part of the building wiring and used for construction personnel shall be GFCI-protected.

Name _____ Date _____

NEC®	Answer	

1. No ____ conductor shall be attached to any terminal or lead so as to reverse designated polarity.

T F 2. The center grip of the operating handle for the switch or CB when in the highest position shall not be more than 6'-7" from the floor or platform.

T F 3. The feeder load for one 10 kVA, one 16 kVA, and one 17 kVA household electric range is 15 kVA.

4. The grounded conductor shall be ____ when SE cable is used in a feeder to supply a subpanel located in the same building.

5. Conduits, including fittings, entering a switchgear shall not rise more than ____" above the bottom of the enclosure.

T F 6. Branch-circuit conductors for fixed resistance space heaters and motors shall be rated at 125% of the total load of heaters and motors.

7. The rating of any one cord-and-plug-connected appliance used on a 30 A branch circuit shall not exceed ____ A.
　　A. 21　　　　　　　C. 27
　　B. 24　　　　　　　D. 30

8. The minimum radius of a conduit bend made with other than a one shot or full shoe bender enclosing THHN Cu conductors is ____" for 2" RMC.

9. A 20 A circuit in a dwelling unit supplies only fixed resistance space heaters. This circuit may be loaded to a maximum value of ____ A.
　　A. 12　　　　　　　C. 18
　　B. 16　　　　　　　D. 20

T F 10. Heating panels shall not be extended beyond partitions or run above walls.

11. The branch-circuit load for a 5 kVA wall-mounted oven and a 7 kVA counter-mounted cooktop shall have the capacity for ____ kVA.
　　A. 2　　　　　　　　C. 8
　　B. 7　　　　　　　　D. 9.2

12. The general rule requires that raceways be installed ____ between junction boxes prior to the installation of conductors.

13. ____ circuits in dwelling units shall supply only loads within that dwelling unit or loads associated only with that dwelling unit.

14. For other than a totally enclosed switchboard, a space of not less than ____' shall be provided between the top of the switchboard and any combustible ceiling if no shielding is provided.

T F 15. Heating cable shall not be used with metal lath ceilings.

16. ____ ____ shall be formed on individual service drop conductors to prevent the entrance of moisture.

_____   _____   **17.** No. 8 THWN conductors supply a central air conditioner, consisting of a sealed (hermetic) motor-compressor. The branch circuit may be loaded to a maximum value of ___ A.
            A. 30   C. 36
            B. 32   D. 50

_____   _____   **18.** In general, in all dwelling units, at least ___ wall switched-control lighting outlet(s) shall be installed in every habitable room other than kitchens and bathrooms.

_____   _____   **19.** Conduits with No. 4 and larger conductors shall be provided with an identified ___ for protection of conductors.

_____   _____   **20.** Type AC cable ___ in walls and other concealed spaces does not require support.

_____   _____   **21.** Conductors normally used to carry current shall be ___ unless otherwise provided in the NEC®.

_____   _____   **22.** The total rating of fastened-in-place appliances connected to a 20 A branch circuit that also supplies lighting units shall not exceed ___ A.
            A. 10   C. 16
            B. 12   D. 20

_____   _____   **23.** Interior metal water piping that is located more than ___' from the point of entrance to the building shall not be used as part of the GES or as a conductor to interconnect electrodes that are part of the GES.

_____   T   F   **24.** Heating cables in a concrete floor shall be placed on at least 2' centers from adjacent runs of cable.

_____   _____   **25.** A 4″ PVC conduit installed with threaded couplings in a straight run shall be supported every ___'.

_____   _____   **26.** A 6″ PVC conduit shall be supported at ___' intervals.

_____   _____   **27.** In general, MC cables shall be supported at ___' intervals.

_____   T   F   **28.** Embedded heating cables are permitted to be spliced only where necessary and if done by approved means.

_____   _____   **29.** The minimum radius of the curve of the inner edge of any bend in AC cable shall not be less than ___ times the diameter of the cable.

_____   _____   **30.** Six current-carrying conductors routed through the same raceway shall be derated ___%.

_____   _____   **31.** The total load on overcurrent protection devices supplying continuous duty loads shall not exceed more than ___% of their rating unless the assembly is listed for operation at 100% of its rating.

_____   _____   **32.** A No. ___ is the smallest THWN neutral feeder conductor permitted to supply two 240 V, 12 kVA household ranges.
            A. 4   C. 8
            B. 6   D. 10

_____   _____   **33.** In walls or ceilings of concrete, tile, or other noncombustible material, boxes shall be installed so that the front edge of the box will not be set back of the finished surface more than ___″.

_____   T   F   **34.** The nonheating leads of heating cables installed in concrete may be protected by EMT.

_____   _____   **35.** The grounded circuit conductor can be used as an equipment grounding conductor and circuit conductor on the ___ side of the service equipment.

_____   T   F   **36.** A separable connector will meet NEC® requirements for a disconnecting means for a hardwired wall-mounted oven.

Name _____ Date _____

NEC®	Answer

_____ T F **1.** In a 1000 sq ft structural addition to a dwelling unit, the "VA-per-square-foot" method per sections 220.12 and 220.14 is not used.

_____ _____ **2.** Heating elements in electrode space-heating equipment that exceed ___ A in a heating unit shall be subdivided.

_____ _____ **3.** The neutral conductor of a 3-wire branch circuit to a household electric range shall not be smaller than No. ___ THW.
  A. 4   C. 8
  B. 6   D. 10

_____ _____ **4.** Unbroken lengths of EMT up to ___' are not required to be supported where structural members do not readily allow fastening.

_____ T F **5.** The overcurrent protection device for a branch circuit supplying power to a 120 gal. fixed water heater can be rated less than 125% of the unit's nameplate rating.

_____ _____ **6.** EBJs shall not be connected to a terminal bar provided for grounded conductors unless the terminal bar is ___ for the purpose and otherwise permitted.

_____ _____ **7.** Frames of electric clothes dryers may be grounded by connection to the grounded circuit conductor only when this copper conductor is not smaller than ___ and it is an existing branch circuit.
  A. No. 8   C. No. 12
  B. No. 10   D. the other conductors

_____ T F **8.** Unit switches with a marked OFF position can be used as a disconnecting means for a fixed appliance where all conductors are disconnected simultaneously and other means for disconnection are also provided.

_____ _____ **9.** Lighting track shall be calculated at ___ VA for each 2' or fraction thereof.

_____ T F **10.** Infrared lamps rated over 300 W shall not be used as screw-shell lampholders unless the lampholders are identified for the use.

_____ _____ **11.** The nameplate on heating equipment shall be located to be ___ or easily accessible after installation.

_____ _____ **12.** Flexible cords identified for hard or extra-hard usage shall be permitted to be used for ___ wiring on branch circuits.

_____ _____ **13.** A heavy-duty, admedium lampholder shall have a rating of not less than ___W.
  A. 150   C. 600
  B. 300   D. 660

_____ _____ **14.** In general, all conductors, and the grounded conductor and the EGC when used, shall be routed in the ___ cable, raceway, trench, or conduit system.

_____ _____ **15.** A clearance of not less than ___″ shall be maintained between the conductors of a concealed knob-and-tube wiring system.

_____ T F **16.** In multifamily dwellings, the disconnecting means for a permanently connected appliance rated at not over 300 VA is permitted to be the branch circuit OCPD in the service equipment for the occupancy.

_____ _____ **17.** Heating elements of cables shall have a separation of at least ___″ from the edge of outlet boxes used for mounting surface lighting fixtures.

_____ _____ **18.** It is recommended that the voltage drop on a feeder circuit be held to ___%.
  A. 2 C. 5
  B. 3 D. 7

_____ T F **19.** A No. 4 grounded conductor in a 277/480 V system installed in a raceway, can be identified by a continuous white or gray finish.

_____ _____ **20.** ___ breakers shall not be installed in panelboards.

_____ _____ **21.** A circuit operating for ___ hours or more is considered a continuous load.

_____ _____ **22.** Receptacle outlets shall be installed for each kitchen countertop that is 12″ or wider so that no point along the wall line is more than ___″ from a receptacle outlet.
  A. 18 C. 36
  B. 24 D. 48

_____ _____ **23.** The exposed blades of knife switches shall be de-energized when in the ___ position.

_____ T F **24.** A 20 A branch circuit supplying air conditioning equipment shall not also supply lighting outlets.

_____ _____ **25.** Heavy-duty lighting track is identified for use with circuits exceeding ___ A.

_____ T F **26.** A lockable branch-circuit overcurrent protection device may serve as the disconnecting means for a motor-driven, permanently-connected appliance rated over ⅛ HP.

_____ _____ **27.** Angle connectors used on flexible metal conduit shall not be ___.

_____ _____ **28.** Neon systems having an open circuit voltage exceeding ___ V shall not be installed in dwelling units.

_____ _____ **29.** A kitchen waste disposal installed in a dwelling unit shall be equipped with a cord that does not exceed ___″ in length.

_____ T F **30.** The total load of four or more fixed appliances in a dwelling unit can have an 80% demand factor automatically applied.

_____ _____ **31.** Outlet boxes designed to support ceiling fans that weigh more than ___ lb shall be marked to indicate the maximum weight to be supported.

_____ _____ **32.** At least one receptacle outlet shall be installed above a show window for each ___′ or major fraction thereof.
  A. 9 C. 18
  B. 12 D. 24

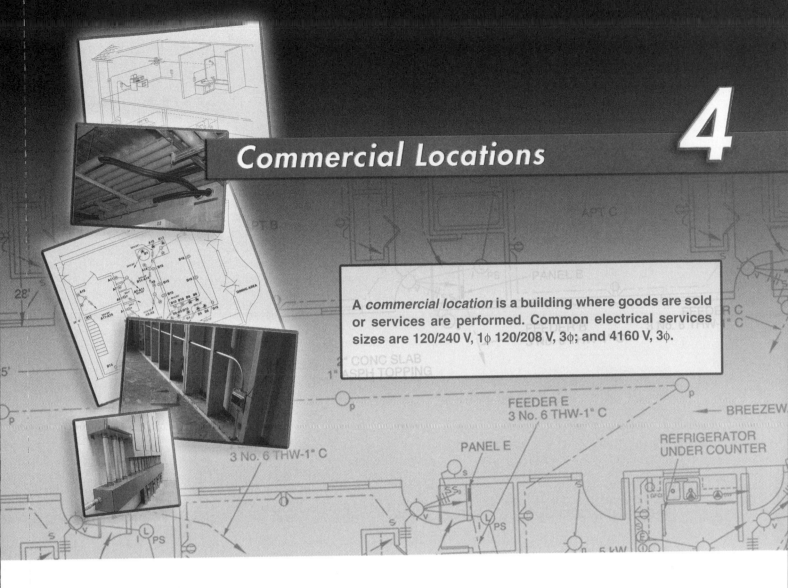

Commercial Locations

4

A *commercial location* is a building where goods are sold or services are performed. Common electrical services sizes are 120/240 V, 1ϕ 120/208 V, 3ϕ; and 4160 V, 3ϕ.

FEEDER E
3 No. 6 THW-1" C

BREEZEW

PANEL E

REFRIGERATOR
UNDER COUNTER

3 No. 6 THW-1" C

READING PRINTS

In preparing to handle a large installation, the electrical supervisor glances through the complete set of plans to obtain an overview of the entire project. The electrical plans are surveyed to determine the extent of the work, kinds of circuits and equipment, and locations of main elements. The electrical specifications are read, and the plans are referred to from time to time as necessary.

The electrical supervisor must refer to the plot, architectural, structural, framing, mechanical, and electrical plans to observe all the details essential to the work. Following are excerpts from electrical specifications for a store building, along with notes that might be taken from the architectural plans.

EXCERPTS FROM SPECIFICATIONS

Note: The NEC® references indicate where the main topic is found in the National Electrical Code®. Always follow manufacturer's directions.

29.1 Main Service – Article 230

(a) The main service shall consist of a 4-wire, 120/208 V, 3ϕ secondary system from utility company utility hole on Sheppard Avenue to the main switchboard.

(b) All underground conduit outside the building shall be buried at least 18″ below grade and shall be painted with bitumastic paint.

(c) All service conductors shall be insulated, including the neutral conductor.

29.4 Conductors – Article 310

(a) Conductors shall be THW unless otherwise specified or required by applicable provisions of the NEC®. All conductors are to be copper unless otherwise noted.

(b) No conductors for lighting, plug receptacles, or power shall be smaller than No. 12 AWG Cu.

(c) No block and tackle or other mechanical means shall be used to draw conductors smaller than No. 2 into raceways. Powdered soapstone shall be the only lubricant permitted.

(d) The feeder voltage drop from the service location to any distribution panel shall not exceed 3%.

29.5 Wiring Methods and Materials – NEC® Chapter 3

(c) The approximate locations of all outlets are marked on the drawings. The contractor shall check all such measurements and, in case of doubt, consult the architect as to exact locations so that boxes will be properly centered with acoustical tile, interior trim, paneling, etc.

29.6 Equipment for General Use – NEC® Chapter 4

(b) Lock-type wall switches shall be installed at points where a switch is marked SK on drawing. These wall switches control lights in the Fitting Room on the first floor.

(m) Two special plug receptacles on the east wall of the Alteration Department shall be equipped with signal lights. These outlets are noted on the Basement Floor Plan.

29.7 Telephone and Loudspeaker Systems – 640, 800, and 810

(a) Telephone outlets shall be located approximately as shown.

29.8 Miscellaneous Equipment – Article 422

(j) Toilet exhaust fans shall be Nuway, 130 VA, Type C2. See Basement Floor Plan.

ELECTRICAL PLANS

Before undertaking a detailed analysis of the store building installation, notice that four drawings are included. The Basement, First Floor, and Second Floor floor plans are included with a special drawing that shows a Feeder Riser Diagram, Switchboard Elevation, Electrical Details, and Fixture Symbols. Discussion could be started with the Basement plan, then carried on to the First Floor and Second Floor plans in turn.

Or, the First Floor might be dealt with, then Second Floor, and finally Basement.

A different order is followed, however, starting with the Second Floor, working downward to the First Floor, and lastly the Basement. This procedure simplifies explanation. The Second Floor is the least complicated area and makes a good starting point. The First Floor plan is more involved. The Basement plan is examined last because it covers service requirements.

SECOND FLOOR PLAN

Two lighting panels are used for this floor, Panel E at the north wall, and Panel D at the south wall. Each of these panels has 13 circuits. Conduit runs between lighting outlets are indicated by solid lines showing them to be concealed in the hung ceiling. All conduit is RMC unless otherwise noted. Arrows indicating homeruns to the panelboards show the number of conductors in conduit with hash marks. For example, Circuits 1, 2, and 3 to Panel E have four conductors in ¾" EMT.

Plug receptacle runs are indicated by alternate long and short dashed lines showing that they are to be run in the poured concrete slab. Telephone runs are indicated by lines marked – T – T –. Speaker system lines are marked – S – S –. Telephone outlets are indicated by standard symbols in the shape of a triangle. Speaker outlets are indicated by the letter S inside a square.

Conduit riser indications are shown in the south wall. At the left or west side, the first such riser is marked DN (DOWN) to show that a conduit run from the stair outlet leads to the floor below. The next two risers are marked UP, and since no conduit runs lead to them from the second floor, it is apparent that they originate below and continue upward toward the roof.

The final riser is marked DN. It is run of conduit leading downward from a lighting outlet at the head of the stairs. Two risers in the east wall of the passenger elevator shaft are both shown as DN. They are run in conduit leading downward from a lighting outlet at the head of the stairs. One of these risers is marked T to indicate a telephone run, the other S to indicate a speaker run. The PBX telephone board and the sound amplifier are in the small room to the right of the elevator shaft.

In determining the load on the VA-per-square-foot basis, 220.12 requires that the floor area be computed from outside dimensions of the building. The measurements are 72′ × 96′, which equals 6912 sq ft (72′ × 96′ = 6912 sq ft). From this amount, it is permissible to deduct 520 sq ft, which is the area of the office space and the area of the two elevators (6912′ − 520′ = 6392 sq ft). The elevator shaft areas and the office space area do not come under the heading of general lighting in Table 220.12.

Table 220.12 lists the unit load for a store as 3 VA per sq ft. NEC® 210.20(A) states that this value shall be increased by 25% where the load on a branch circuit is continuous (load is expected to continue for 3 hours or more). The unit load in the present case is 1.25 × 3 VA, or 3.75 VA per sq ft. The minimum provision for the lighting load is 6392 sq ft × 3.75 VA, or 23,970 VA. The lighting load may also be found by multiplying the sq ft area by 3 VA per Table 220.12 and then multiplying by 125% per 210.20(A). For example, 6392 sq ft × 3 VA = 19,176 VA × 125% = 23,970 VA. Panels D and E must have a sufficient number of lighting circuits to furnish this amount of general illumination.

The Specifications provide that all lighting circuits shall be run with No. 12 THW Cu wire, protected by 20 A CBs. NEC® 210.3 states that branch circuits shall be classified according to the rating of the overcurrent protection devices. Because the loads are continuous, these 20 A circuits, each of which can supply 2400 VA of power (20 A × 120 V = 2400 VA), can be loaded to only 80% of this value, or 1920 VA (2400 VA × 0.80 = 1920 VA) per 210.20(A). A *continuous load* is a load that, in normal operation, lasts for three hours or more.

Calculations for Panel E

The loads on the circuit are:

Circuit	Load*	Circuit	Load*
1	1400	8	1536
2	1536	9	972
3	972	10	900
4	1400	11	720
5	1536	12	540
6	972	13	1080
7	1400		

* in VA

The fixture list, fixture specifications, and the print show the load on Circuit 1 as seven 200 VA incandescent units, totaling 1400 VA. Continuing to Circuit 2, 220.18(B) requires that the load on a lighting fixture that has a ballast shall be calculated according to the ampere rating of the ballast and not the VA of the lamps. The current used in the two ballasts that are part of each fixture on the circuit is 3.2 A or 384 VA (3.2 A × 120 V = 384 VA).

The total VA for Circuit 2, which supplies four such units, is 1536 VA (384 VA × 4 = 1536 VA). Circuit 3 has two of these units and a third with a ballast rating of 1.7 A. The load is 972 VA for all units in Circuit 3 (384 VA × 2 = 768 VA; 1.7 A × 120 V = 204 VA; 768 VA + 204 VA = 972 VA). The load for this group of three circuits is 3908 VA (1400 VA + 1536 VA + 972 VA = 3908 VA).

Loads on the two groups 4, 5, 6 and 7, 8, 9 are the same as the loads for groups 1, 2, and 3, so the total lighting power to be supplied by Panel E amounts to 11,724 VA (3 × 3908 VA = 11,724 VA). Table 220.42 lists a demand factor of 100% for lighting loads of this nature.

In addition to the lighting outlets, a number of wall receptacles and floor outlets are supplied from the panel. The Specifications require that No. 12 THW Cu wire be used for these circuits. NEC® 220.14(I) designates a load of 180 VA or 1½ A (180 VA ÷ 120 V = 1.5 A) for each of such other outlets. The eighteen plug receptacles connected to this panel represent a current input of 27 A × 120 V, or 3240 VA. *Note:* 18 × 1.5 A = 27 A.

The load imposed on the feeder is 11,724 VA + 3240 VA = 14,964 VA. At this point, determine if the feeder will carry its required proportion of minimum lighting load. The floor plan shows that circuits from Panel E provide general illumination for an area of 3456 sq ft (36′ × 96′ = 3456 sq ft).

Since the unit load cannot be less than 3.75 VA per sq ft, the amount of general lighting burden allotted to Panel E cannot be less than 12,960 VA (3456 sq ft × 3.75 VA = 12,960 VA). This value is used in the feeder calculations because it is larger than the actual connected load of lighting circuits.

The minimum value of lighting VA does not have to be used, but must be present in circuit and feeder capacity. The corrected load on the feeder then is 16,200 VA (12,960 VA + 3240 VA = 16,200 VA).

With a balanced, 4-wire, 120/208 V, 3φ system, the current per conductor may be determined by dividing VA by 360 (1.73 × 208 V = 360). The current in the feeder to Panel E is 45 A (16,200 VA ÷ 360 V = 45 A). Table 310.15(B)(16) lists the nearest size of THW Cu conductor to carry 45 A as No. 8, which is rated at 50 A if there are not more than three current-carrying conductors in a raceway.

Section 310.15(B)(2)(a) states that where the number of conductors in a raceway exceeds three, the allowable ampacity of each conductor shall be reduced as shown in Table 310.15(B)(3)(a). Since there are four conductors, it seems that the capacity of the No. 8 wire is reduced to 80% of 50 A, or 40 A. Section 310.15(B)(5), however, states that a neutral conductor that carries only the unbalanced current from other conductors shall not be counted in determining ampacities where there are more than three conductors in a raceway.

Such a neutral conductor is included in the feeder under consideration, and the carrying capacity of the No. 8 wire will not be reduced in value. Nevertheless, the rule may be altered later on because the neutral conductor of a 4-wire, 3φ wye system that supplies circuits for electric discharge lamps carries some third harmonic current even when the load is perfectly balanced.

The *American Electricians' Handbook* and other engineering data (See Ch 9, Table 8) lists the resistance of No. 6 Cu as 0.491 W per 1000′. When floor heights and distances on the architectural plans are measured, allowing a sufficient length of conductor for connections at either end of the run, the length of single conductor is found to be 104′ from Panel E to the switchboard in the basement.

The preferred method for calculating voltage drop (VD) for a 3φ circuit and % voltage drop (%VD) is by using the formulas:

$$VD = \frac{2 \times R \times L \times I \times 0.866}{1000}$$

where

VD = voltage drop (in V)

V_l = voltage loss (in V)

V_s = supply voltage (in V)

R = resistance of conductor (in Ω)

L = length of conductor (in feet)

I = current (in A)

$0.866 = \dfrac{\sqrt{3}}{2}$ (used for 3φ only)

$1000 = 1000′$ or less of conductor

$$VD = \frac{2 \times R \times L \times I \times 0.866}{1000}$$

$$VD = \frac{2 \times 0.491 \times 104 \times 45 \times 0.866}{1000}$$

$$VD = \frac{3979.93}{1000}$$

$$VD = 3.9799 = \mathbf{3.98\ V}$$

$$\%VD = \frac{V_l}{V_s} \times 100$$

$$\%VD = \frac{3.98}{208} \times 100$$

$$\%VD = 0.01913 \times 100$$

$$\%VD = 1.913 = \mathbf{1.91\%}$$

Note 2 following 215.2 recommends that the voltage drop for feeders should be limited to 3% of the phase voltage at the farthest outlet. The recommendation for voltage drop for both feeders and branch circuits is 5%. In this calculation, the recommended voltage drop should not exceed 6.24 V (208 V × 3% = 6.24 V). The 3.98 V is well below the 3% recommendation. The feeder to Panel E consists of four No. 6 THW Cu conductors in 1″ RMC.

Calculations for Panel D

The loads on the circuits are:

Circuit	Load*	Circuit	Load*
1	1200	8	1152
2	1920	9	720
3	1200	10	1080
4	1920	11	1080
5	800	12	720
6	1152	13	720
7	1152		

* in VA

The circuits devoted to general lighting are 1, 2, 3, 4, 5, and 6. The general lighting circuits require a total of 8192 VA (1200 VA + 1920 VA + 1200 VA + 1920 VA + 800 VA + 1152 VA = 8192 VA).

Measurement shows the floor area supplied with general illumination from this panel to be 2496 sq ft. The minimum VA required is 9360 VA (2496 sq ft × 3.75 VA = 9360 VA). This value, being somewhat higher than the actual load on the circuits, is used in the feeder calculations.

The office has an area of 440 sq ft, requiring 1650 VA of illumination at the 3.75 VA rate (440 sq ft × 3.75 VA = 1650 VA). Table 220.12 lists office buildings at a minimum unit load of 3.5 VA per sq ft. For a continuous load, this minimum would have to be increased to 4.4 VA (1.25 × 3.5 VA = 4.375 = 4.4 VA).

Although this provision does not apply strictly to the one small office space here, it does indicate that office buildings and office areas need somewhat more lighting intensity than sales areas. Using 4.4 VA per sq ft, this area requires 1936 VA (440 sq ft × 4.4 VA = 1936 VA). The lighting load connected to Circuits 7 and 8 amounts to 2304 VA. Because the connected load of 2880 VA (2304 VA × 1.25 = 2880 VA) is larger than that required by Table 220.12 and 210.20(A), the higher value of 2880 VA is used.

Plug receptacle circuits 9, 10, 11, 12, and 13 take 4320 VA. The total load on the feeder is 16,560 VA (9360 VA + 2880 VA + 4320 VA = 16,560 VA). The current is 46 A (16,430 VA ÷ 360 = 46 A), for which the minimum size of THW conductors is No. 8.

The length of single conductor between Panel D and the switchboard is approximately 83′. The voltage drop for a No. 8 stranded conductor is too high. Use of No. 4 conductor results in an acceptable drop. Appd C, Table C8 shows that four No. 4 THW Cu conductors may be installed in 1¼″ RMC. Ch 9, Table 8 lists the resistance of No. 4 Cu as 0.308 Ω per 1000′.

$$VD = \frac{2 \times R \times L \times I \times 0.866}{1000}$$

$$VD = \frac{2 \times 0.308 \times 83 \times 46 \times 0.866}{1000}$$

$$VD = \frac{2036.73}{1000}$$

$$VD = 2.036 = \textbf{2.04 V}$$

$$\%VD = \frac{V_l}{V_s} \times 100$$

$$\%VD = \frac{2.04}{208} \times 100$$

$$\%VD = 0.0098 \times 100$$

$$\% VD = \textbf{0.98\% VD}$$

FIRST FLOOR PLAN

Three lighting panels are shown on this floor, Panel B in the north wall, and Panels A and C in the south wall. Panel C is primarily confined to supplying outlets in the entrance, canopy, and show window areas. Panels A and B are confined to the general store illumination. The wiring plan on this floor is similar to that on the second floor.

A single riser is indicated in the north wall. The feeder conduit is associated with Panel E on the second floor. The riser near the left end of the south wall, marked UP, is the same one shown on the second floor plan, marked DN.

The riser next to Panel A is a feeder conduit to Panel D, above. The next two risers, also marked UP, are explained on the Basement Floor Plan. The third riser, marked UP 10, is the other end of the conduit supplying the lighting unit at the head of the stairs on the second floor. The two risers in the east wall of the passenger elevator shaft are shown on this floor plan.

Calculations for Panel B

The loads on the circuits are:

Circuit	Load*	Circuit	Load*
1	1800	8	1920
2	1920	9	1556
3	1556	10	720
4	1800	11	1080
5	1920	12	1260
6	1556	13	1080
7	1800		

* in VA

The circuits devoted to general lighting are 1, 2, 3, 4, 5, 6, 7, 8, and 9. The general lighting circuits use 15,828 VA (1800 VA + 1920 VA + 1556 VA + 1800 VA + 1920 VA + 1556 VA + 1800 VA + 1920 VA + 1556 VA = 15,828 VA).

Measurement shows the floor area lighted by these circuits to be 3344 sq ft. The minimum acceptable VA for this area is 12,540 VA (3344 sq ft × 3.75 VA = 12,540 VA). Since the connected load is greater than this amount, the higher value is used in the feeder calculations.

Plug receptacle circuits 10, 11, 12, and 13 require 4140 VA. The load on the feeder is 19,968 VA (15,828 VA + 4140 VA = 19,968 VA). The current is approximately

56 A (19,968 VA ÷ 360 = 55.47 = 56 A). The minimum size of THW Cu conductor per Table 310.15(B)(16) is No. 6.

The length of single conductor between Panel B and the switchboard is approximately 92′. The voltage drop in this length of No. 6 conductor equals 4.38 V. The resistance for a No. 6 Cu conductor is 0.491 Ω per Ch 9, Table 8.

$$VD = \frac{2 \times R \times L \times I \times 0.866}{1000}$$

$$VD = \frac{2 \times 0.491 \times 92 \times 56 \times 0.866}{1000}$$

$$VD = \frac{4381.32}{1000}$$

$$VD = \mathbf{4.38\ V}$$

$$\%VD = \frac{V_l}{V_s} \times 100$$

$$\%VD = \frac{4.38}{208} \times 100$$

$$\%VD = 0.021 \times 100$$

$$\%VD = \mathbf{2.1\%}$$

The feeder to Panel B consists of four No. 6 THW Cu conductors in 1″ RMC.

Calculations for Panel A

The loads on the circuits are:

Circuit	Load*	Circuit	Load*
1	1800	8	1920
2	1920	9	1556
3	1556	10	720
4	1800	11	1080
5	1920	12	1260
6	1556	13	1080
7	1800		

* in VA

The circuits concerned with general lighting are 1, 2, 3, 4, 5, 6, 7, and 8. The general lighting circuits have a connected load of 11,852 VA (300 VA + 1000 VA + 1800 VA + 1920 VA + 1556 VA + 1800 VA + 1920 VA + 1556 VA = 11,852 VA).

Measurement shows the floor area lighted by these circuits as 3080 sq ft, for which the minimum VA must be not less than 11,550 VA (3080 sq ft × 3.75 VA = 11,550 VA).

The connected load, being somewhat higher than this value, is used in feeder calculations.

Plug receptacle circuits 9, 10, and 11 take 2880 VA. The total load on the feeder is 14,732 VA (11,852 VA + 2880 VA = 14,732 VA). The current is approximately 41 A (14,732 VA ÷ 360 = 40.9 = 41 A) and calls for a minimum size of No. 8 Cu THW Cu conductor per Table 310.16.

The length of single conductor from Panel A to the switchboard is about 73′. Voltage drop in this No. 8 conductor is 4 V. The resistance for a No. 8 Cu conductor is 0.778 Ω per Ch 9, Table 8.

$$VD = \frac{2 \times R \times L \times I \times 0.866}{1000}$$

$$VD = \frac{2 \times 0.778 \times 73 \times 41 \times 0.866}{1000}$$

$$VD = \frac{4033.05}{1000}$$

$$VD = 4.03\ V = \mathbf{4\ V}$$

$$\%VD = \frac{V_l}{V_s} \times 100$$

$$\%VD = \frac{4}{208} \times 100$$

$$\%VD = 0.019 \times 100$$

$$\%VD = \mathbf{1.9\%}$$

The feeder to Panel A consists of four No. 8 THW Cu conductors in 1″ RMC.

Calculations for Panel C

The loads on the circuits are:

Circuit	Load*	Circuit	Load*
1	1200	8	1500
2	1000	9	1000
3	1500	10	300
4	1000	11	1800
5	1500	12	864
6	1200	13	900
7	1000		

* in VA

Except for front stair lighting, Circuit 10, and cove lighting, Circuit 12, this panelboard feeds only show window, entrance, and canopy areas.

The circuits for lighting the show windows are 1, 2, 3, 4, 5, 6, 7, 8, and 9. They require a total of 10,900 VA (1200 VA + 1000 VA + 1500 VA + 1000 VA + 1500 VA + 1200 VA + 1000 VA + 1500 VA + 1000 VA = 10,900 VA).

NEC® 220.43(A) provides that a minimum of 200 VA per linear foot be assigned for show window lighting. The two show windows are each 22′ long, so that a minimum of 8800 VA (200 VA × 22′ × 2 = 8800 VA) shall be used in the feeder calculations. Here, the greater value of the 10,900 VA connected load is used.

The total load on Panel C is equal to 14,764 VA. The current is approximately 41 A (14,764 VA ÷ 360 = 41.01 = 41 A). No. 8 THW Cu is required per Table 310.15(B)(16). The conductor is 73′ in length. Since Panel C is directly over the switchboard, voltage drop in the feeder conductors is negligible, and four No. 8 THW Cu conductors in 1″ RMC are satisfactory.

Since the Specifications state that the basement ceiling is unfinished, and since floors are of poured, reinforced concrete construction, it is necessary to set basement lighting outlet boxes and conduit on the first floor deck at the same time that first floor plug receptacle boxes are spotted there. Plans and Specifications should always be checked with a view to detect these situations.

BASEMENT FLOOR PLAN

The basement lighting panel is included in the main switchboard. A power panel, located in the boiler room, furnishes current to pump motors there and to a motor that operates the package elevator unit.

The basement contains a shipping and receiving space, an alteration department, a large stockroom area, employee toilets, public toilets, and a small office. Both employee toilets and public toilets have exhaust fans connected directly to respective lighting circuits so they are in operation at all times while light switches are turned ON. The Specifications furnish details regarding the fans, stating that they consume 130 VA each. Plug receptacle circuit 13, in the alteration department, has two special outlets described in the Specifications.

A pair of risers in the north wall are feeders to Panels B and E. In the south wall, risers at the left are associated with Panels D and A. Toward the other end, a conduit rises to the power center on the roof. A feeder conduit to Panel C is next in line, followed by one to an electric sign on the roof. The DN marking in the east wall of the freight elevator shaft refers to a conduit leading to the pit light, and the identical marking in the passenger elevator wall shows a run to its pit light.

Calculations for Basement Panel

The loads on the circuits are:

Circuit	Load*	Circuit	Load*
1	1200	8	1360
2	870	9	1100
3	1224	10	1464
4	1624	11	1620
5	1220	12	1260
6	1220	13	1500
7	1220	14	1260

* in VA

The total load is 18,142 VA (1200 VA + 870 VA + 1224 VA + 1624 VA + 1220 VA + 1220 VA + 1220 VA + 1360 VA + 1100 VA + 1464 VA + 1620 VA + 1260 VA + 1500 VA + 1260 VA = 18,142 VA). The current is approximately 50 A (18,142 VA ÷ 360 = 50.39 = 50 A) and a No. 6 conductor is large enough to supply the panel per Table 310.15(B)(16).

Service Calculation

The service load may be obtained by adding all the feeder requirements, including boiler room power, sign, and roof power. A load of 4800 VA is to be allowed for boiler room power, 8640 VA for the sign, and 36,000 VA for roof power.

SERVICE CALCULATION	
Panel	**Load***
E	16,200
D	16,560
B	19,968
A	14,732
C	14,764
Basement Lighting	18,142
Sign	8640
Boiler Room Power	4800
Roof Power	36,000
Total VA	149,806

* in VA

Starting with Panel E and continuing in the order of calculations, the total load is 16,200 VA + 16,560 VA + 19,968 VA + 14,732 VA + 14,764 VA + 18,142 VA + 8640 VA + 4800 VA + 36,000 VA, or 149,806 VA.

The total load for the service calculation is 149,806 VA. The current is 416 A (149,806 VA ÷ 360 = 416.13 = 416 A).

$$I = \frac{VA}{V \times \sqrt{3}}$$

$$I = \frac{149,806}{360}$$

$$I = 416 \text{ A}$$

Note: When making calculations for balanced 4-wire, 120/208 V, 3ϕ systems, divide the total VA by 360 (208 × 1.732 = 360) in order to obtain current per phase wire.

Table 310.15(B)(16) shows that a 600 kcmil THW conductor will carry 420 A. Two smaller conductors are more easily handled than a single large one. NEC® 310.10(H) permits conductors of No. 1/0 and larger to be run in parallel, provided they are of the same length, conductor material, circular.mil area, insulation type, and are terminated in the same manner. Where parallel conductors are run in separate raceways or cables, the raceways or cables shall have the same physical characteristics.

Two parallel conductors on each leg of the circuit would each carry one-half of the total current, or 208 A. With all conductors in the same conduit, a total of six current-carrying wires, the rating of each is reduced to 80% of the value indicated in Table 310.15(B)(16) per Table 310.15(B)(3)(a). A No. 4/0 THW conductor has an ampacity of 230 A where there are not more than three current-carrying conductors in a raceway. For six conductors, each shall have a normal ampacity equal to 208 A divided by 0.8, or 260 A. A 300 kcmil THW conductor would be required in this case. But if two sets of No. 4/0 conductors are installed in separate conduits, the full ampacity of each conductor is restored, and there is considerable saving of copper.

NEC® 220.61(A) states that the neutral feeder shall be large enough to carry the maximum unbalanced load. The maximum unbalanced load is the load that can be connected between the neutral and any ungrounded conductor, which means that loads that have no neutral component, such as power loads confined to the 3ϕ wires, are subtracted from the total load in order to determine this value of current. The total load in the present instance is 416 A and the 3ϕ power load 114 A, (boiler room power plus roof power). The maximum unbalanced load on the neutral is 302 A (416 A – 114 A = 302 A).

NEC® 220.61(B) provides that a demand factor of 70% may be applied to that portion of the neutral load in excess of 200 A. It appears that the neutral load could be reduced to 200 A + 102 A (0.7 × 102 A = 71.4 A = 72 A) or 272 A. NEC® 220.61(C) states that no reduction of neutral capacity shall be allowed for that part of the load that consists of electric discharge lighting. Since electric discharge lighting in the present example accounts for at least 150 A, there is no excess over 200 A to which a 70% factor might be applied. The neutral load, then, is calculated on the basis of 302 A.

NEC® 300.3(B) requires that a grounded conductor be run in the same raceway with each set of phase conductors. Either of the two neutral wires in this case carry one-half (151 A) of the unbalanced current (302 A ÷ 2 = 151 A). The next largest conductor size listed in Table 310.16 is No. 2/0 THW, which has an ampacity of 175 A.

NEC® 230.30, Ex. permits an uninsulated copper grounded conductor to be installed underground in duct or conduit. It may appear that a larger ampacity might be assigned to an uninsulated conductor, so that a size smaller than No. 2/0 THW could be used. But 310.15(B)(4) states that where bare conductors are used with insulated conductors, their allowable ampacity shall be limited to that permitted for an insulated conductor of the same size. The present Specifications require that an insulated neutral conductor be used. The service consists of two sets of three No. 4/0 THW conductors and one set of No. 2/0 THW conductors in parallel.

Service Conduit Size

When all conductors in a raceway are of the same size, Annex C, Tables C1 – C12 determine the size of conduit required. Each table includes all sizes of conductors with all of the different types of conductor insulation listed. Each of the tables in Annex C is for a specific type of raceway. For different sizes of wire in a conduit, calculations are based upon Chapter 9, Table 1 and Tables 4 through 8.

Table 4 gives dimensions and percent of area of conduit and tubing. The other tables give dimensions for the various types of conductor coverings including rubber, thermoplastic, lead-covered, asbestos-varnished-cambric, cross-linked polyethylene, and bare.

In the store service, each conduit has three No 4/0 THW conductors and one No. 2/0 THW conductor. Table 5 gives the cross-sectional area of a No. 4/0 THW conductor as 0.3718 sq in. and a No. 2/0 THW conductor as 0.2624 sq in. The area of all four conductors equals 1.3778 sq in. (3 × 0.3718 sq in. + 0.2624 sq in. = 1.3778 sq in.). Table 1 allows 40% of the conduit cross-sectional area to be used when there are three or more conductors of different sizes. Cross-sectional area of the conduit, therefore, is equal to 3.4445 sq in. (1.3778 sq in. ÷ 0.4 = 3.4445 sq in.). According to Table 4, the required size of conduit is 2½″.

480 V Lighting System

A method of lighting that has become increasingly popular in commercial and industrial establishments is the four-wire, 277/480 V, 3ϕ electric discharge system. Lighting units are connected between the neutral conductor and phase wires, imposing 277 V on ballast assemblies. Switching may be at the panelboard if CBs are listed and marked SWD or HID per 240.83(D). Where wall switches are used and the voltage between adjacent switches exceeds 300 VA, an identified and securely installed barrier is required in outlet boxes between switches per 404.8(B).

Aluminum Wire and Conduit

In recent years, aluminum wire and conduit have competed with copper wire and steel conduit. Since aluminum conduit is made in the same standard trade sizes as steel, there are no NEC® problems. Aluminum conductors, however, vary with respect to copper in both allowable ampacity and ohms of resistance because ampacity for a given size is less than that of

copper and resistance is higher. The sizes of wire and conduit required are calculated when THW Al feeder and service conductors are substituted for THW Cu conductors in the store building example.

The formula for calculating voltage drop for aluminum conductors is the same as the formula used for copper conductors. The resistivity (R) for copper and aluminum conductors is given in Chapter 9, Table 8.

The formulas for calculating voltage drop (VD) and %VD are:

$$VD = \frac{2 \times R \times L \times I \times 0.866}{1000}$$

where

VD = voltage drop (in V)

R = resistance of conductor (in Ω)

L = length of conductor (in ft)

I = current (in A)

0.866 = (used for 3ϕ only)

1000 = 1000′ or less of conductor

$$\%VD = \frac{V_l}{V_s} \times 100$$

where

%VD = percentage of voltage drop (in %)

V_l = voltage loss (in V)

V_s = voltage supply (in V)

100 = constant

The VD and %VD has been determined for each of the panels using Cu conductors. The formulas may be used to find VD and %VD for aluminum conductors. Ch 9, Table 8 gives the resistance values and circular mils used in the voltage drop formulas.

COPPER					
Panel	R	L	I	Conductors	Conduit
A	0.778	73	41	4 No. 8 THW	1″
B	0.491	92	56	4 No. 6 THW	1″
C	0.778	73	41	4 No. 8 THW	1″
D	0.308	83	46	4 No. 4 THW	1¼″
E	0.491	104	45	4 No. 6 THW	1″

ALUMINUM					
Panel	R	L	I	Conductors	Conduit
A	0.808	73	41	4 No. 6 THW	1″
B	0.508	92	56	4 No. 4 THW	1¼″
C	0.808	73	41	4 No. 6 THW	1″
D	0.808	83	46	4 No. 2 THW	1¼″
E	0.508	104	45	4 No. 4 THW	1″

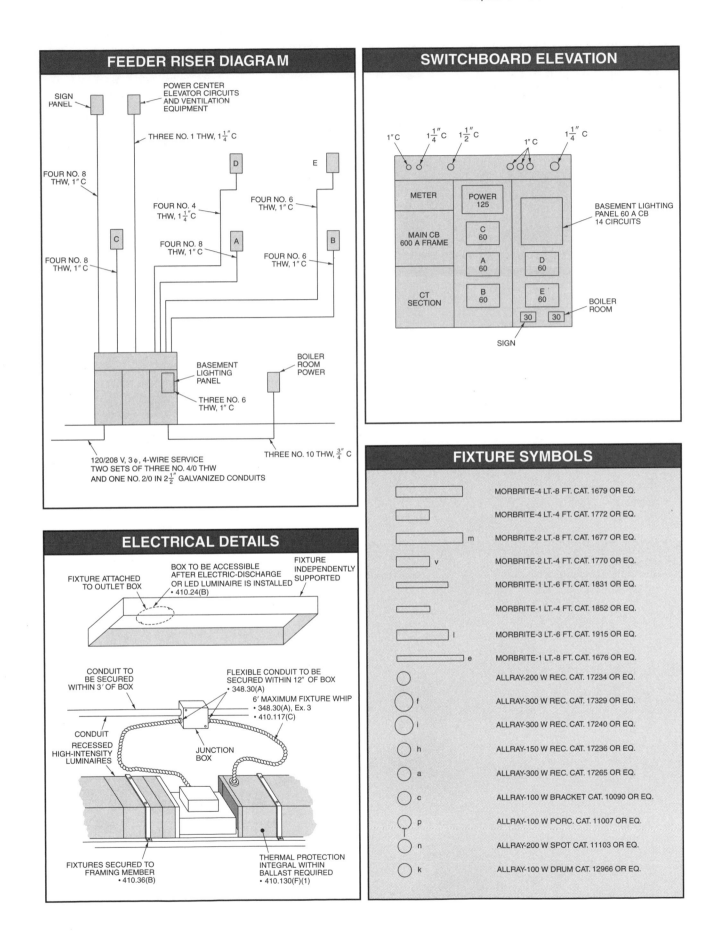

FEEDER RISER DIAGRAM

SIGN PANEL

POWER CENTER ELEVATOR CIRCUITS AND VENTILATION EQUIPMENT

THREE NO. 1 THW, $1\frac{1}{4}''$ C

FOUR NO. 8 THW, 1" C

D

E

FOUR NO. 4 THW, $1\frac{1}{4}''$ C

FOUR NO. 6 THW, 1" C

C

FOUR NO. 8 THW, 1" C

A

FOUR NO. 8 THW, 1" C

FOUR NO. 6 THW, 1" C

B

BASEMENT LIGHTING PANEL

BOILER ROOM POWER

THREE NO. 6 THW, 1" C

THREE NO. 10 THW, $\frac{3}{4}''$ C

120/208 V, 3 φ, 4-WIRE SERVICE
TWO SETS OF THREE NO. 4/0 THW
AND ONE NO. 2/0 IN $2\frac{1}{2}''$ GALVANIZED CONDUITS

SWITCHBOARD ELEVATION

1" C $1\frac{1}{4}''$ C $1\frac{1}{2}''$ C   1" C $1\frac{1}{4}''$ C

METER

POWER 125

BASEMENT LIGHTING PANEL 60 A CB 14 CIRCUITS

MAIN CB 600 A FRAME

C 60

A 60

D 60

CT SECTION

B 60

E 60

30   30

BOILER ROOM

SIGN

ELECTRICAL DETAILS

FIXTURE ATTACHED TO OUTLET BOX

BOX TO BE ACCESSIBLE AFTER ELECTRIC-DISCHARGE OR LED LUMINAIRE IS INSTALLED
• 410.24(B)

FIXTURE INDEPENDENTLY SUPPORTED

CONDUIT TO BE SECURED WITHIN 3′ OF BOX

FLEXIBLE CONDUIT TO BE SECURED WITHIN 12″ OF BOX
• 348.30(A)

6′ MAXIMUM FIXTURE WHIP
• 348.30(A), Ex. 3
• 410.117(C)

CONDUIT

RECESSED HIGH-INTENSITY LUMINAIRES

JUNCTION BOX

FIXTURES SECURED TO FRAMING MEMBER
• 410.36(B)

THERMAL PROTECTION INTEGRAL WITHIN BALLAST REQUIRED
• 410.130(F)(1)

FIXTURE SYMBOLS

	MORBRITE-4 LT.-8 FT. CAT. 1679 OR EQ.
	MORBRITE-4 LT.-4 FT. CAT. 1772 OR EQ.
m	MORBRITE-2 LT.-8 FT. CAT. 1677 OR EQ.
v	MORBRITE-2 LT.-4 FT. CAT. 1770 OR EQ.
	MORBRITE-1 LT.-6 FT. CAT. 1831 OR EQ.
	MORBRITE-1 LT.-4 FT. CAT. 1852 OR EQ.
l	MORBRITE-3 LT.-6 FT. CAT. 1915 OR EQ.
e	MORBRITE-1 LT.-8 FT. CAT. 1676 OR EQ.
	ALLRAY-200 W REC. CAT. 17234 OR EQ.
f	ALLRAY-300 W REC. CAT. 17329 OR EQ.
i	ALLRAY-300 W REC. CAT. 17240 OR EQ.
h	ALLRAY-150 W REC. CAT. 17236 OR EQ.
a	ALLRAY-300 W REC. CAT. 17265 OR EQ.
c	ALLRAY-100 W BRACKET CAT. 10090 OR EQ.
p	ALLRAY-100 W PORC. CAT. 11007 OR EQ.
n	ALLRAY-200 W SPOT CAT. 11103 OR EQ.
k	ALLRAY-100 W DRUM CAT. 12966 OR EQ.

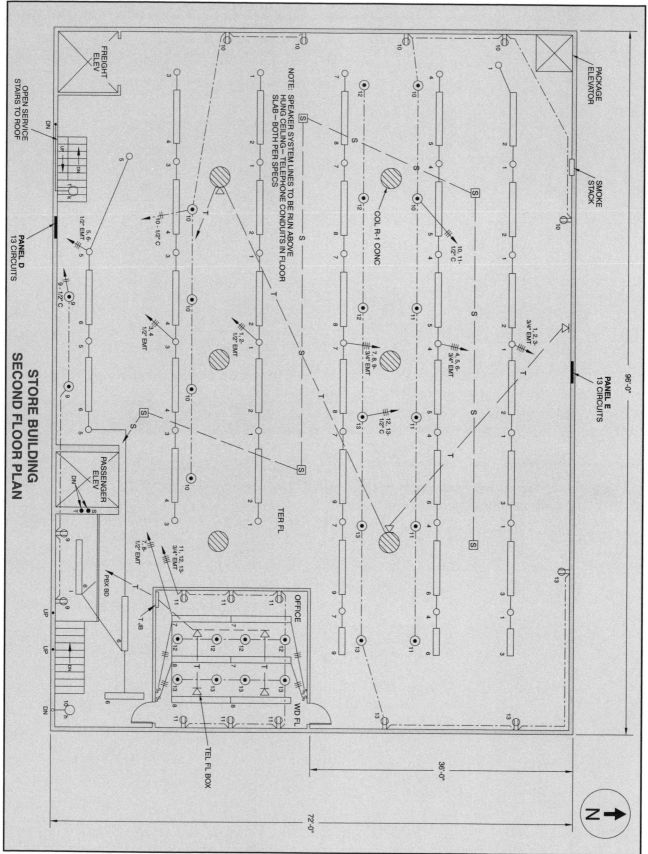

STORE BUILDING
SECOND FLOOR PLAN

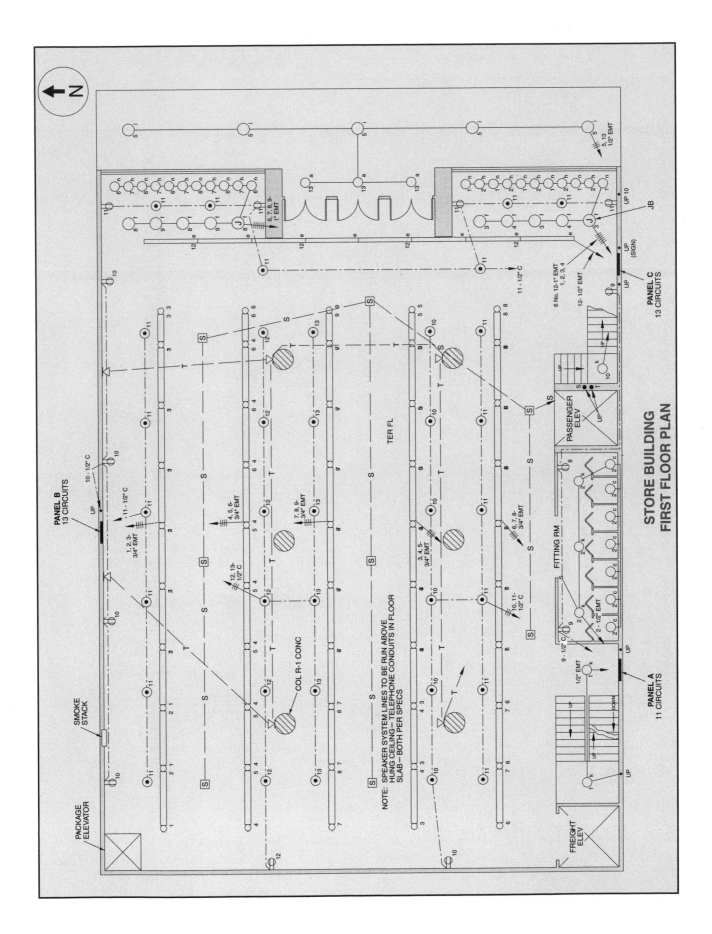

STORE BUILDING
FIRST FLOOR PLAN

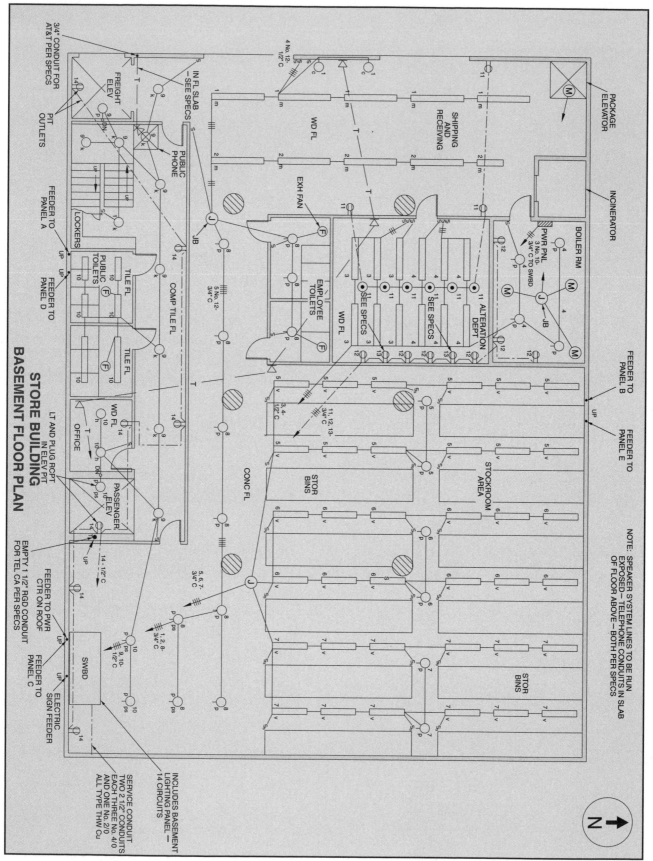

STORE BUILDING
BASEMENT FLOOR PLAN

SERVICES

Building—A structure that stands alone or is cut off from adjoining structures by fire walls with all openings therein protected by approved fire doors.

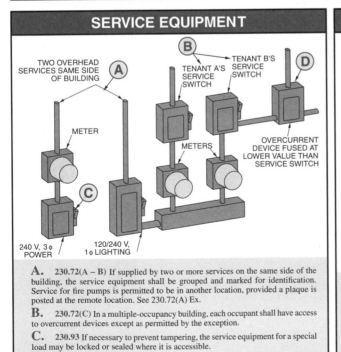

SERVICE HEAD
SERVICE WIRES IN CONDUIT
SERVICE-DROP CONDUCTORS
DRIP LOOPS
TO SERVICE-ENTRANCE EQUIPMENT
SIDEWALK HANDHOLE

A. **225.31** Conductors supplying each building shall be provided with readily accessible means for disconnecting ungrounded conductors from the source of supply. The disconnecting means for garages, etc., for residential use may be 3- and 4-way snap switches. See 225.36, Ex.

B. **230.3** Service conductors shall not feed one building or structure through another building or structure.

C. **230.6(1 – 5)** Conductors installed under at least 2″ of concrete beneath a building, enclosed by concrete or brick at least 2″ thick, in a transformer vault, or where traveling through a building eave per 230.24 shall be considered outside a building or under not less than 18″ of earth.

D. **230.43** Wiring methods for service conductors are limited to: (1) open wiring on insulators; (2) IGS cable; (3) RMC; (4) IMC; (5) EMT; (6) ENT; (7)

SE cable; (8) wireways; (9) busways; (10) auxiliary gutters; (11) PVC; (12) cablebus; (13) MC cable; (14) MI cable; (15) FMC or LFMC, 6′ maximum length, with EBJ; (16) LFNC; (17) HDPE; (18) NUCC; (19) RTRC.

E. **230.46** Service-entrance conductors shall only be spliced or tapped per 110.14, 300.5(E), 300.13 and 300.15.

F. **230.54(F)** Drip loops are required to prevent moisture and are used to connect service-entrance conductors to service-drop or overhead service conductors below level of service head.

G. **250.32(A – B) (D)** Where a grounded system serves two or more buildings from the same service, each building shall have a neutral to its own grounding electrode except if the second building has a single-branch circuit and no equipment that requires grounding or an EGC is run to the second building.

SERVICE EQUIPMENT

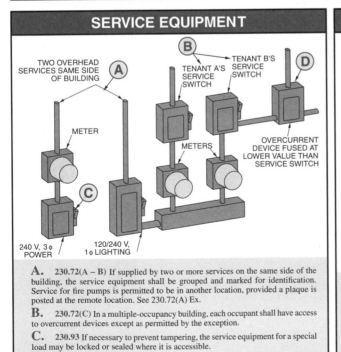

TWO OVERHEAD SERVICES SAME SIDE OF BUILDING
METER
TENANT A'S SERVICE SWITCH
TENANT B'S SERVICE SWITCH
METERS
OVERCURRENT DEVICE FUSED AT LOWER VALUE THAN SERVICE SWITCH
240 V, 3 ø POWER
120/240 V, 1 ø LIGHTING

A. **230.72(A – B)** If supplied by two or more services on the same side of the building, the service equipment shall be grouped and marked for identification. Service for fire pumps is permitted to be in another location, provided a plaque is posted at the remote location. See 230.72(A) Ex.

B. **230.72(C)** In a multiple-occupancy building, each occupant shall have access to overcurrent devices except as permitted by the exception.

C. **230.93** If necessary to prevent tampering, the service equipment for a special load may be locked or sealed where it is accessible.

D. **230.92** If service equipment is locked or inaccessible, overcurrent devices for branch-circuits or feeders of a lower rating than the service OCPD must be installed in an accessible location.

SERVICE DISCONNECTING MEANS

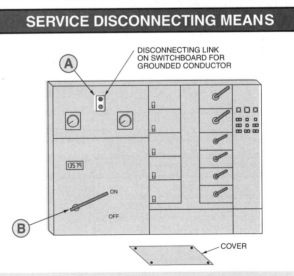

DISCONNECTING LINK ON SWITCHBOARD FOR GROUNDED CONDUCTOR
ON
OFF
COVER

A. **230.75** If the switch or CB does not interrupt the grounded conductor, other means of disconnection shall be provided in the cabinet or switchboard.

B. **230.76** The service disconnecting means shall be operable by hand and shall consist of either a manually-operable switch or CB or a power-operated switch or CB also operable by hand.

408.3(C) Switchboard or panelboards used as service equipment shall have a main bonding jumper or a means within the service disconnect section for connecting the supply neutral to the metal frame and sized per 250.28(D).

CABLE LIMITERS

240.2: Current-Limiting Overcurrent Protective Device—An overcurrent device that, when interrupting currents in current-limiting range, reduces current in faulted circuit to a value less than obtainable in same circuit if replaced with a solid conductor of comparable impedance.

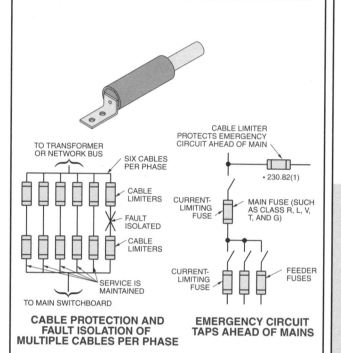

CABLE PROTECTION AND FAULT ISOLATION OF MULTIPLE CABLES PER PHASE

EMERGENCY CIRCUIT TAPS AHEAD OF MAINS

230.82 In general, equipment shall not be installed on the line or supply side of service disconnecting means. See Special Applications (1 – 9).

230.91 The service overcurrent device shall be immediately adjacent to the service disconnecting means or an integral part of it.

240.40 In circuits over 150 V to ground, a disconnecting means shall be installed on the line side of all fuses, where accessible to non-qualified personnel. See Special Applications for (1) current-limiting devices and (2) group operations in 430.112, Ex. and 424.22(C).

240.60(B) Current-limiting fuseholders shall be designed so that they only accept current-limiting fuses.

GROUNDING SEPARATELY DERIVED AC SYSTEMS — 250.30

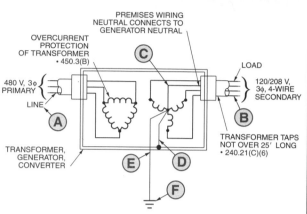

A.   **240.4** Overcurrent protection of conductors shall be in accordance with their ampacities per 310.15 unless otherwise permitted in 240.4 (A-G).

B.   **240.21** Overcurrent protection shall be connected in each ungrounded circuit conductor at the point of supply.

C.   **250.20(D)** A premises wiring system with power derived from a generator and having no direct electrical connection to supply conductors originating in another system shall be grounded per 250.30(A).

D.   **250.28** The main bonding jumper is sized per 250.28(D) for derived phase conductors. The MBJ connects the circuit conductor of the system that is required to be grounded to the system noncurrent-carrying equipment enclosures except as permitted in 250.24(A)(3). This connection is to be made on the supply side of separately derived system and ahead of any system disconnect or overcurrent device.

E.   **250.30(A)(3)** The GEC is sized per 250.66 for derived phase conductors of a separately derived system. The GEC connects the circuit conductor of the system that is required to be grounded to the grounding electrode. Except as permitted by 250.30(A), 250.32, or 250.142, this connection is to be made at the source of separately derived system and ahead of any disconnect means or overcurrent device.

F.   **250.30(A)(4)** The grounding electrode shall be the nearest of the following: (1) metal water pipe per 250.52(A)(1); (2) structural metal of a building per 250.52(A)(2); or (3) other electrodes when (1) and (2) are not present per 250.30(A)(4) Ex.1.

ABC FIRE EXTINGUISHERS

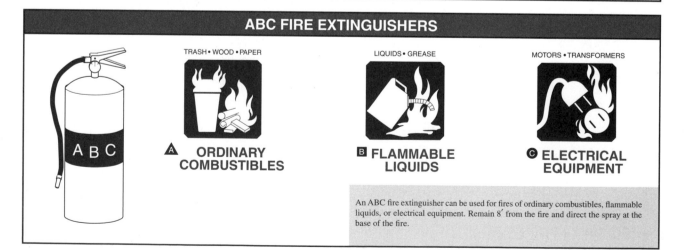

An ABC fire extinguisher can be used for fires of ordinary combustibles, flammable liquids, or electrical equipment. Remain 8′ from the fire and direct the spray at the base of the fire.

GROUND-FAULT PROTECTION OF EQUIPMENT — 230.95

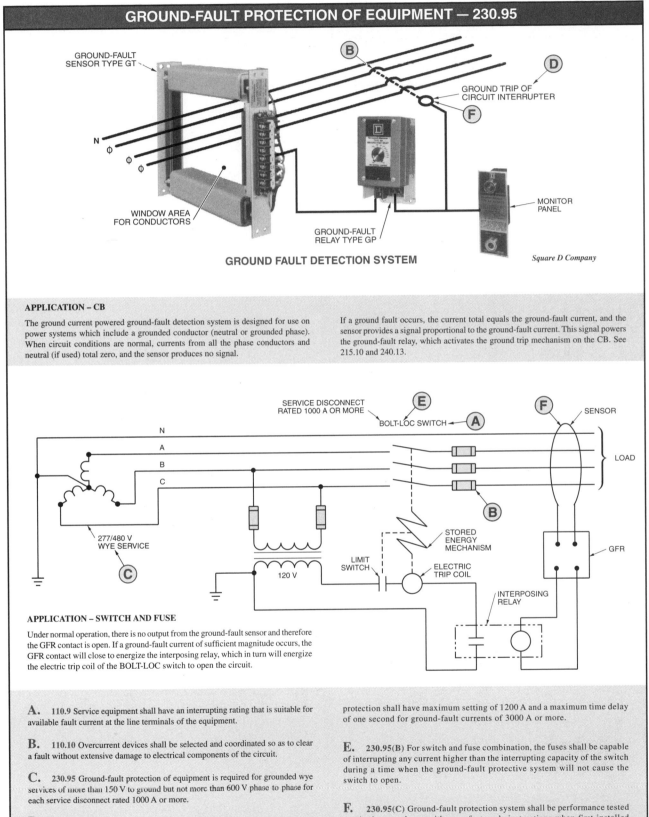

GROUND-FAULT
SENSOR TYPE GT

N
φ
φ
φ

WINDOW AREA
FOR CONDUCTORS

GROUND-FAULT
RELAY TYPE GP

GROUND TRIP OF
CIRCUIT INTERRUPTER

MONITOR
PANEL

GROUND FAULT DETECTION SYSTEM

Square D Company

APPLICATION – CB

The ground current powered ground-fault detection system is designed for use on power systems which include a grounded conductor (neutral or grounded phase). When circuit conditions are normal, currents from all the phase conductors and neutral (if used) total zero, and the sensor produces no signal.

If a ground fault occurs, the current total equals the ground-fault current, and the sensor provides a signal proportional to the ground-fault current. This signal powers the ground-fault relay, which activates the ground trip mechanism on the CB. See 215.10 and 240.13.

SERVICE DISCONNECT
RATED 1000 A OR MORE

BOLT-LOC SWITCH

SENSOR

N
A
B
C

LOAD

277/480 V
WYE SERVICE

120 V

LIMIT
SWITCH

STORED
ENERGY
MECHANISM

ELECTRIC
TRIP COIL

GFR

INTERPOSING
RELAY

APPLICATION – SWITCH AND FUSE

Under normal operation, there is no output from the ground-fault sensor and therefore the GFR contact is open. If a ground-fault current of sufficient magnitude occurs, the GFR contact will close to energize the interposing relay, which in turn will energize the electric trip coil of the BOLT-LOC switch to open the circuit.

A.   **110.9** Service equipment shall have an interrupting rating that is suitable for available fault current at the line terminals of the equipment.

B.   **110.10** Overcurrent devices shall be selected and coordinated so as to clear a fault without extensive damage to electrical components of the circuit.

C.   **230.95** Ground-fault protection of equipment is required for grounded wye services of more than 150 V to ground but not more than 600 V phase to phase for each service disconnect rated 1000 A or more.

D.   **230.95(A)** Operation of ground-fault protection shall cause service disconnect to open all ungrounded wires of faulted circuit. Ground-fault

protection shall have maximum setting of 1200 A and a maximum time delay of one second for ground-fault currents of 3000 A or more.

E.   **230.95(B)** For switch and fuse combination, the fuses shall be capable of interrupting any current higher than the interrupting capacity of the switch during a time when the ground-fault protective system will not cause the switch to open.

F.   **230.95(C)** Ground-fault protection system shall be performance tested on-site in accordance with manufacturer's instructions when first installed and a record of this test shall be made available to the AHJ.

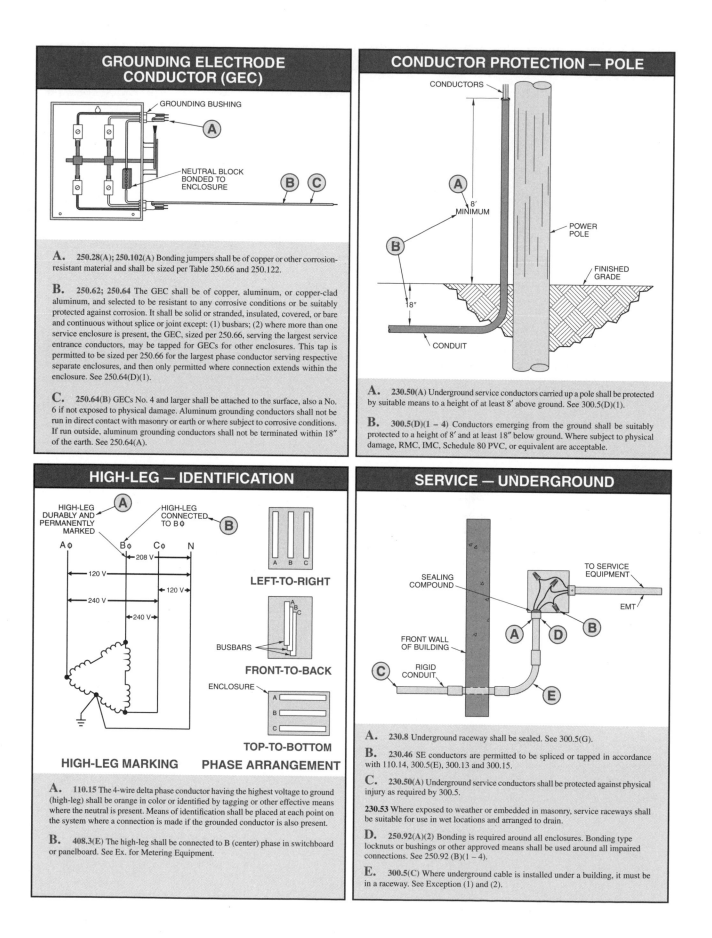

GROUNDING ELECTRODE CONDUCTOR (GEC)

GROUNDING BUSHING
(A)

NEUTRAL BLOCK BONDED TO ENCLOSURE

(B) (C)

A. **250.28(A); 250.102(A)** Bonding jumpers shall be of copper or other corrosion-resistant material and shall be sized per Table 250.66 and 250.122.

B. **250.62; 250.64** The GEC shall be of copper, aluminum, or copper-clad aluminum, and selected to be resistant to any corrosive conditions or be suitably protected against corrosion. It shall be solid or stranded, insulated, covered, or bare and continuous without splice or joint except: (1) busbars; (2) where more than one service enclosure is present, the GEC, sized per 250.66, serving the largest service entrance conductors, may be tapped for GECs for other enclosures. This tap is permitted to be sized per 250.66 for the largest phase conductor serving respective separate enclosures, and then only permitted where connection extends within the enclosure. See 250.64(D)(1).

C. **250.64(B)** GECs No. 4 and larger shall be attached to the surface, also a No. 6 if not exposed to physical damage. Aluminum grounding conductors shall not be run in direct contact with masonry or earth or where subject to corrosive conditions. If run outside, aluminum grounding conductors shall not be terminated within 18″ of the earth. See 250.64(A).

CONDUCTOR PROTECTION — POLE

CONDUCTORS

(A)
8′ MINIMUM

(B)

POWER POLE

FINISHED GRADE

18″

CONDUIT

A. **230.50(A)** Underground service conductors carried up a pole shall be protected by suitable means to a height of at least 8′ above ground. See 300.5(D)(1).

B. **300.5(D)(1 – 4)** Conductors emerging from the ground shall be suitably protected to a height of 8′ and at least 18″ below ground. Where subject to physical damage, RMC, IMC, Schedule 80 PVC, or equivalent are acceptable.

HIGH-LEG — IDENTIFICATION

HIGH-LEG DURABLY AND PERMANENTLY MARKED
(A)
HIGH-LEG CONNECTED TO B φ
(B)

Aφ   Bφ   Cφ   N
208 V
120 V
120 V
240 V
240 V

HIGH-LEG MARKING

A B C
LEFT-TO-RIGHT

A
B
C
BUSBARS
FRONT-TO-BACK

ENCLOSURE
A
B
C
TOP-TO-BOTTOM

PHASE ARRANGEMENT

A. **110.15** The 4-wire delta phase conductor having the highest voltage to ground (high-leg) shall be orange in color or identified by tagging or other effective means where the neutral is present. Means of identification shall be placed at each point on the system where a connection is made if the grounded conductor is also present.

B. **408.3(E)** The high-leg shall be connected to B (center) phase in switchboard or panelboard. See Ex. for Metering Equipment.

SERVICE — UNDERGROUND

SEALING COMPOUND
TO SERVICE EQUIPMENT
EMT
FRONT WALL OF BUILDING
(A) (D) (B)
(C)
RIGID CONDUIT
(E)

A. **230.8** Underground raceway shall be sealed. See 300.5(G).

B. **230.46** SE conductors are permitted to be spliced or tapped in accordance with 110.14, 300.5(E), 300.13 and 300.15.

C. **230.50(A)** Underground service conductors shall be protected against physical injury as required by 300.5.

230.53 Where exposed to weather or embedded in masonry, service raceways shall be suitable for use in wet locations and arranged to drain.

D. **250.92(A)(2)** Bonding is required around all enclosures. Bonding type locknuts or bushings or other approved means shall be used around all impaired connections. See 250.92 (B)(1 – 4).

E. **300.5(C)** Where underground cable is installed under a building, it must be in a raceway. See Exception (1) and (2).

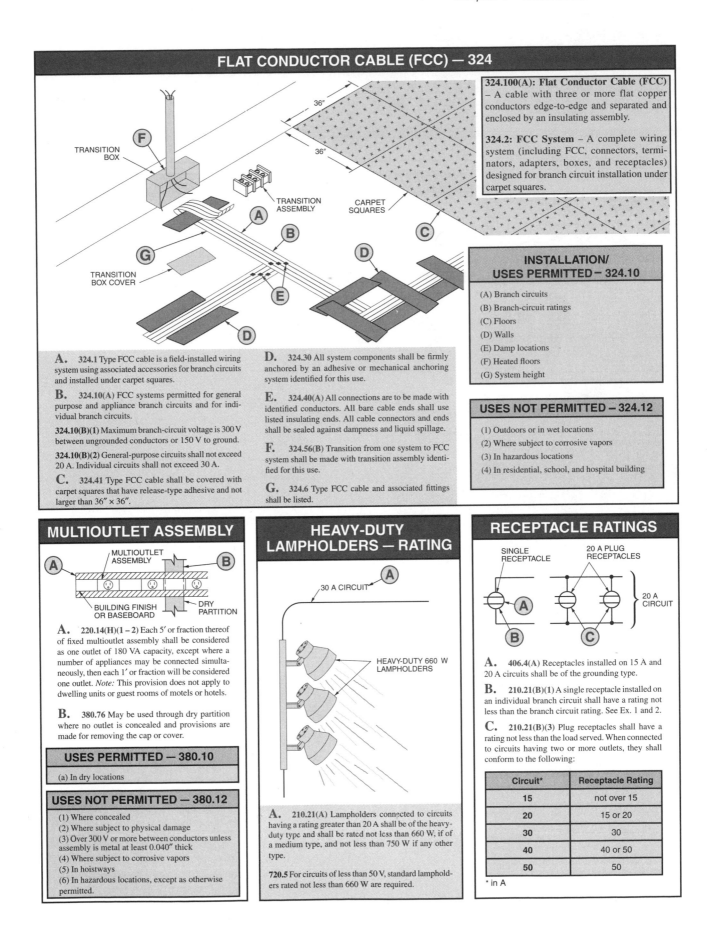

FLAT CONDUCTOR CABLE (FCC) — 324

TRANSITION BOX

TRANSITION ASSEMBLY

CARPET SQUARES

TRANSITION BOX COVER

36″

36″

324.100(A): Flat Conductor Cable (FCC) – A cable with three or more flat copper conductors edge-to-edge and separated and enclosed by an insulating assembly.

324.2: FCC System – A complete wiring system (including FCC, connectors, terminators, adapters, boxes, and receptacles) designed for branch circuit installation under carpet squares.

A. **324.1** Type FCC cable is a field-installed wiring system using associated accessories for branch circuits and installed under carpet squares.

B. **324.10(A)** FCC systems permitted for general purpose and appliance branch circuits and for individual branch circuits.

324.10(B)(1) Maximum branch-circuit voltage is 300 V between ungrounded conductors or 150 V to ground.

324.10(B)(2) General-purpose circuits shall not exceed 20 A. Individual circuits shall not exceed 30 A.

C. **324.41** Type FCC cable shall be covered with carpet squares that have release-type adhesive and not larger than 36″ × 36″.

D. **324.30** All system components shall be firmly anchored by an adhesive or mechanical anchoring system identified for this use.

E. **324.40(A)** All connections are to be made with identified conductors. All bare cable ends shall use listed insulating ends. All cable connectors and ends shall be sealed against dampness and liquid spillage.

F. **324.56(B)** Transition from one system to FCC system shall be made with transition assembly identified for this use.

G. **324.6** Type FCC cable and associated fittings shall be listed.

INSTALLATION/ USES PERMITTED – 324.10

(A) Branch circuits

(B) Branch-circuit ratings

(C) Floors

(D) Walls

(E) Damp locations

(F) Heated floors

(G) System height

USES NOT PERMITTED – 324.12

(1) Outdoors or in wet locations

(2) Where subject to corrosive vapors

(3) In hazardous locations

(4) In residential, school, and hospital building

MULTIOUTLET ASSEMBLY

MULTIOUTLET ASSEMBLY

BUILDING FINISH OR BASEBOARD

DRY PARTITION

A. **220.14(H)(1 – 2)** Each 5′ or fraction thereof of fixed multioutlet assembly shall be considered as one outlet of 180 VA capacity, except where a number of appliances may be connected simultaneously, then each 1′ or fraction will be considered one outlet. *Note:* This provision does not apply to dwelling units or guest rooms of motels or hotels.

B. **380.76** May be used through dry partition where no outlet is concealed and provisions are made for removing the cap or cover.

USES PERMITTED — 380.10

(a) In dry locations

USES NOT PERMITTED — 380.12

(1) Where concealed
(2) Where subject to physical damage
(3) Over 300 V or more between conductors unless assembly is metal at least 0.040″ thick
(4) Where subject to corrosive vapors
(5) In hoistways
(6) In hazardous locations, except as otherwise permitted.

HEAVY-DUTY LAMPHOLDERS — RATING

30 A CIRCUIT

HEAVY-DUTY 660 W LAMPHOLDERS

A. **210.21(A)** Lampholders connected to circuits having a rating greater than 20 A shall be of the heavy-duty type and shall be rated not less than 660 W, if of a medium type, and not less than 750 W if any other type.

720.5 For circuits of less than 50 V, standard lampholders rated not less than 660 W are required.

RECEPTACLE RATINGS

SINGLE RECEPTACLE

20 A PLUG RECEPTACLES

20 A CIRCUIT

A. **406.4(A)** Receptacles installed on 15 A and 20 A circuits shall be of the grounding type.

B. **210.21(B)(1)** A single receptacle installed on an individual branch circuit shall have a rating not less than the branch circuit rating. See Ex. 1 and 2.

C. **210.21(B)(3)** Plug receptacles shall have a rating not less than the load served. When connected to circuits having two or more outlets, they shall conform to the following:

Circuit*	Receptacle Rating
15	not over 15
20	15 or 20
30	30
40	40 or 50
50	50

* in A

OVERCURRENT DEVICES AND AVAILABLE FAULT CURRENT

Non-Selective System – A fault on an individual branch circuit not only opens the branch-circuit overcurrent protective device, but also opens the feeder overcurrent protective device.

Selective System – A fault on an individual branch circuit opens only the branch-circuit overcurrent protective device and does not affect the feeder overcurrent protective device.

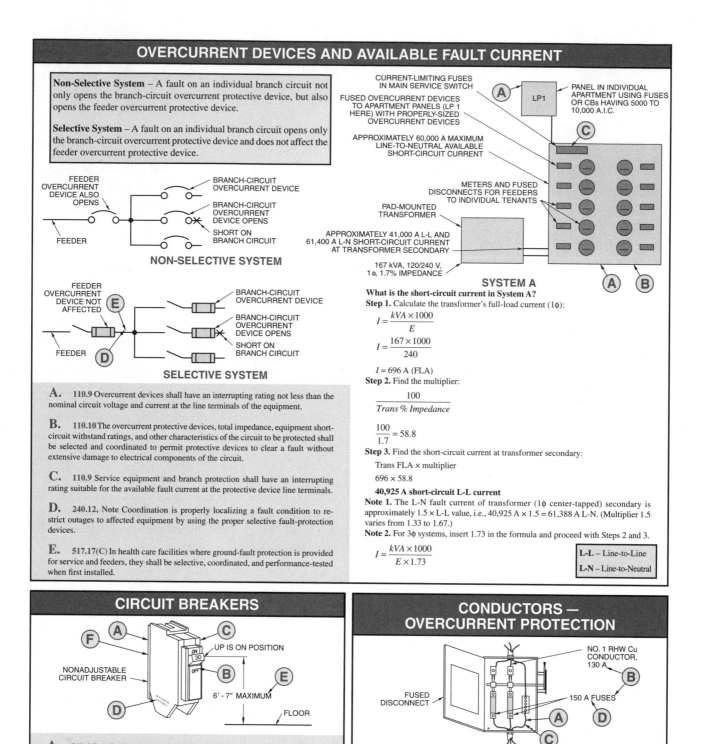

FEEDER OVERCURRENT DEVICE ALSO OPENS

BRANCH-CIRCUIT OVERCURRENT DEVICE

BRANCH-CIRCUIT OVERCURRENT DEVICE OPENS

SHORT ON BRANCH CIRCUIT

FEEDER

NON-SELECTIVE SYSTEM

FEEDER OVERCURRENT DEVICE NOT AFFECTED

BRANCH-CIRCUIT OVERCURRENT DEVICE

BRANCH-CIRCUIT OVERCURRENT DEVICE OPENS

SHORT ON BRANCH CIRCUIT

FEEDER

SELECTIVE SYSTEM

CURRENT-LIMITING FUSES IN MAIN SERVICE SWITCH

FUSED OVERCURRENT DEVICES TO APARTMENT PANELS (LP 1 HERE) WITH PROPERLY-SIZED OVERCURRENT DEVICES

APPROXIMATELY 60,000 A MAXIMUM LINE-TO-NEUTRAL AVAILABLE SHORT-CIRCUIT CURRENT

PANEL IN INDIVIDUAL APARTMENT USING FUSES OR CBs HAVING 5000 TO 10,000 A.I.C.

METERS AND FUSED DISCONNECTS FOR FEEDERS TO INDIVIDUAL TENANTS

PAD-MOUNTED TRANSFORMER

APPROXIMATELY 41,000 A L-L AND 61,400 A L-N SHORT-CIRCUIT CURRENT AT TRANSFORMER SECONDARY

167 kVA, 120/240 V, 1φ, 1.7% IMPEDANCE

SYSTEM A

What is the short-circuit current in System A?

Step 1. Calculate the transformer's full-load current (1φ):

$$I = \frac{kVA \times 1000}{E}$$

$$I = \frac{167 \times 1000}{240}$$

$I = 696$ A (FLA)

Step 2. Find the multiplier:

$$\frac{100}{Trans\ \%\ Impedance}$$

$$\frac{100}{1.7} = 58.8$$

Step 3. Find the short-circuit current at transformer secondary:

Trans FLA × multiplier

696 × 58.8

40,925 A short-circuit L-L current

Note 1. The L-N fault current of transformer (1φ center-tapped) secondary is approximately 1.5 × L-L value, i.e., 40,925 A × 1.5 = 61,388 A L-N. (Multiplier 1.5 varies from 1.33 to 1.67.)

Note 2. For 3φ systems, insert 1.73 in the formula and proceed with Steps 2 and 3.

$$I = \frac{kVA \times 1000}{E \times 1.73}$$

L-L – Line-to-Line

L-N – Line-to-Neutral

A. 110.9 Overcurrent devices shall have an interrupting rating not less than the nominal circuit voltage and current at the line terminals of the equipment.

B. 110.10 The overcurrent protective devices, total impedance, equipment short-circuit withstand ratings, and other characteristics of the circuit to be protected shall be selected and coordinated to permit protective devices to clear a fault without extensive damage to electrical components of the circuit.

C. 110.9 Service equipment and branch protection shall have an interrupting rating suitable for the available fault current at the protective device line terminals.

D. 240.12, Note Coordination is properly localizing a fault condition to restrict outages to affected equipment by using the proper selective fault-protection devices.

E. 517.17(C) In health care facilities where ground-fault protection is provided for service and feeders, they shall be selective, coordinated, and performance-tested when first installed.

CIRCUIT BREAKERS

UP IS ON POSITION

NONADJUSTABLE CIRCUIT BREAKER

6' - 7" MAXIMUM

FLOOR

A. 240.4(B – C) If the allowable ampacity of a conductor does not correspond with a standard rating of an overcurrent device, the next larger rating may be used, provided that the current is not greater than 800 A.

B. 240.81 CBs shall be clearly marked ON or OFF. Up is ON position.

C. 240.82 A CB shall be designed so that alteration of its operating characteristics requires dismantling the device.

D. 240.83(D) Where CBs are used to switch 120 V and 277 V fluorescent lighting circuits, they shall be marked SWD or HID and listed for such switching duty.

E. 404.8(A) Center of operating handle shall not be over 6'-7" from floor.

F. 404.11 CBs used as a switch shall conform to 240.81 and 240.83.

CONDUCTORS — OVERCURRENT PROTECTION

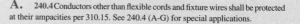

NO. 1 RHW Cu CONDUCTOR, 130 A

FUSED DISCONNECT

150 A FUSES

A. 240.4 Conductors other than flexible cords and fixture wires shall be protected at their ampacities per 310.15. See 240.4 (A-G) for special applications.

B. 240.4(B – C) If the allowable ampacity of a conductor does not correspond with a standard rating of an overcurrent device, the next larger rating may be used, provided that the current is not greater than 800 A and the conductor is not part of a multioutlet branch circuit supplying receptacles for cord- and plug-connected portable loads.

C. 210.19(A)(1) Branch-circuit conductors shall have an ampacity not less than the branch circuit and not less than the maximum load served.

D. 240.21 Overcurrent protection is required where conductors receive their supply, except as permitted in 240.21(A-H).

CORDS AND FIXTURE WIRE — OVERCURRENT PROTECTION

APPLIANCE

FLEXIBLE CORD

240.5(A) Flexible cords (including tinsel) shall be protected at their ampacities per Tables 400.5(A)(1) and (2).

240.5(B)(1) Flexible cord or tinsel cord is approved for use with a specific listed appliance or portable lamp. It shall be considered to be protected when applied within the appliance or portable lamp listing requirements.

240.5(B)(2) Fixture wire connected to 120 V or more standard branch circuit shall be protected as follows:

Cord	Circuit
No. 18 (to 50′)	20
No. 16 (to 100′)	20
No. 14 +	20 or 30
No. 12 +	40 or 50

TEMPORARY INSTALLATIONS

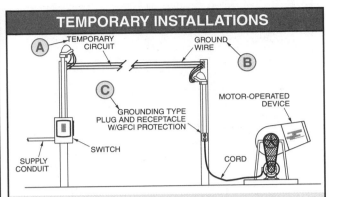

A. 590.4(A – J) Services shall conform to 230. Feeders and branch circuits shall be protected at rated ampacity. Receptacles shall be the grounding type and connected to a grounding conductor. Receptacles are not permitted on lighting branch circuits. Suitable disconnecting switches or plug connectors required to disconnect all ungrounded conductors of each temporary circuit. Lamps shall be protected by a suitable fixture or lampholder with guard. A box is not required for splices or junction connections when using cable assemblies, cords, or cables. See 110.14(B) and 400.9.

B. Grounding shall be per Article 250.

C. 590.6 All temporary wiring installations shall comply with GFCI protection per 590.6(A) or by use of an assured equipment grounding program for other outlets per 590.6(B)(2).

THREADED BOX SUPPORTS

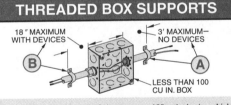

A. 314.23(E) Threaded boxes or fittings not over 100 cu in. in size which contain no devices are considered fastened if two or more conduits are threaded into the box and are firmly supported within 3′ of the box. See Exception.

B. 314.23(F) Threaded boxes or fittings not over 100 cu in. which contain devices are considered supported if two or more conduits are threaded into the box and are firmly secured within 18″ of the box. See Ex No.1 and No. 2.

SNAP SWITCHES

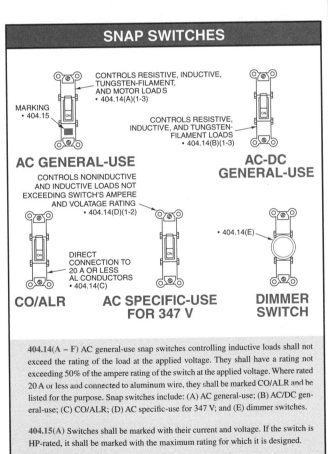

404.14(A – F) AC general-use snap switches controlling inductive loads shall not exceed the rating of the load at the applied voltage. They shall have a rating not exceeding 50% of the ampere rating of the switch at the applied voltage. Where rated 20 A or less and connected to aluminum wire, they shall be marked CO/ALR and be listed for the purpose. Snap switches include: (A) AC general-use; (B) AC/DC general-use; (C) CO/ALR; (D) AC specific-use for 347 V; and (E) dimmer switches.

404.15(A) Switches shall be marked with their current and voltage. If the switch is HP-rated, it shall be marked with the maximum rating for which it is designed.

SWITCHBOARDS

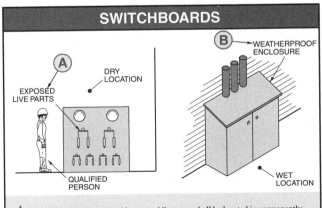

A. 408.20 Switchboards with exposed live parts shall be located in permanently dry locations accessible only to qualified persons.

B. 408.16 A switchboard in a wet location or outside a building shall have a weatherproof enclosure per 312.2.

PANELBOARDS — 408 PART III

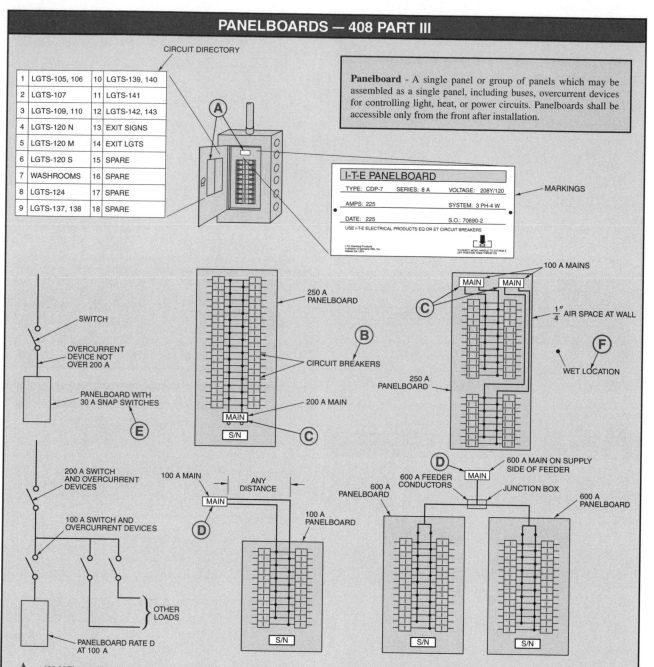

CIRCUIT DIRECTORY

1	LGTS-105, 106	10	LGTS-139, 140
2	LGTS-107	11	LGTS-141
3	LGTS-109, 110	12	LGTS-142, 143
4	LGTS-120 N	13	EXIT SIGNS
5	LGTS-120 M	14	EXIT LGTS
6	LGTS-120 S	15	SPARE
7	WASHROOMS	16	SPARE
8	LGTS-124	17	SPARE
9	LGTS-137, 138	18	SPARE

Panelboard - A single panel or group of panels which may be assembled as a single panel, including buses, overcurrent devices for controlling light, heat, or power circuits. Panelboards shall be accessible only from the front after installation.

I-T-E PANELBOARD

TYPE: CDP-7   SERIES: 8 A   VOLTAGE: 208Y/120

AMPS: 225   SYSTEM: 3 PH-4 W

DATE: 225   S.O.: 70690-2

USE I-T-E ELECTRICAL PRODUCTS EQ OR ET CIRCUIT BREAKERS

I-T-E Electrical Products
A division of Siemens-Allis, Inc
Atlanta GA, USA

TO RESET, MOVE HANDLE TO EXTREME
OFF POSITION THEN THROW ON

MARKINGS

SWITCH

OVERCURRENT DEVICE NOT OVER 200 A

PANELBOARD WITH 30 A SNAP SWITCHES

250 A PANELBOARD

CIRCUIT BREAKERS

200 A MAIN

MAIN

S/N

100 A MAINS

MAIN   MAIN

1/4" AIR SPACE AT WALL

250 A PANELBOARD

WET LOCATION

MAIN

600 A MAIN ON SUPPLY SIDE OF FEEDER

600 A FEEDER CONDUCTORS

600 A PANELBOARD

JUNCTION BOX

600 A PANELBOARD

S/N

S/N

200 A SWITCH AND OVERCURRENT DEVICES

100 A SWITCH AND OVERCURRENT DEVICES

100 A MAIN

ANY DISTANCE

MAIN

100 A PANELBOARD

S/N

OTHER LOADS

PANELBOARD RATE D AT 100 A

A. 408.30 The panelboard rating shall not be less than the minimum feeder capacity per Article 220. They shall be marked with the voltage, current rating, number of phases, and manufacturer's name or trademark. A circuit directory shall be provided on the face or inside the door. See 110.22.

B. 240.80 Circuit breakers shall be trip free and capable of being closed and opened by manual operation.

C. 408.36 Panelboards shall be protected by an OCPD having a rating not greater than that of the panelboard. The OCPD must be located within or at any point on the supply side of the panelboard. See Exceptions.

D. 408.36 Ex. 1 Individual protection of the panelboard is not required where the panelboard is used as service equipment with multiple disconnecting means as per 230.71.

408.36 Ex. 2 Individual protection of the panelboard is not required where the panelboard is protected on the supply side by two sets of fuses or two main CBs

having a combined rating not greater than that of the panelboard. In this case, the panelboard shall not contain more than 42 OCPDs.

E. 408.36(A) Panelboards with snap switches rated 30 A or less shall have overcurrent protection not in excess of 200 A.

210.20(A) Where a branch circuit supplies continuous loads, the rating of the OCPD shall not be less than the noncontinuous load plus 125% of the continuous load.

F. 408.37 Panelboards in damp or wet locations shall comply with 312.2.

408.38 Panelboards shall be mounted in enclosures designed for the purpose and shall be dead-front.

408.39 Except as permitted for services, panelboard fuses shall be installed on the load side of switches.

408.40 Metal panelboard cabinets and frames shall be grounded per Section 408.40 and Article 250.

FLEXIBLE CORDS AND CABLES — 400

USES PERMITTED — 400.7(A)

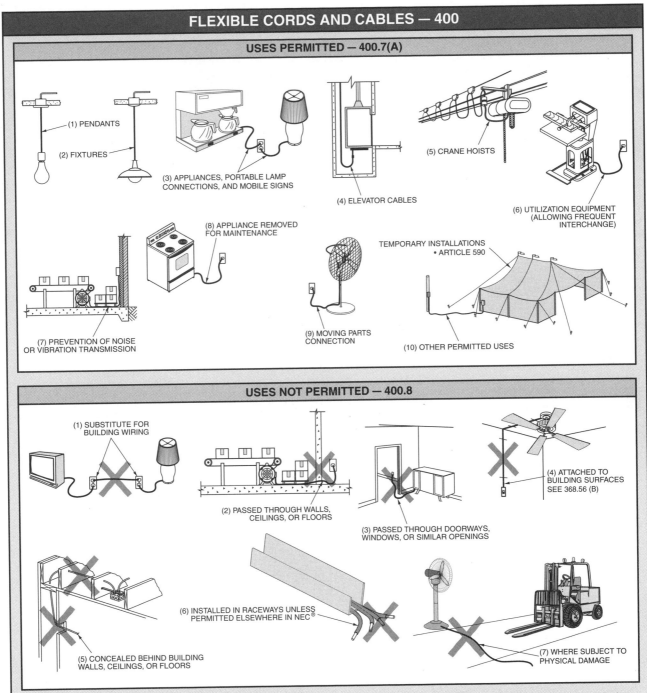

(1) PENDANTS

(2) FIXTURES

(3) APPLIANCES, PORTABLE LAMP CONNECTIONS, AND MOBILE SIGNS

(4) ELEVATOR CABLES

(5) CRANE HOISTS

(6) UTILIZATION EQUIPMENT (ALLOWING FREQUENT INTERCHANGE)

(8) APPLIANCE REMOVED FOR MAINTENANCE

TEMPORARY INSTALLATIONS
• ARTICLE 590

(7) PREVENTION OF NOISE OR VIBRATION TRANSMISSION

(9) MOVING PARTS CONNECTION

(10) OTHER PERMITTED USES

USES NOT PERMITTED — 400.8

(1) SUBSTITUTE FOR BUILDING WIRING

(2) PASSED THROUGH WALLS, CEILINGS, OR FLOORS

(3) PASSED THROUGH DOORWAYS, WINDOWS, OR SIMILAR OPENINGS

(4) ATTACHED TO BUILDING SURFACES SEE 368.56 (B)

(6) INSTALLED IN RACEWAYS UNLESS PERMITTED ELSEWHERE IN NEC®

(5) CONCEALED BEHIND BUILDING WALLS, CEILINGS, OR FLOORS

(7) WHERE SUBJECT TO PHYSICAL DAMAGE

400.6(A) Types S, SC, SCE, SCT, SE, SEO, SEOO, SJ, SJE, SJEO, SJEOO, SJO, SJT, SJTO, SJTOO, SO, SOO, ST, STO, STOO, SEW, SEOW, SEOOW, SJEW, SJEOW, SJEOOW, SJOW, SJTW, SJTOW, SJTOOW, SOW, SOOW, STW, STOW, and STOOW flexible cords and G, G–GC, PPE, and W flexible cables shall be marked on the surface every 24″ with the type, size, and number of conductors.

400.7(A) Flexible cords and cables shall be used for: (1) pendants; (2) fixtures; (3) appliances or portable lamp connections; (4) elevator cables; (5) cranes and hoists; (6) stationary equipment (allowing frequent interchange); (7) prevention of noise or vibration transmission; (8) appliance removed for maintenance; (9) moving parts connections; or (10) other permitted uses.

400.8 Flexible cords and cables shall not be used: (1) as a substitute for building wiring; (2) where passed through walls, ceilings, or floors; (3) where passed through doorways, windows, or similar openings; (4) where attached to building surfaces; (5) where concealed behind building walls, ceilings, or floors; (6) where installed in raceways unless permitted elsewhere in the NEC®; or (7) where subject to physical damage.

400.9 Flexible cords shall be used only in continuous lengths, not spliced or tapped except junior hard-service and hard-service flexible cord No. 14 and larger if spliced per 110.14(B).

400.11 Flexible cords used in show windows and show cases shall, in general, be Types S, SE, SEO, SEOO, SJ, SJE, SJEO, SJEOO, SJO, SJOO, SJT, SJTO, SJTOO, SO, SOO, ST, STO, STOO, SEW, SEOW, SEEOW, SJEW, SJEOW, SJEOOW, SJOW, SJOOW, SJTW, SJTOW, SJTOOW, SOW, SOOW, STW, STOW, and STOOW.

400.21(A) Flexible cords or cables shall not be smaller than the sizes in Table 400.4. See Exception for Type G-GC cables.

FEEDER — 10′ FEEDER TAP RULE

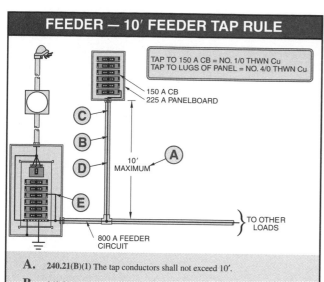

TAP TO 150 A CB = NO. 1/0 THWN Cu
TAP TO LUGS OF PANEL = NO. 4/0 THWN Cu

150 A CB
225 A PANELBOARD

C
B
A
D
10′
MAXIMUM
E

TO OTHER
LOADS

800 A FEEDER
CIRCUIT

A. 240.21(B)(1) The tap conductors shall not exceed 10′.

B. 240.21(B)(1)(1) The tap conductor's ampacity shall not be less than the combined load and not less than the rating of the device or OCPD supplied.

C. 240.21(B)(1)(2) The tap conductors shall not extend beyond the panelboard, etc. supplied.

D. 240.21(B)(1)(3) The tap conductors shall be protected in a raceway.

E. 240.21(B)(1)(4) The ampacity of the tap conductors is not less than one-tenth of the rating of the OCPD protecting the feeder.

FEEDER — 25′ FEEDER TAP RULE

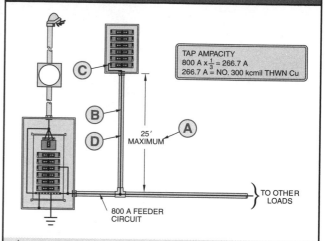

TAP AMPACITY
$800 \text{ A} \times \frac{1}{3} = 266.7 \text{ A}$
266.7 A = NO. 300 kcmil THWN Cu

C
B
A
D
25′
MAXIMUM

TO OTHER
LOADS

800 A FEEDER
CIRCUIT

A. 240.21(B)(2) The tap conductor shall not exceed 25′.

B. 240.21(B)(2)(1) The tap conductor's ampacity shall not be less than one-third of the rating of the feeder OCPD.

C. 240.21(B)(2)(2) The tap conductors shall terminate in a single OCPD which limits the load to the ampacity of the tap conductors.

D. 240.21(B)(2)(3) The tap conductors shall be protected from physical damage by enclosure in an approved raceway or other approved means.

25′ TRANSFORMER FEEDER TAP RULE

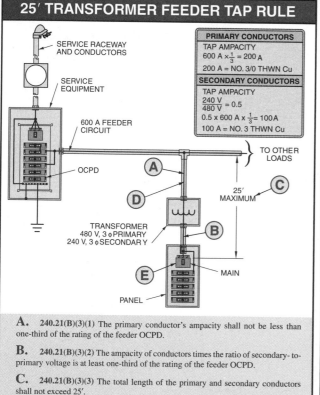

SERVICE RACEWAY
AND CONDUCTORS

SERVICE
EQUIPMENT

PRIMARY CONDUCTORS	
TAP AMPACITY	
$600 \text{ A} \times \frac{1}{3} = 200 \text{ A}$	
200 A = NO. 3/0 THWN Cu	
SECONDARY CONDUCTORS	
TAP AMPACITY	
$\frac{240 \text{ V}}{480 \text{ V}} = 0.5$	
$0.5 \times 600 \text{ A} \times \frac{1}{3} = 100\text{A}$	
100 A = NO. 3 THWN Cu	

600 A FEEDER
CIRCUIT

OCPD

TO OTHER
LOADS

A
D
C
25′
MAXIMUM

TRANSFORMER
480 V, 3 φ PRIMARY
240 V, 3 φ SECONDARY

B

E
MAIN

PANEL

A. 240.21(B)(3)(1) The primary conductor's ampacity shall not be less than one-third of the rating of the feeder OCPD.

B. 240.21(B)(3)(2) The ampacity of conductors times the ratio of secondary-to-primary voltage is at least one-third of the rating of the feeder OCPD.

C. 240.21(B)(3)(3) The total length of the primary and secondary conductors shall not exceed 25′.

D. 240.21(B)(3)(4) The tap conductors shall be protected in an approved raceway from physical damage or by other approved means.

E. 240.21(B)(3)(5) The secondary conductors shall terminate in a single OCPD which limits the load current to no more than the conductor's ampacity.

OVER 25′ FEEDER TAP RULE

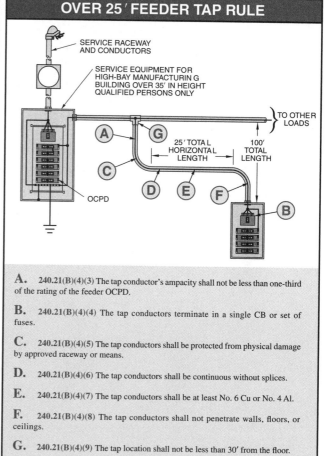

SERVICE RACEWAY
AND CONDUCTORS

SERVICE EQUIPMENT FOR
HIGH-BAY MANUFACTURING
BUILDING OVER 35′ IN HEIGHT
QUALIFIED PERSONS ONLY

TO OTHER
LOADS

A
G
C
25′ TOTAL
HORIZONTAL
LENGTH
100′
TOTAL
LENGTH

OCPD

D
E
F
B

A. 240.21(B)(4)(3) The tap conductor's ampacity shall not be less than one-third of the rating of the feeder OCPD.

B. 240.21(B)(4)(4) The tap conductors terminate in a single CB or set of fuses.

C. 240.21(B)(4)(5) The tap conductors shall be protected from physical damage by approved raceway or means.

D. 240.21(B)(4)(6) The tap conductors shall be continuous without splices.

E. 240.21(B)(4)(7) The tap conductors shall be at least No. 6 Cu or No. 4 Al.

F. 240.21(B)(4)(8) The tap conductors shall not penetrate walls, floors, or ceilings.

G. 240.21(B)(4)(9) The tap location shall not be less than 30′ from the floor.

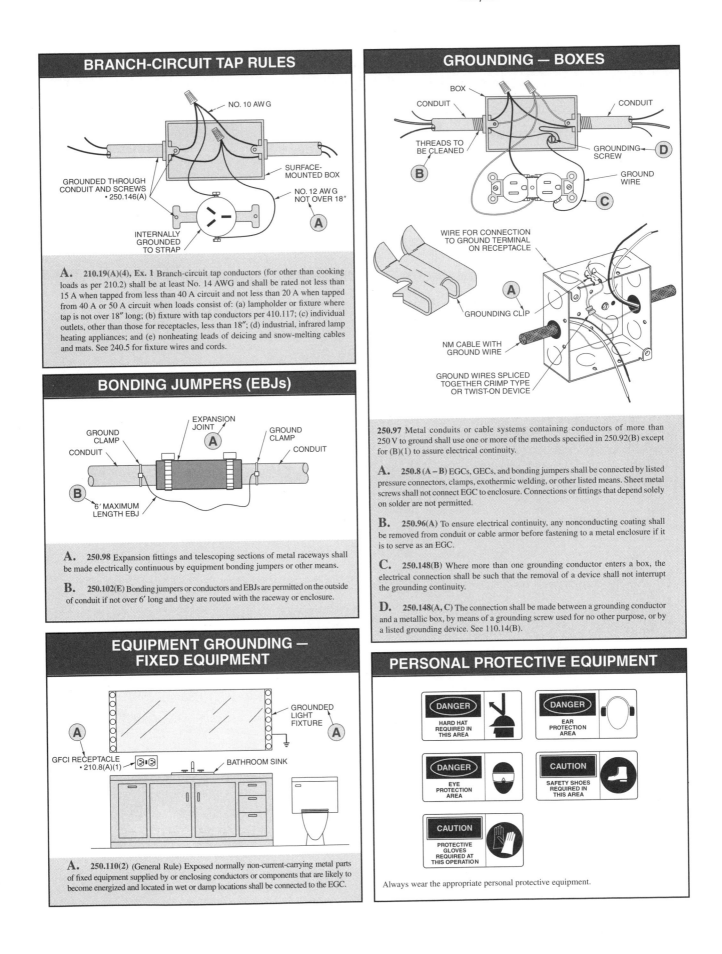

BRANCH-CIRCUIT TAP RULES

NO. 10 AWG

GROUNDED THROUGH CONDUIT AND SCREWS • 250.146(A)

SURFACE-MOUNTED BOX

NO. 12 AWG NOT OVER 18"

INTERNALLY GROUNDED TO STRAP

A. **210.19(A)(4), Ex. 1** Branch-circuit tap conductors (for other than cooking loads as per 210.2) shall be at least No. 14 AWG and shall be rated not less than 15 A when tapped from less than 40 A circuit and not less than 20 A when tapped from 40 A or 50 A circuit when loads consist of: (a) lampholder or fixture where tap is not over 18" long; (b) fixture with tap conductors per 410.117; (c) individual outlets, other than those for receptacles, less than 18"; (d) industrial, infrared lamp heating appliances; and (e) nonheating leads of deicing and snow-melting cables and mats. See 240.5 for fixture wires and cords.

BONDING JUMPERS (EBJs)

EXPANSION JOINT

GROUND CLAMP

CONDUIT

GROUND CLAMP

CONDUIT

6' MAXIMUM LENGTH EBJ

A. **250.98** Expansion fittings and telescoping sections of metal raceways shall be made electrically continuous by equipment bonding jumpers or other means.

B. **250.102(E)** Bonding jumpers or conductors and EBJs are permitted on the outside of conduit if not over 6' long and they are routed with the raceway or enclosure.

EQUIPMENT GROUNDING — FIXED EQUIPMENT

GROUNDED LIGHT FIXTURE

GFCI RECEPTACLE • 210.8(A)(1)

BATHROOM SINK

A. **250.110(2)** (General Rule) Exposed normally non-current-carrying metal parts of fixed equipment supplied by or enclosing conductors or components that are likely to become energized and located in wet or damp locations shall be connected to the EGC.

GROUNDING — BOXES

BOX

CONDUIT

CONDUIT

THREADS TO BE CLEANED

GROUNDING SCREW

GROUND WIRE

WIRE FOR CONNECTION TO GROUND TERMINAL ON RECEPTACLE

GROUNDING CLIP

NM CABLE WITH GROUND WIRE

GROUND WIRES SPLICED TOGETHER CRIMP TYPE OR TWIST-ON DEVICE

250.97 Metal conduits or cable systems containing conductors of more than 250 V to ground shall use one or more of the methods specified in 250.92(B) except for (B)(1) to assure electrical continuity.

A. **250.8 (A – B)** EGCs, GECs, and bonding jumpers shall be connected by listed pressure connectors, clamps, exothermic welding, or other listed means. Sheet metal screws shall not connect EGC to enclosure. Connections or fittings that depend solely on solder are not permitted.

B. **250.96(A)** To ensure electrical continuity, any nonconducting coating shall be removed from conduit or cable armor before fastening to a metal enclosure if it is to serve as an EGC.

C. **250.148(B)** Where more than one grounding conductor enters a box, the electrical connection shall be such that the removal of a device shall not interrupt the grounding continuity.

D. **250.148(A, C)** The connection shall be made between a grounding conductor and a metallic box, by means of a grounding screw used for no other purpose, or by a listed grounding device. See 110.14(B).

PERSONAL PROTECTIVE EQUIPMENT

DANGER HARD HAT REQUIRED IN THIS AREA

DANGER EAR PROTECTION AREA

DANGER EYE PROTECTION AREA

CAUTION SAFETY SHOES REQUIRED IN THIS AREA

CAUTION PROTECTIVE GLOVES REQUIRED AT THIS OPERATION

Always wear the appropriate personal protective equipment.

SUPPORTING CONDUCTORS IN VERTICAL RACEWAYS — 300.19

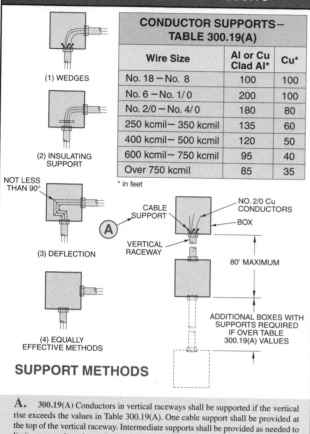

CONDUCTOR SUPPORTS— TABLE 300.19(A)		
Wire Size	**Al or Cu Clad Al***	**Cu***
No. 18 — No. 8	100	100
No. 6 — No. 1/0	200	100
No. 2/0 — No. 4/0	180	80
250 kcmil — 350 kcmil	135	60
400 kcmil — 500 kcmil	120	50
600 kcmil — 750 kcmil	95	40
Over 750 kcmil	85	35

* in feet

(1) WEDGES

(2) INSULATING SUPPORT

(3) DEFLECTION

NOT LESS THAN 90°

(A)

CABLE SUPPORT

VERTICAL RACEWAY

NO. 2/0 Cu CONDUCTORS

BOX

80' MAXIMUM

ADDITIONAL BOXES WITH SUPPORTS REQUIRED IF OVER TABLE 300.19(A) VALUES

(4) EQUALLY EFFECTIVE METHODS

SUPPORT METHODS

A. 300.19(A) Conductors in vertical raceways shall be supported if the vertical rise exceeds the values in Table 300.19(A). One cable support shall be provided at the top of the vertical raceway. Intermediate supports shall be provided as needed to limit supported conductor length to values in Table 300.19(A).

300.19(C) Support methods for conductors shall be by one of the following:

(1) Wedges inserted in the top ends of the raceways

(2) Boxes at required intervals with insulating supports installed

(3) In junction boxes by deflecting cables not less than 90° and carrying them horizontally not less than twice the cable diameter. Support spacing is reduced 20% for this method.

(4) Equally effective methods

WIRING WITHIN FLUORESCENT FIXTURES

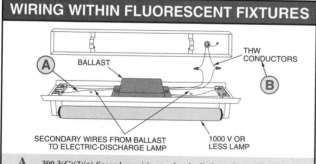

(A)

BALLAST

THW CONDUCTORS

(B)

SECONDARY WIRES FROM BALLAST TO ELECTRIC-DISCHARGE LAMP

1000 V OR LESS LAMP

A. 300.3(C)(2)(a) Secondary wiring to electric-discharge lamps of 1000 V or less may occupy the same enclosure as branch-circuit conductors if insulated for the secondary voltage of the ballast.

B. 410.68 Branch-circuit conductors within 3″ of a ballast, LED driver, power supply, or transformer shall have a temperature rating of at least 90°C, unless they supply a fixture that is marked as being suitable for a different insulation temperature.

RACEWAYS EXPOSED TO DIFFERENT TEMPERATURES

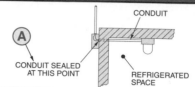

(A)

CONDUIT

CONDUIT SEALED AT THIS POINT

REFRIGERATED SPACE

A. 300.7(A) To prevent circulation of air through interior conduit runs whose ends are exposed to widely different temperatures, raceways, or sleeves shall be sealed off with an approved material.

300.7(B) Expansion joints shall be provided where necessary to compensate for thermal movement.

CONTINUITY — RACEWAYS AND BOXES

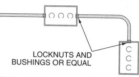

LOCKNUTS AND BUSHINGS OR EQUAL

250.96(A) Where a metal raceway serves as an equipment grounding conductor, it shall be electrically continuous or bonded to assure ability to safely conduct possible fault currents.

250.102(E)(1)(2) Bonding jumpers and EBJs shall be inside the conduit with circuit wires except if they are not over 6′, then they may be outside the conduit if routed with the conduit.

300.10 Metal raceways, cable armor, and enclosures shall be metallically joined throughout the whole installation to provide electrical continuity. See Ex. No. 1 and No. 2.

AIR PLENUMS — WIRING METHODS

INTERMEDIATE RMC, EMT, MI, TYPE MC CABLE, IMC, OR ALS CABLE

DUCTS OR PLENUMS SPECIFICALLY FABRICATED FOR ENVIRONMENTAL AIR

300.22(A) No wiring of any kind shall be installed in ducts used to transport dust, loose stock, flammable vapors, or commercial cooking.

300.22(B) Where necessary to run a wiring system through fabricated ducts or plenums, types MI or MC cables, EMT, FMC, IMC, or RMC may be used. Physically adjustable equipment and devices in ducts may be wired with FMC in lengths not exceeding 4′.

300.22(C)(1) Wiring methods in hollow spaces above lowered ceilings that are also used for supply or return plenums for environmental air and are not specifically fabricated ducts are limited to types MI, MC, AC cables, and other factory-assembled multiconductor control or power cable specifically listed for the use. Other cables and conductors shall be in EMT, IMC, RMC, FMC, or where accessible, surface metal raceway, metal wireway, or solid-bottom metal cable tray with solid metal cover. See 300.22(C)(2)(a-b).

300.22(C), Ex. Stud or joist space in dwelling units where used as air handling permits normal wiring to pass through if perpendicular to space.

300.22(C)(1) Listed prefabricated cable assemblies of manufactured metallic wiring systems where listed for this use shall be permitted.

424.58 Electric duct heaters shall be identified as suitable for the installation.

424.65 The electric duct heater controller shall be accessible and the disconnecting means shall be within sight from the controller or as permitted by 424.19(A).

UNDERFLOOR RACEWAYS — 390

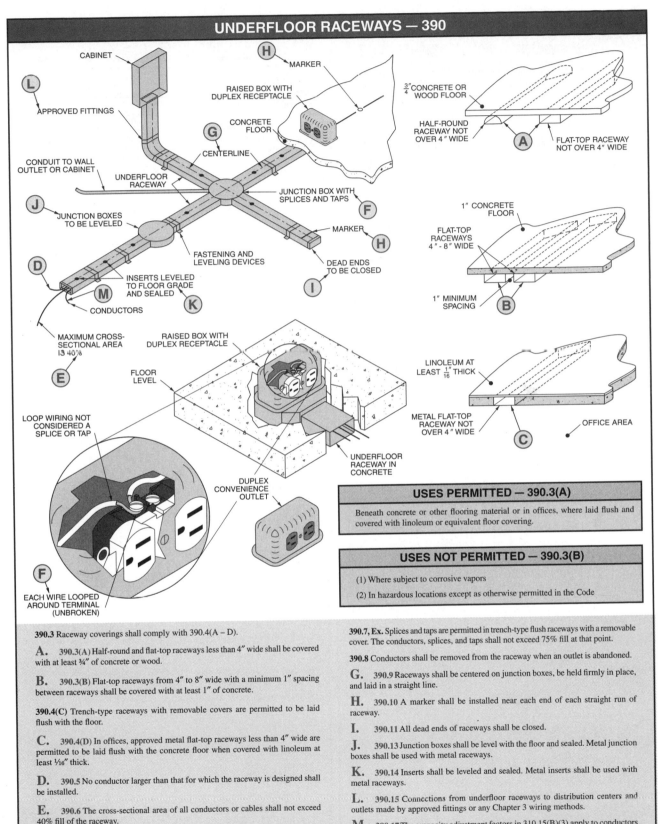

USES PERMITTED — 390.3(A)

Beneath concrete or other flooring material or in offices, where laid flush and covered with linoleum or equivalent floor covering.

USES NOT PERMITTED — 390.3(B)

(1) Where subject to corrosive vapors

(2) In hazardous locations except as otherwise permitted in the Code

---

390.3 Raceway coverings shall comply with 390.4(A – D).

A. 390.3(A) Half-round and flat-top raceways less than 4″ wide shall be covered with at least ¾ of concrete or wood.

B. 390.3(B) Flat-top raceways from 4″ to 8″ wide with a minimum 1″ spacing between raceways shall be covered with at least 1″ of concrete.

390.4(C) Trench-type raceways with removable covers are permitted to be laid flush with the floor.

C. 390.4(D) In offices, approved metal flat-top raceways less than 4″ wide are permitted to be laid flush with the concrete floor when covered with linoleum at least ¹⁄₁₆″ thick.

D. 390.5 No conductor larger than that for which the raceway is designed shall be installed.

E. 390.6 The cross-sectional area of all conductors or cables shall not exceed 40% fill of the raceway.

F. 390.7 Splices and taps are permitted only in junction boxes.

390.7, Ex. Splices and taps are permitted in trench-type flush raceways with a removable cover. The conductors, splices, and taps shall not exceed 75% fill at that point.

390.8 Conductors shall be removed from the raceway when an outlet is abandoned.

G. 390.9 Raceways shall be centered on junction boxes, be held firmly in place, and laid in a straight line.

H. 390.10 A marker shall be installed near each end of each straight run of raceway.

I. 390.11 All dead ends of raceways shall be closed.

J. 390.13 Junction boxes shall be level with the floor and sealed. Metal junction boxes shall be used with metal raceways.

K. 390.14 Inserts shall be leveled and sealed. Metal inserts shall be used with metal raceways.

L. 390.15 Connections from underfloor raceways to distribution centers and outlets made by approved fittings or any Chapter 3 wiring methods.

M. 390.17 The ampacity adjustment factors in 310.15(B)(3) apply to conductors installed in underfloor raceway.

CELLULAR METAL FLOOR RACEWAYS — 374

374.2: Cellular Metal Floor Raceway – The hollow spaces of cellular metal floors, with fittings, approved as enclosures for electric conductors.

Cell – A single, enclosed tubular space in a cellular metal floor with the axis being parallel to the axis of the metal floor member.

Header – A transverse raceway for electric conductors that provides access to particular cells.

CELLULAR CONCRETE FLOOR RACEWAYS — 372

372.2: Cell – A single, enclosed, tubular space in a precast cellular concrete slab floor with the direction of the cell being parallel to the direction of the floor member.

Header – A transverse metal raceway for electric conductors with access to predetermined cells, permitting the installation of electric conductors from a distribution center to the floor cell.

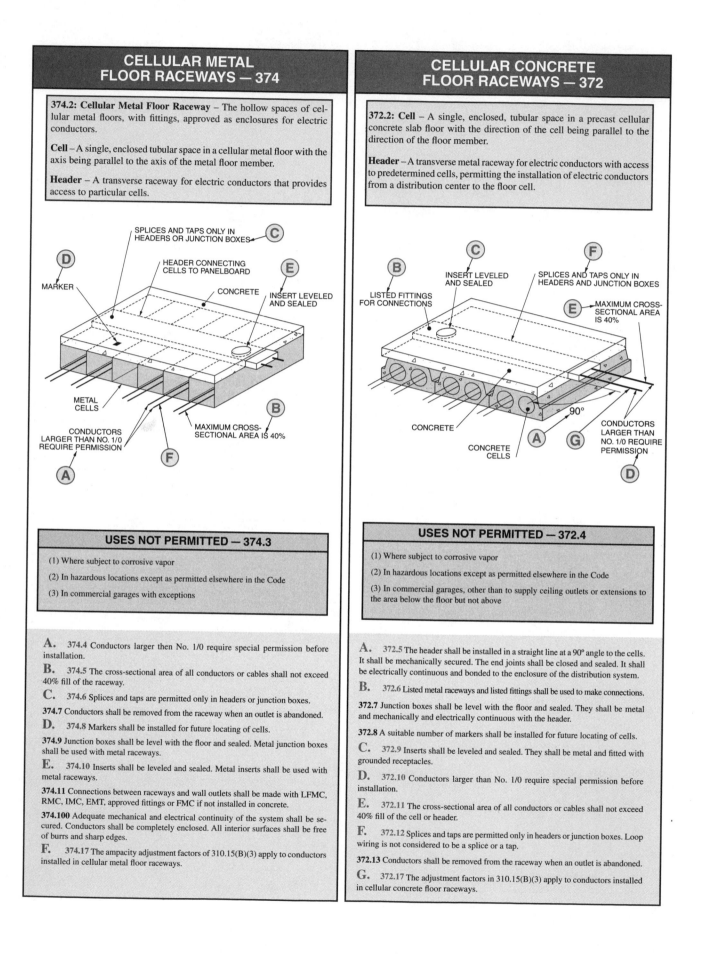

USES NOT PERMITTED — 374.3

(1) Where subject to corrosive vapor

(2) In hazardous locations except as permitted elsewhere in the Code

(3) In commercial garages with exceptions

USES NOT PERMITTED — 372.4

(1) Where subject to corrosive vapor

(2) In hazardous locations except as permitted elsewhere in the Code

(3) In commercial garages, other than to supply ceiling outlets or extensions to the area below the floor but not above

A. 374.4 Conductors larger then No. 1/0 require special permission before installation.

B. 374.5 The cross-sectional area of all conductors or cables shall not exceed 40% fill of the raceway.

C. 374.6 Splices and taps are permitted only in headers or junction boxes.

374.7 Conductors shall be removed from the raceway when an outlet is abandoned.

D. 374.8 Markers shall be installed for future locating of cells.

374.9 Junction boxes shall be level with the floor and sealed. Metal junction boxes shall be used with metal raceways.

E. 374.10 Inserts shall be leveled and sealed. Metal inserts shall be used with metal raceways.

374.11 Connections between raceways and wall outlets shall be made with LFMC, RMC, IMC, EMT, approved fittings or FMC if not installed in concrete.

374.100 Adequate mechanical and electrical continuity of the system shall be secured. Conductors shall be completely enclosed. All interior surfaces shall be free of burrs and sharp edges.

F. 374.17 The ampacity adjustment factors of 310.15(B)(3) apply to conductors installed in cellular metal floor raceways.

A. 372.5 The header shall be installed in a straight line at a 90° angle to the cells. It shall be mechanically secured. The end joints shall be closed and sealed. It shall be electrically continuous and bonded to the enclosure of the distribution system.

B. 372.6 Listed metal raceways and listed fittings shall be used to make connections.

372.7 Junction boxes shall be level with the floor and sealed. They shall be metal and mechanically and electrically continuous with the header.

372.8 A suitable number of markers shall be installed for future locating of cells.

C. 372.9 Inserts shall be leveled and sealed. They shall be metal and fitted with grounded receptacles.

D. 372.10 Conductors larger than No. 1/0 require special permission before installation.

E. 372.11 The cross-sectional area of all conductors or cables shall not exceed 40% fill of the cell or header.

F. 372.12 Splices and taps are permitted only in headers or junction boxes. Loop wiring is not considered to be a splice or a tap.

372.13 Conductors shall be removed from the raceway when an outlet is abandoned.

G. 372.17 The adjustment factors in 310.15(B)(3) apply to conductors installed in cellular concrete floor raceways.

WIREWAYS

376.2: Metal Wireway – sheet metal trough with a hinged or removable cover that houses and protects wires and cables laid in place after the wireway has been installed as a complete system.

378.2: Nonmetallic Wireway – flame-retardant nonmetallic trough with a removable cover that houses and protects wires and cables laid in place after the wireway has been installed as a complete system.

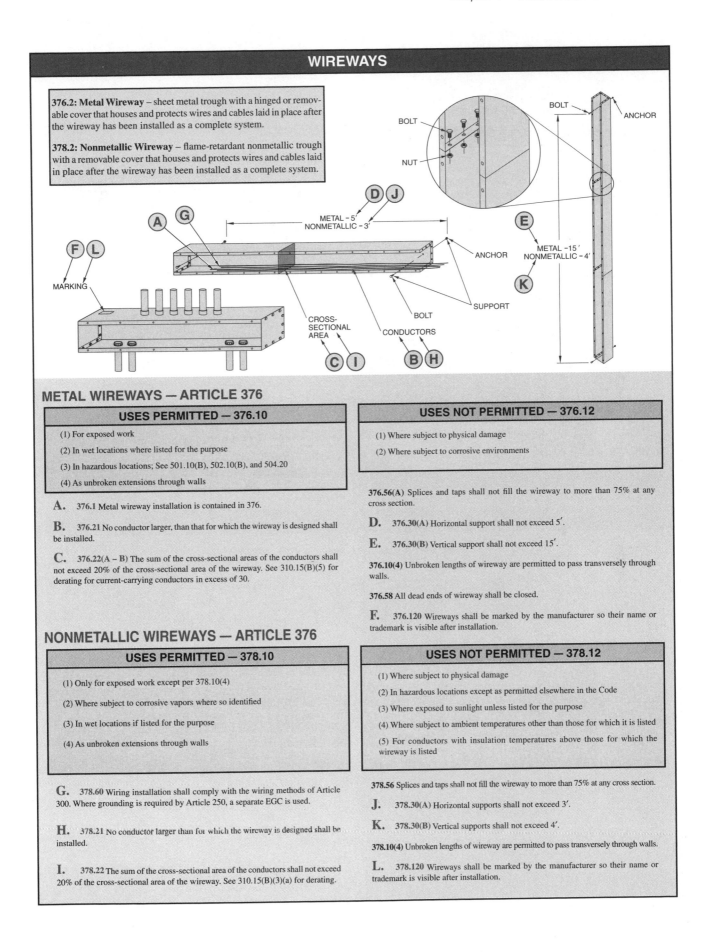

METAL WIREWAYS — ARTICLE 376

USES PERMITTED — 376.10

(1) For exposed work

(2) In wet locations where listed for the purpose

(3) In hazardous locations; See 501.10(B), 502.10(B), and 504.20

(4) As unbroken extensions through walls

A. 376.1 Metal wireway installation is contained in 376.

B. 376.21 No conductor larger, than that for which the wireway is designed shall be installed.

C. 376.22(A – B) The sum of the cross-sectional areas of the conductors shall not exceed 20% of the cross-sectional area of the wireway. See 310.15(B)(5) for derating for current-carrying conductors in excess of 30.

NONMETALLIC WIREWAYS — ARTICLE 376

USES PERMITTED — 378.10

(1) Only for exposed work except per 378.10(4)

(2) Where subject to corrosive vapors where so identified

(3) In wet locations if listed for the purpose

(4) As unbroken extensions through walls

G. 378.60 Wiring installation shall comply with the wiring methods of Article 300. Where grounding is required by Article 250, a separate EGC is used.

H. 378.21 No conductor larger than for which the wireway is designed shall be installed.

I. 378.22 The sum of the cross-sectional area of the conductors shall not exceed 20% of the cross-sectional area of the wireway. See 310.15(B)(3)(a) for derating.

USES NOT PERMITTED — 376.12

(1) Where subject to physical damage

(2) Where subject to corrosive environments

376.56(A) Splices and taps shall not fill the wireway to more than 75% at any cross section.

D. 376.30(A) Horizontal support shall not exceed 5′.

E. 376.30(B) Vertical support shall not exceed 15′.

376.10(4) Unbroken lengths of wireway are permitted to pass transversely through walls.

376.58 All dead ends of wireway shall be closed.

F. 376.120 Wireways shall be marked by the manufacturer so their name or trademark is visible after installation.

USES NOT PERMITTED — 378.12

(1) Where subject to physical damage

(2) In hazardous locations except as permitted elsewhere in the Code

(3) Where exposed to sunlight unless listed for the purpose

(4) Where subject to ambient temperatures other than those for which it is listed

(5) For conductors with insulation temperatures above those for which the wireway is listed

378.56 Splices and taps shall not fill the wireway to more than 75% at any cross section.

J. 378.30(A) Horizontal supports shall not exceed 3′.

K. 378.30(B) Vertical supports shall not exceed 4′.

378.10(4) Unbroken lengths of wireway are permitted to pass transversely through walls.

L. 378.120 Wireways shall be marked by the manufacturer so their name or trademark is visible after installation.

AUXILIARY GUTTERS

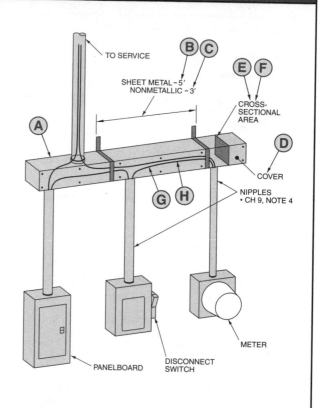

> **Auxiliary Gutter** - A metal or nonmetallic enclosure with a hinged or removable cover designed to supplement wiring space for conductors or busbars.

A.   366.12(2) Auxiliary gutters shall not extend over 30′ beyond the equipment supplemented.

B.   366.30(A) Sheet metal auxiliary gutter supporting and securing shall not exceed 5′ throughout their entire length.

C.   366.30(B) Nonmetallic auxiliary gutter supporting and securing shall not exceed 3′ and at each end or joint. In no case shall the distance between supports exceed 10′.

D.   366.100(D) Covers shall be securely fastened to the gutter.

E.   366.22(A) Sheet metal auxiliary gutters shall not contain over 30 current-carrying conductors at a cross section without derating per 310.15(B)(3)(a). The cross-sectional area of the conductors shall not exceed 20% of the cross-sectional gutter area.

F.   366.22(B) The cross-sectional area of conductors in a nonmetallic auxiliary gutter shall not exceed 20% of the cross-sectional gutter area.

G.   366.23(A) Thirty or less current-carrying conductors in a sheet metal auxiliary gutter shall not be derated.

H.   366.23(B) All current-carrying conductors in a nonmetallic auxiliary gutter shall be derated per 310.15(B)(3)(a).

366.56(A) Splices and taps shall not fill the gutter to more than 75% of its cross-sectional area.

366.100(A) Electrical and mechanical continuity shall be secured by the construction and installation of the gutter.

CONDUIT BODIES, PULL, AND JUNCTION BOXES

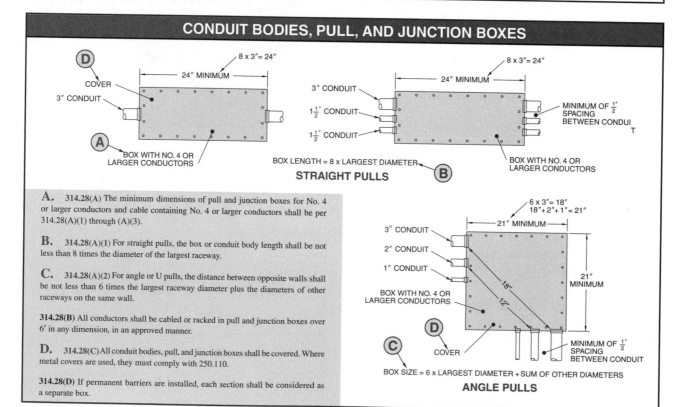

STRAIGHT PULLS

ANGLE PULLS

A.   314.28(A) The minimum dimensions of pull and junction boxes for No. 4 or larger conductors and cable containing No. 4 or larger conductors shall be per 314.28(A)(1) through (A)(3).

B.   314.28(A)(1) For straight pulls, the box or conduit body length shall be not less than 8 times the diameter of the largest raceway.

C.   314.28(A)(2) For angle or U pulls, the distance between opposite walls shall be not less than 6 times the largest raceway diameter plus the diameters of other raceways on the same wall.

314.28(B) All conductors shall be cabled or racked in pull and junction boxes over 6′ in any dimension, in an approved manner.

D.   314.28(C) All conduit bodies, pull, and junction boxes shall be covered. Where metal covers are used, they must comply with 250.110.

314.28(D) If permanent barriers are installed, each section shall be considered as a separate box.

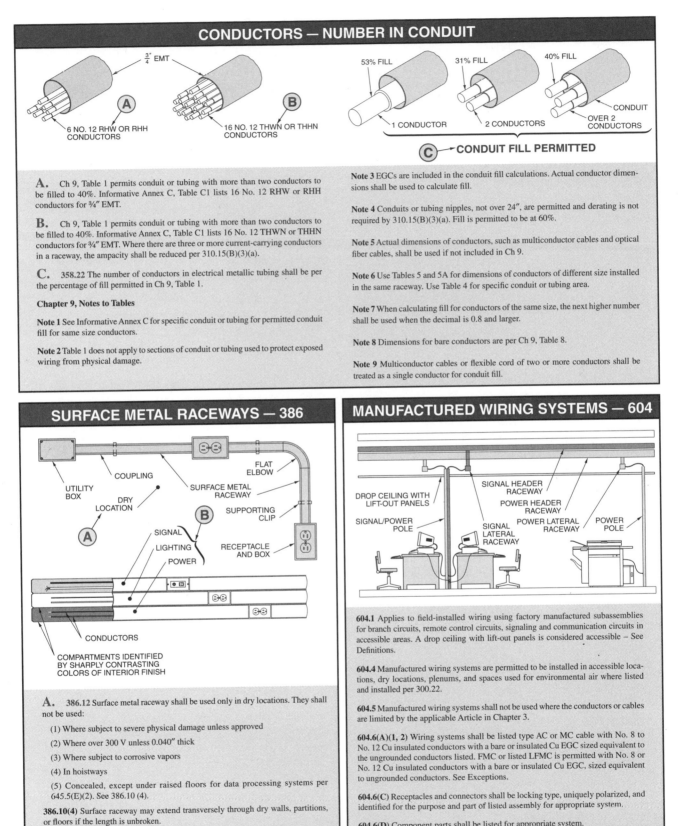

CONDUCTORS — NUMBER IN CONDUIT

A. Ch 9, Table 1 permits conduit or tubing with more than two conductors to be filled to 40%. Informative Annex C, Table C1 lists 16 No. 12 RHW or RHH conductors for ¾" EMT.

B. Ch 9, Table 1 permits conduit or tubing with more than two conductors to be filled to 40%. Informative Annex C, Table C1 lists 16 No. 12 THWN or THHN conductors for ¾" EMT. Where there are three or more current-carrying conductors in a raceway, the ampacity shall be reduced per 310.15(B)(3)(a).

C. 358.22 The number of conductors in electrical metallic tubing shall be per the percentage of fill permitted in Ch 9, Table 1.

Chapter 9, Notes to Tables

Note 1 See Informative Annex C for specific conduit or tubing for permitted conduit fill for same size conductors.

Note 2 Table 1 does not apply to sections of conduit or tubing used to protect exposed wiring from physical damage.

Note 3 EGCs are included in the conduit fill calculations. Actual conductor dimensions shall be used to calculate fill.

Note 4 Conduits or tubing nipples, not over 24", are permitted and derating is not required by 310.15(B)(3)(a). Fill is permitted to be at 60%.

Note 5 Actual dimensions of conductors, such as multiconductor cables and optical fiber cables, shall be used if not included in Ch 9.

Note 6 Use Tables 5 and 5A for dimensions of conductors of different size installed in the same raceway. Use Table 4 for specific conduit or tubing area.

Note 7 When calculating fill for conductors of the same size, the next higher number shall be used when the decimal is 0.8 and larger.

Note 8 Dimensions for bare conductors are per Ch 9, Table 8.

Note 9 Multiconductor cables or flexible cord of two or more conductors shall be treated as a single conductor for conduit fill.

SURFACE METAL RACEWAYS — 386

A. 386.12 Surface metal raceway shall be used only in dry locations. They shall not be used:

(1) Where subject to severe physical damage unless approved

(2) Where over 300 V unless 0.040" thick

(3) Where subject to corrosive vapors

(4) In hoistways

(5) Concealed, except under raised floors for data processing systems per 645.5(E)(2). See 386.10 (4).

386.10(4) Surface raceway may extend transversely through dry walls, partitions, or floors if the length is unbroken.

B. 386.70 Combination metallic raceways are permitted for signal, power, and lighting circuits if in different compartments and identified throughout the system by sharply contrasting colors of the interior finish.

MANUFACTURED WIRING SYSTEMS — 604

604.1 Applies to field-installed wiring using factory manufactured subassemblies for branch circuits, remote control circuits, signaling and communication circuits in accessible areas. A drop ceiling with lift-out panels is considered accessible – See Definitions.

604.4 Manufactured wiring systems are permitted to be installed in accessible locations, dry locations, plenums, and spaces used for environmental air where listed and installed per 300.22.

604.5 Manufactured wiring systems shall not be used where the conductors or cables are limited by the applicable Article in Chapter 3.

604.6(A)(1, 2) Wiring systems shall be listed type AC or MC cable with No. 8 to No. 12 Cu insulated conductors with a bare or insulated Cu EGC sized equivalent to the ungrounded conductors listed. FMC or listed LFMC is permitted with No. 8 or No. 12 Cu insulated conductors with a bare or insulated Cu EGC, sized equivalent to ungrounded conductors. See Exceptions.

604.6(C) Receptacles and connectors shall be locking type, uniquely polarized, and identified for the purpose and part of listed assembly for appropriate system.

604.6(D) Component parts shall be listed for appropriate system.

604.7 Wiring systems shall be secured and supported per the applicable cable or conduit article.

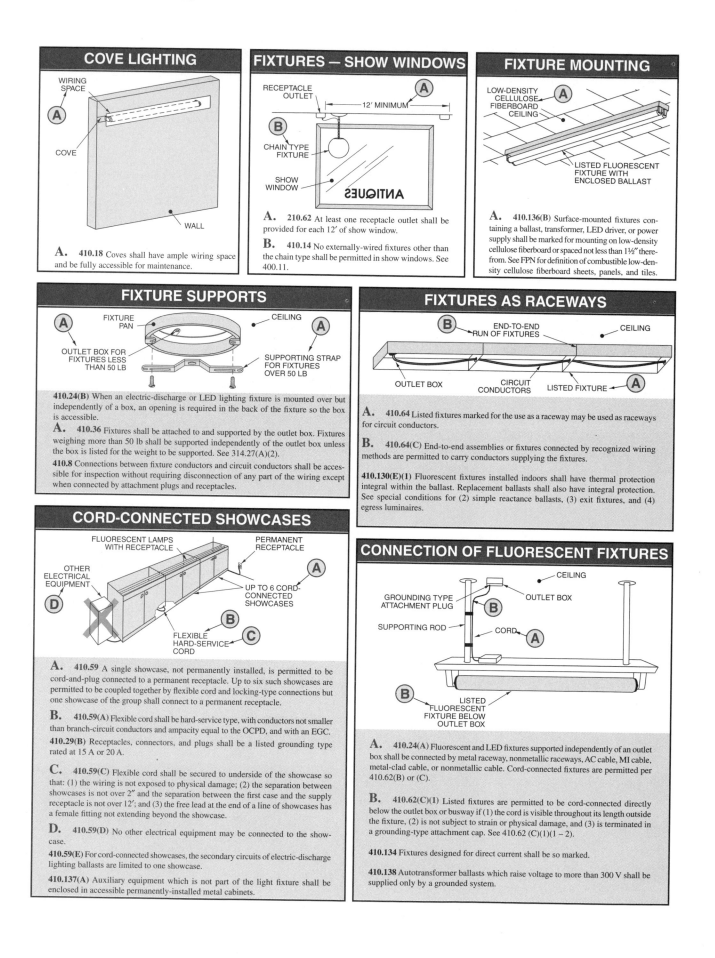

COVE LIGHTING

A. 410.18 Coves shall have ample wiring space and be fully accessible for maintenance.

FIXTURES — SHOW WINDOWS

A. 210.62 At least one receptacle outlet shall be provided for each 12′ of show window.

B. 410.14 No externally-wired fixtures other than the chain type shall be permitted in show windows. See 400.11.

FIXTURE MOUNTING

A. 410.136(B) Surface-mounted fixtures containing a ballast, transformer, LED driver, or power supply shall be marked for mounting on low-density cellulose fiberboard or spaced not less than 1½″ therefrom. See FPN for definition of combustible low-density cellulose fiberboard sheets, panels, and tiles.

FIXTURE SUPPORTS

410.24(B) When an electric-discharge or LED lighting fixture is mounted over but independently of a box, an opening is required in the back of the fixture so the box is accessible.

A. 410.36 Fixtures shall be attached to and supported by the outlet box. Fixtures weighing more than 50 lb shall be supported independently of the outlet box unless the box is listed for the weight to be supported. See 314.27(A)(2).

410.8 Connections between fixture conductors and circuit conductors shall be accessible for inspection without requiring disconnection of any part of the wiring except when connected by attachment plugs and receptacles.

FIXTURES AS RACEWAYS

A. 410.64 Listed fixtures marked for the use as a raceway may be used as raceways for circuit conductors.

B. 410.64(C) End-to-end assemblies or fixtures connected by recognized wiring methods are permitted to carry conductors supplying the fixtures.

410.130(E)(1) Fluorescent fixtures installed indoors shall have thermal protection integral within the ballast. Replacement ballasts shall also have integral protection. See special conditions for (2) simple reactance ballasts, (3) exit fixtures, and (4) egress luminaires.

CORD-CONNECTED SHOWCASES

A. 410.59 A single showcase, not permanently installed, is permitted to be cord-and-plug connected to a permanent receptacle. Up to six such showcases are permitted to be coupled together by flexible cord and locking-type connections but one showcase of the group shall connect to a permanent receptacle.

B. 410.59(A) Flexible cord shall be hard-service type, with conductors not smaller than branch-circuit conductors and ampacity equal to the OCPD, and with an EGC. **410.29(B)** Receptacles, connectors, and plugs shall be a listed grounding type rated at 15 A or 20 A.

C. 410.59(C) Flexible cord shall be secured to underside of the showcase so that: (1) the wiring is not exposed to physical damage; (2) the separation between showcases is not over 2″ and the separation between the first case and the supply receptacle is not over 12′; and (3) the free lead at the end of a line of showcases has a female fitting not extending beyond the showcase.

D. 410.59(D) No other electrical equipment may be connected to the showcase.

410.59(E) For cord-connected showcases, the secondary circuits of electric-discharge lighting ballasts are limited to one showcase.

410.137(A) Auxiliary equipment which is not part of the light fixture shall be enclosed in accessible permanently-installed metal cabinets.

CONNECTION OF FLUORESCENT FIXTURES

A. 410.24(A) Fluorescent and LED fixtures supported independently of an outlet box shall be connected by metal raceway, nonmetallic raceways, AC cable, MI cable, metal-clad cable, or nonmetallic cable. Cord-connected fixtures are permitted per 410.62(B) or (C).

B. 410.62(C)(1) Listed fixtures are permitted to be cord-connected directly below the outlet box or busway if (1) the cord is visible throughout its length outside the fixture, (2) is not subject to strain or physical damage, and (3) is terminated in a grounding-type attachment cap. See 410.62 (C)(1)(1 – 2).

410.134 Fixtures designed for direct current shall be so marked.

410.138 Autotransformer ballasts which raise voltage to more than 300 V shall be supplied only by a grounded system.

SCREW-SHELL LAMPHOLDERS

A. 410.90 Screw-shell lampholders shall not be installed for use as plug receptacles. Where supplied by a circuit having a grounded conductor, the conductor shall connect to the screw-shell.

410.96 Lampholders installed in wet locations shall be listed for use in wet locations.

410.97 Lampholders shall be constructed, installed, or equipped with shades or guards so that combustible material is not subject to temperatures in excess of 90°C.

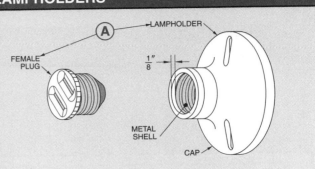

ELECTRIC DISCHARGE LIGHTING — OVER 1000 V

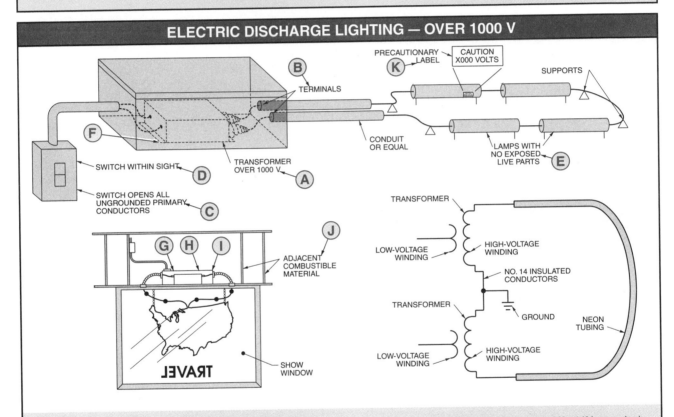

410.42(A) Where a circuit operates at over 150 V to ground, all exposed metal parts shall be grounded except tie wires, screws, clips, etc. at least 1½" from lamp terminals.

A. 410.140(B) Equipment with open-circuit voltage exceeding 1000 V shall not be installed in or on dwellings.

B. 410.140(C) The terminal of an electric-discharge lamp connected to a circuit of over 1000 V is considered a live part.

C. 410.141(A) Fixtures or lamp installations shall be connected singly or in groups by an externally-operable switch or CB that opens all ungrounded primary conductors.

D. 410.141(B) Switch or CB shall be within sight of fixture or lamps or else capable of being locked in open position. Portable means for adding a lock shall not be permitted.

E. 410.142 Lamps and lampholders are to be designed so that there will be no exposed live parts when lamps are inserted or removed.

F. 410.143(B – C) The secondary circuit voltage shall not be over 15,000 V under any load condition. The secondary current rating shall not be over 150 mA if

the open-circuit voltage is over 7500 V. It shall not be over 300 mA if the open-circuit voltage is 7500 V or less.

G. 410.143(A) Transformers shall be enclosed, identified for the use and listed.

410.143(D) High-voltage transformer secondary windings shall not be connected in series or in parallel.

H. 410.144(A) Transformers shall be accessible after installation.

I. 410.144(B) Transformers shall be installed as near to the lamps as practicable.

J. 410.144(C) Transformers shall be located so that adjacent combustible materials are not subjected to temperature in excess of 90° (194°F).

K. 410.146 Each fixture or circuit of tubing over 1000 V shall have a precautionary label. The voltage indicated is the rated open-circuit voltage.

TAPS FROM BRANCH CIRCUITS

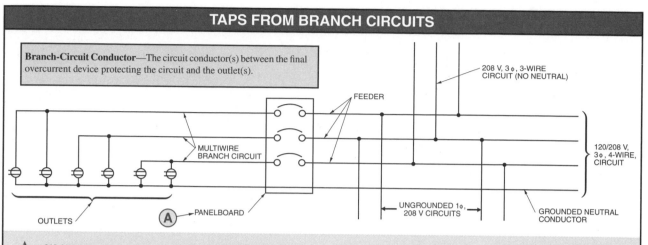

Branch-Circuit Conductor—The circuit conductor(s) between the final overcurrent device protecting the circuit and the outlet(s).

FEEDER

208 V, 3φ, 3-WIRE CIRCUIT (NO NEUTRAL)

MULTIWIRE BRANCH CIRCUIT

120/208 V, 3φ, 4-WIRE, CIRCUIT

OUTLETS

(A) PANELBOARD

UNGROUNDED 1φ, 208 V CIRCUITS

GROUNDED NEUTRAL CONDUCTOR

A. **210.4(A)** All multiwire branch-circuit conductors shall originate in same panelboard or distribution equipment.

210.4(B) Each multiwire branch circuit shall be provided with a means to simultaneously disconnect all ungrounded conductors at the point where the circuit originates.

210.4(C) Multiwire branch circuits are permitted by Article 210 and shall supply only line-to-neutral load, except: (1) where they supply only one piece of utilization equipment; (2) where branch-circuit overcurrent protective device simultaneously opens all ungrounded conductors of the multiwire circuit.

210.5(C)(1) If more than one nominal voltage system exists in a building, each ungrounded system conductor of a multiwire branch circuit shall be identified by phase and system at all termination, connection, and splice points.

210.10 Ungrounded conductors for feeders and branch circuits may be tapped from ungrounded wires of circuits which have identified neutrals. Switching devices in these circuits shall have a pole in each conductor. For branch circuits, all poles shall open simultaneously where they serve as the disconnecting means for appliances. See 422.31(B), 410.93, 410.104(B), 424.20, 426.51, 430.85, and 430.103.

215.7 Two-wire DC circuits and AC circuits with two or more ungrounded conductors are permitted to be tapped from the ungrounded conductors of circuits with a grounded neutral. Switching devices in each tapped circuit shall have a pole in each ungrounded conductor.

FEEDER AND BRANCH-CIRCUIT LOAD CALCULATIONS

215.2, Note 2 The voltage drop for power, heating, or lighting feeders is sized to not exceed 3%. The voltage drop for the feeder plus the branch circuit is sized to not exceed 5%. See 210.19(A)(1), Note 4.

210.19(A)(1) The branch-circuit rating shall be not less than the noncontinuous load plus 125% of the continuous load. The minimum branch-circuit conductor size, without correction factors, shall be equal to or greater than the noncontinuous load plus 125% of the continuous load.

Table 220.42, Note Demand factors in Table 220.42 do not apply to computed loads on feeders to areas in hospitals, hotels, and motels where the entire lighting load is used generally at one time.

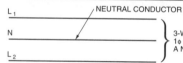

L_1

NEUTRAL CONDUCTOR

N

L_2

3-WIRE DC OR 1φ AC CIRCUIT WITH A NEUTRAL CONDUCTOR

220.61 The feeder neutral is sized for maximum unbalance between neutral and phase conductors, except the neutral for household ranges, ovens, and dryers is permitted to be reduced to 70% of the load as calculated per Table 220.54 and Table 220.55. A further demand of 70% is permitted for the portion of the maximum unbalanced load in excess of 200 A. This also applies to 4-wire, 3φ and 5-wire, 2φ systems. This 70% demand factor does not apply for that portion of the feeder load which consists of nonlinear loads.

COMMON NEUTRAL FOR FEEDERS

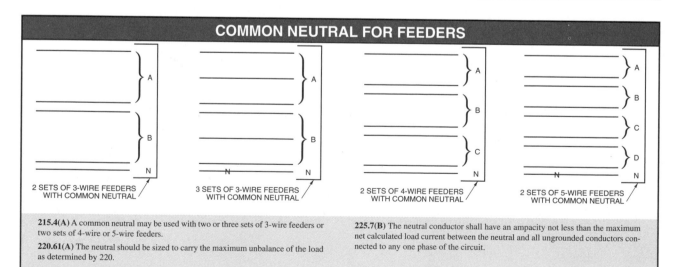

2 SETS OF 3-WIRE FEEDERS WITH COMMON NEUTRAL

3 SETS OF 3-WIRE FEEDERS WITH COMMON NEUTRAL

2 SETS OF 4-WIRE FEEDERS WITH COMMON NEUTRAL

2 SETS OF 5-WIRE FEEDERS WITH COMMON NEUTRAL

215.4(A) A common neutral may be used with two or three sets of 3-wire feeders or two sets of 4-wire or 5-wire feeders.

220.61(A) The neutral should be sized to carry the maximum unbalance of the load as determined by 220.

225.7(B) The neutral conductor shall have an ampacity not less than the maximum net calculated load current between the neutral and all ungrounded conductors connected to any one phase of the circuit.

MANUFACTURED BUILDINGS

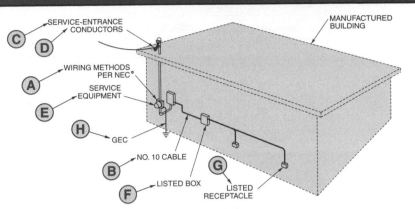

545.2: Manufactured Building—A building of closed construction, normally factory-made or assembled for installation or assembly and installation on building site. This does not include mobile homes, recreational vehicles, park trailers, or manufactured homes.

A. 545.4(A) All wiring methods in the NEC® are permitted for use, and other wiring systems and fittings specifically listed for use with manufactured buildings are also permitted. Wiring devices with integral enclosures are permitted if a sufficient length of conductor is provided for replacement of devices.

B. 545.4(B) No. 10 and smaller concealed cables are permitted to be installed without supports, such as staples or straps, as required in normal construction, if cables are secured at cabinets, boxes, and fittings and protected from physical damage.

C. 545.5 Provisions shall be made to route the service conductors, feeders and branch-circuit conductors to the service or building disconnecting means.

D. 545.6 Service-entrance conductors are to be installed after erection at the building site, except where the point of attachment is known before manufacture.

E. 545.7 Service equipment shall be located at nearest, readily-accessible point, either inside or outside of the building per 230.70.

545.8 Protection of equipment and conductors that are exposed during manufacturing, transit, and erection at building site is required.

F. 545.9(A) Boxes other than those in Table 314.16(A) are permitted if listed, identified, and tested.

545.9(B) Boxes under 100 cu in. for mounting in closed construction shall have anchors or clamps for rigid and secure installation.

G. 545.10 A switch or receptacle with an integral enclosure is permitted when tested, identified, and listed.

545.11 Bonding and grounding of prewired panels and building components shall conform to 250, Parts V, VI, and VII.

H. 545.12 Means shall be provided to route the GEC from the service equipment to the point of attachment to the grounding electrode.

545.13 Where on-site interconnection of modules or other building components is necessary, fittings and connectors when tested, identified, and listed are permitted to be used. These fittings are to be equal to the wiring method and may be concealed at the time of on-site assembly.

AGRICULTURAL BUILDINGS

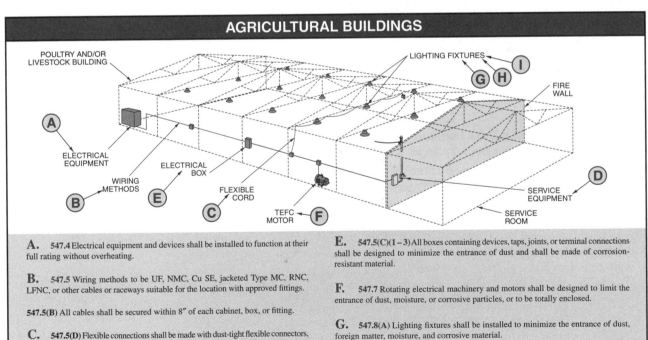

A. 547.4 Electrical equipment and devices shall be installed to function at their full rating without overheating.

B. 547.5 Wiring methods to be UF, NMC, Cu SE, jacketed Type MC, RNC, LFNC, or other cables or raceways suitable for the location with approved fittings.

547.5(B) All cables shall be secured within 8″ of each cabinet, box, or fitting.

C. 547.5(D) Flexible connections shall be made with dust-tight flexible connectors, LTFMC, LFNC, or hard usage flexible cord, with listed and identified fittings.

D. 547.6 Devices, such as switches, fuses, CBs, controllers, pushbuttons, relays, etc., shall be installed in dustproof, waterproof, watertight, or corrosion-resistant enclosures designed to keep out dust, water, and corrosive elements, and shall have telescoping or close-fitting covers. See 547.5(C)(1 – 3).

E. 547.5(C)(1 – 3) All boxes containing devices, taps, joints, or terminal connections shall be designed to minimize the entrance of dust and shall be made of corrosion-resistant material.

F. 547.7 Rotating electrical machinery and motors shall be designed to limit the entrance of dust, moisture, or corrosive particles, or to be totally enclosed.

G. 547.8(A) Lighting fixtures shall be installed to minimize the entrance of dust, foreign matter, moisture, and corrosive material.

H. 547.8(B) Lighting fixtures exposed to physical damage shall have suitable guards.

I. 547.8(C) Lighting fixtures, if exposed to water from condensation and/or cleansing water or solution, shall be listed as suitable for use in wet locations.

SOLAR PHOTOVOLTAIC SYSTEMS — 690

A. 690.2 Array-An assembly of modules or panels mounted on a support structure to form a direct current, power-producing unit.

690.2 Interactive System-A solar PV system that operates in parallel with and is permitted to deliver power to a normal utility service source connected to the same load.

690.2 Inverter-Equipment used to change voltage level or waveform, or both, of electrical energy.

B. 690.2 Module-A complete, environmentally protected assembly of solar cells, exclusive of tracking, designed to generate DC power under sunlight.

C. 690.2 Panel-A collection of modules, secured together, wired, and designed to provide a field-installable unit.

690.2 Photovoltaic Output Circuit-Conductors between photovoltaic source circuits and the inverter or DC utilization equipment.

690.2 Photovoltaic Power Source-One or more arrays which generate DC power at system voltage and current.

690.2 Photovoltaic Source Circuit-Conductors between modules and from modules to common junction point(s) of the DC system.

D. 690.2 Solar Cell-Basic photovoltaic device that generates electricity when exposed to light.

690.2 Solar Photovoltaic System-A solar system used to convert solar energy into electrical energy suitable for connection to a utilization load.

690, Part VIII covers storage batteries. Article 690.71 deals with installation, 690.72 deals with charge control.

690.1 Solar photovoltaic systems may interact with other electric power production sources or may stand alone. Either AC or DC output for utilization is permitted.

E. 690.5 Dwelling units with roof-mounted grounded DC photovoltaic arrays shall be provided with ground-fault protection to reduce fire hazards.

F. 690.7(D) In one-family and two-family dwellings, live parts in source and output circuits over 150 V to ground, while energized, shall not be accessible to other than qualified persons.

690.7(D), Note See 110.27 for guarding of live parts and 210.6 for voltage to ground and between conductors.

690.15 Means shall be provided to disconnect solar photovoltaic equipment from all ungrounded conductors of all sources.

G. 690.18 A means shall be provided to disable an array or portions of an array to protect persons against shock while installing, servicing, or replacing an array.

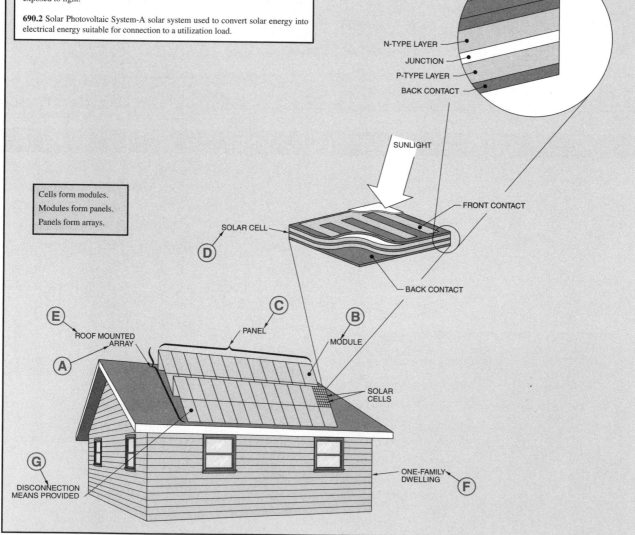

Cells form modules.
Modules form panels.
Panels form arrays.

Name _____ Date _____

NEC®	Answer	
_____	T ⓕ	**1.** In a fuseholder, the screw shell shall be connected to the line side of the circuit.
_____	_____	**2.** Header ducts used with cellular concrete raceways shall be constructed of ___.
_____	T F	**3.** Taps not over 18″ long for individual lampholders may be smaller than the branch circuit conductors.
_____	*capacity*	**4.** Equipment intended to interrupt current at fault levels shall have a(n) ___ rating sufficient for the nominal circuit voltage and current that is available at the line terminals of the equipment.
_____	*Covered*	**5.** Dead ends of underfloor raceways shall be ___.
_____	*C*	**6.** One of the approved methods of identifying the high-leg of a 4-wire, delta-connected secondary system is by using a ___ cable on the outer finish of the conductor wire.

  A. white   C. red
  B. green   D. none of the above

| _____ | T F | **7.** Openings around electrical penetrations through fire-resistant-rated walls shall be firestopped using approved methods. |
| _____ | *200* | **8.** The minimum bench-circuit load for show-window lighting shall be ___ VA per linear foot. |

  A. 100   C. 300
  B. 200   D. 500

_____	*2″*	**9.** Service conductors of 600 V or less buried under ___″ or more of concrete beneath a building or structure are considered outside the building.
_____	_____	**10.** Conductors from discontinued outlets in an underfloor raceway installation shall be ___ from the raceway.
_____	_____	**11.** Raceways installed between areas differing widely in temperature and where condensation is likely shall be sealed with a(n) ___ material.
_____	_____	**12.** ___ are required at or near the end of each run of underfloor raceway to locate the last insert.
_____	_____	**13.** The total load on a branch circuit shall not exceed its ___ rating.
_____	*B*	**14.** The general lighting unit load for a store is listed as ___ VA per sq ft.

  A. 2   C. 3
  B. 2.5   D. 3.75

| _____ | _____ | **15.** If the allowable current-carrying capacity of a conductor does not correspond to the rating of a standard size overcurrent protection device, the next larger size may be used provided the current does not exceed ___ A. |
| _____ | _____ | **16.** Temporary wiring methods shall be acceptable only if ___ based on the conditions of use and any special requirements of the temporary installation. |

_____ _____ **17.** Threaded boxes not over 100 cu in. in size that contain devices need no additional support if two or more conduits, which are properly supported within ___" of the box, are threaded into the box from the same side.

_____ _____ **18.** Where permissible, the demand factor applied to that portion of an unbalanced neutral feeder load in excess of 200 A is ___%.

    A. 40   C. 70

    B. 60   D. 80

_____ _____ **19.** In other than dwellings, each 5' or fraction of multioutlet assembly shall normally be considered as a load of ___ VA.

_____ T   F **20.** The neutral feeder conductor shall be the same size as the ungrounded conductors for a resistance load.

_____ _____ **21.** Neon installed on dwelling units shall not have an open-circuit voltage that exceeds ___ V.

_____ _____ **22.** Metal auxiliary gutters shall be supported and secured throughout their entire length at intervals not exceeding ___'.

_____ _____ **23.** The unit load for a hall in other than a one-family dwelling or an apartment is listed as ___ VA per square foot.

_____ T   F **24.** Lighting outlets installed in other than dwelling units shall not be permitted to be connected to 50 A branch circuits.

_____ _____ **25.** A CB having an interrupting rating other than ___ A shall be marked with the rating of a standard size overcurrent device.

_____ _____ **26.** A(n) ___ or terminal fitting is required where underground cable leaves conduit providing physical protection.

_____ _____ **27.** Bonding shall be provided where necessary to ensure electrical ___ and the capacity to safely conduct any fault current which is likely to be imposed.

_____ _____ **28.** A minimum size of No. ___ Cu EGC is required for a 30 A automatic overcurrent device ahead of the equipment.

_____ _____ **29.** CBs used to switch 120/277 V fluorescent fixtures shall be listed for the purpose and marked ___ or HID.

_____ T   F **30.** In interior wiring, a common neutral feeder wire shall not be permitted to be run for two or three sets of 3-wire feeder circuits.

_____ _____ **31.** CBs shall clearly ___ whether in open or closed position.

_____ _____ **32.** A 15 A receptacle on a standard 15 A or 20 A branch circuit shall not supply a portable appliance load in excess of ___ A.

    A. 9   C. 12

    B. 10   D. 15

_____ T   F **33.** In general, flexible cords and cables are permitted to be attached to building surfaces over 6'-7" from the floor or a platform.

_____ _____ **34.** The ___ of an underfloor raceway shall determine the largest conductor that can be installed.

_____ _____ **35.** A unit load for each outlet supplying heavy-duty lampholders that are not used for general illumination in industrial occupancies shall be ___.

    A. 1½ A   C. 180 VA

    B. 5 A   D. 600 VA

Name_____ Date _____

NEC®	Answer	

_____   T   F   **1.** Splices and taps are sometimes permitted in service conductors.

_____   _____   **2.** Auxiliary gutters shall not extend more than ___' beyond the equipment they supplement.

_____   _____   **3.** EBJs are permitted to be routed outside a conduit system if not over ___' in length and they are routed with the raceway.

_____   _____   **4.** The cross-sectional fill of a metal wireway shall not exceed ___% at any one point.

_____   _____   **5.** Every commercial building accessible to pedestrians must have at least one ___ A circuit for a sign outlet.

_____   T   F   **6.** The maximum ampere rating of a 250 V cartridge fuse is 30 A.

_____   _____   **7.** Underground service conductors carried up a pole shall be protected from mechanical injury to a height of at least ___.
     A. 8   C. 12
     B. 9   D. 15

_____   _____   **8.** Metal enclosures for a GEC using a water pipe electrode for a 120/208 V service with three 250 kcmil Cu conductors and a No. 3/0 neutral shall be ___.
     A. rigid conduit only   C. RNC
     B. not less than ¾" in dia.   D. electrically continuous

_____   T   F   **9.** Bonding jumpers required for grounding interior equipment on the load side of the service shall be selected from Table 250.122.

_____   _____   **10.** The area of conductors and splices in an auxiliary gutter shall not exceed ___% of the cross-sectional area.

_____   T   F   **11.** Fuses shall never be field connected in parallel.

_____   _____   **12.** The size of a copper GEC using a water pipe electrode for a 120/208 V service with three 250 kcmil Cu conductors (one per phase) and a No. 3/0 neutral shall be No. ___.
     A. 2   C. 1/0
     B. 4   D. 2/0

_____   _____   **13.** Panelboards equipped with snap switches rated at 30 A or less shall have overcurrent protection not in excess of ___ A.
     A. 100   C. 200
     B. 150   D. 300

_____   T   F   **14.** The service OCPD shall be an integral part of, or shall be located adjacent to, the service disconnecting means.

_____   _____   **15.** The current carried continuously in bare copper bars in metal auxiliary gutters shall not exceed ___ A per sq in.

_____ _____ **16.** No. 1/0 Cu conductors in vertical raceways shall be supported at intervals not exceeding ___'.

    A. 50   C. 100

    B. 75   D. 125

_____ _____ **17.** Three-way and four-way switches shall be wired so that all switching is done in the ___ conductors.

_____ T   F **18.** In general, the disconnecting means for a motor controller shall be within sight of the controller.

_____ _____ **19.** The wiring between 3-way switches and outlets in metal enclosures shall be run with all ___ in the same enclosure.

_____ _____ **20.** The largest standard ampere rating for a fuse is ___ A.

    A. 100   C. 1200

    B. 600   D. 6000

_____ _____ **21.** Screw-shell lampholders shall be installed for use as ___ only.

_____ T   F **22.** Switchboards with exposed live parts shall be located in permanently dry locations accessible only to qualified persons.

_____ _____ **23.** A nipple having a maximum length of 24″ can be filled to ___ of its internal cross-sectional area.

_____ _____ **24.** Transformers for electric-discharge lighting systems of more than 1000 V shall be ___ after installation.

_____ _____ **25.** Sheet metal auxiliary gutters shall be supported and secured throughout their entire length at intervals not exceeding ___'.

_____ _____ **26.** 25' tap taken from a 100 A feeder shall have an ampacity of not less than ___ A.

    A. 15   C. 25

    B. 20   D. 33⅓

_____ _____ **27.** When applying the over 25' feeder tap rule in a high-bay manufacturing building, the horizontal run is limited to ___' or less.

_____ _____ **28.** Tap conductors shall have at least one-third ampacity of supply conductors when applying the ___' tap rule.

_____ _____ **29.** In straight pulls, the length of a box or conduit body shall be not less than ___ times the diameter of the largest conduit.

_____ _____ **30.** When applying the 25' tap rule from the secondary side of the transformer for industrial installations, OCPDs shall be ___.

_____ _____ **31.** Branch-circuit conductors routed within 3″ of a ballast in a fixture shall have insulation rated at least ___°C unless the fixture is marked otherwise.

_____ _____ **32.** In U-pulls, the size of the box or conduit body shall be ___ times the diameter of the largest conduit plus all remaining conduits on the same wall.

_____ _____ **33.** No. 12 conductors enclosed in vertical raceways shall be supported every ___'.

_____ _____ **34.** IMC larger than trade size ___″ shall not be used.

_____ _____ **35.** In a building where water pipe is available, the minimum size of the GEC for an ungrounded system with No. 3/0 Cu service conductors shall be No. ___.

    A. 2 Cu   C. 1/0 Cu

    B. 4 Cu   D. 2/0 Al

Name _____ Date _____

NEC®	Answer	

_____   T   F   **1.** Conduit nipples not over 24″ long may be filled to 60% of their cross-sectional area.

_____   _____   **2.** Nonmetallic wireways shall not be used where exposed to ___ unless listed and marked as suitable for the purpose.

_____   _____   **3.** Coves that enclose lighting units shall have ___ space for proper installation and maintenance.

_____   _____   **4.** The secondary OCPD of a transformer may consist of ___ or less CBs or fuses.
   A. two   C. four
   B. three   D. six

_____   _____   **5.** In cellular concrete floor raceways, conductors larger than No. ___ shall not be installed unless by special permission.

_____   _____   **6.** An autotransformer ballast that raises the voltage above ___ V shall be supplied by a grounded system.

_____   _____   **7.** The largest Type THHN conductor permitted in ⅜″ FMC is No. ___.
   A. 10   C. 14
   B. 12   D. 16

_____   _____   **8.** No wiring method of any kind shall be installed in ___ used with commercial cooking.

_____   T   F   **9.** Conductors in underfloor raceway systems may be spliced or tapped only in junction boxes.

_____   _____   **10.** When applying the over 25′ feeder tap rule for a high-bay manufacturing building, the junction box containing the tap shall be at least ___′ from the floor.

_____   _____   **11.** Underfloor raceways over 4″ but less than 8″ wide, and spaced less than 1″ apart, shall be covered with concrete to a depth of not less than ___″.
   A. ¾   C. 1½
   B. 1   D. 2

_____   T   F   **12.** Unbroken lengths of surface metal raceway may be extended through dry floors.

_____   _____   **13.** In multiwire branch circuits, the continuity of a ___ conductor shall not depend upon device connections.

_____   _____   **14.** If a change occurs in the size of the ungrounded conductor, a similar change is permitted in the ___ conductor.

_____ _____ **15.** The combined cross-sectional area of all conductors or cables in an underfloor raceway shall not exceed ___%.
    A. 20   C. 40
    B. 30   D. 50

_____ _____ **16.** Where more than one voltage system exists in a building, each ungrounded conductor of a branch circuit shall be identified by phase or line and ___.

_____   T   F **17.** A switchboard enclosure in a damp or wet location shall be weatherproof.

_____   T   F **18.** Header ducts used with cellular concrete floor raceways may be run diagonally across the cells, provided they run in straight lines.

_____ _____ **19.** Cord-equipped fixtures can be suspended directly below an outlet box if the cord ___ in a grounding-type attachment plug.

_____ _____ **20.** Underfloor raceways not over 4″ wide shall be covered by concrete or wood to a thickness of not less than ___″.
    A. ¾   C. ½
    B. 1   D. 2

_____ _____ **21.** Electric sign transformers shall have a secondary circuit current rating of not more than ___ mA.

_____ _____ **22.** A feeder neutral conductor on a pure resistive load shall have a demand factor of ___% applied for all loads above 200 A.

_____   T   F **23.** The full cross-sectional area of cellular metal floor raceway may be used if AC cable is used.

_____ _____ **24.** Fixtures that are designed to be supplied by direct current shall be ___ for DC operation.

_____ _____ **25.** Expansion joints in raceway systems shall be installed where necessary to compensate for ___ expansion and contraction.

_____ _____ **26.** The largest size conductor allowed in a cellular metal floor raceway without special permission is ___.
    A. No. 1/0   C. 250 kcmil
    B. No. 3/0   D. 500 kcmil

_____ _____ **27.** Raceway systems that are exposed to widely different temperatures shall be ___ if condensation is known to be a problem.

_____   T   F **28.** Surface metal raceways are permitted for use in dry or wet locations.

_____ _____ **29.** In cellular metal floor raceways, splices and taps shall be made only in ___ access units or junction boxes.

_____ _____ **30.** A fuse or OCPD shall be connected in ___ with each ungrounded conductor.

_____ _____ **31.** The center of the grip of a switch or CB operating handle in its highest position shall not exceed ___ from the floor.

_____ _____ **32.** Each fixture of each secondary circuit of tubing rated over ___ V shall have a warning label.

_____ _____ **33.** An outlet tap for a fixture shall not extend over ___″.

_____   T   F **34.** No conduit shall be run through a duct used for dust removal.

Name_____ Date _____

NEC®	Answer	
_____	_____	**1.** Panelboards with overcurrent protection not above 200 A may contain snap switches of ___ A or less.
_____	T F	**2.** Nonmetallic inserts may be used with metal underfloor raceways.
_____	_____	**3.** A 15 A or 20 A receptacle shall not supply a cord-and-plug-connected load that exceeds ___% of the receptacle rating.

4. If the ampacity of a conductor does not correspond with a standard overcurrent protection device, the next larger size up to ___ A can be used.
  A. 800   C. 1200
  B. 1000   D. 1500

5. No wiring system of any type shall be installed in a duct used for ventilation of commercial cooking equipment.   T F

6. The OCPD of a phase converter is computed at ___% of the 1φ input FLA for variable loads.

7. SE conductors can be spliced where the splice is made per 110.14, 300.5(E), 300.13, and 300.15.   T F

8. The panels of switchboards shall be made of moisture-resistant ___ material.

9. Conductors are considered outside the building when installed in a raceway that is encased in at least 2″ of concrete or brick.   T F

10. Splices and wires within any portion of an auxiliary gutter shall not fill the cross-sectional area of the gutter by more than ___%.
  A. 20   C. 60
  B. 50   D. 75

11. The cross-sectional area of conductors in any section of sheet metal auxiliary gutter shall not exceed 20% of its area.   T F

12. The use of a multioutlet assembly shall be permitted in ___ locations.

13. AC-DC general-use snap switches controlling ordinary inductive loads shall have an ampere rating twice that of the load.   T F

14. Busbars in switchboards may be insulated or bare, provided they are rigidly mounted.   T F

15. SE conductors shall not feed one building or structure by passing through the interior of another building or structure.   T F

16. The general rule requires grounding receptacles to be installed on ___ A and ___ A circuits.
  A. 15; 20   C. 25; 30
  B. 20; 25   D. none of the above

_____   _____   **17.** Field assembled extension cord sets with No. ___ or larger conductors can be connected to a 20 A branch circuit.

_____   T   F   **18.** CBs shall clearly indicate whether they are in the ON position or OFF position.

_____   _____   **19.** Fixture wire used on a parking lot lighting pole can extend up to ___' for No. 18 conductors tapped from a 20 A circuit.

_____   _____   **20.** The load for show windows is computed at a unit load of ___ VA per linear foot.

_____   _____   **21.** A Type AC fixture whip for fixture tap conductors can be used if its length does not exceed ___'.

_____   _____   **22.** All 15 A and 20 A, 120 V receptacles used for temporary power that are not part of the permanent wiring of the building shall be ___-protected.
    A. GFCI   C. GP
    B. GPF   D. none of the above

_____   _____   **23.** A 2″ FMC does not require support for lengths of ___' or less at terminals where flexibility is required.

_____   _____   **24.** Receptacles located in floors shall be installed in floor boxes ___ for such applications.

_____   _____   **25.** Angle-pull dimensional requirements apply to junction boxes only when the size of conductor is equal to or larger than No. ___.
    A. 1/0   C. 4
    B. 3/0   D. 6

_____   _____   **26.** Electric mixers traveling in and out of open-type mixing tanks are permitted to be wired with ___ cord in a Class I, Division 1 location.

_____   _____   **27.** Flexible cords of the hard or junior-hard type, enclosing No. ___ and larger conductors are permitted to be spliced.

_____   _____   **28.** The total load on any OCPD in a panelboard shall not exceed ___% of its rating if in normal use for over 3 hours unless otherwise listed for continuous operation.

_____   _____   **29.** For existing installations in residential occupancies, ___ protection is not required in a panelboard used as service equipment for individual residential occupancy.

_____   _____   **30.** A 120 V CB used to switch fluorescent fixtures shall be listed and marked ___ or HID.
    A. SWD   C. SWF
    B. SWR   D. FLS

_____   _____   **31.** The maximum setting of ground-fault protection of equipment for service equipment shall be ___ A.

_____   _____   **32.** The general rule requires individual conductors used in flexible cords to be not smaller than the sizes in Table ___.

_____   _____   **33.** The maximum number of quarter bends permitted in a run of FMC is ___.

_____   T   F   **34.** The general rule allows overcurrent protection devices to be inaccessible to occupants of a building.

_____   T   F   **35.** Where GFCI protection of personnel for temporary electrical power from existing permanent building is required, listed cord sets incorporating this protection shall be permitted.

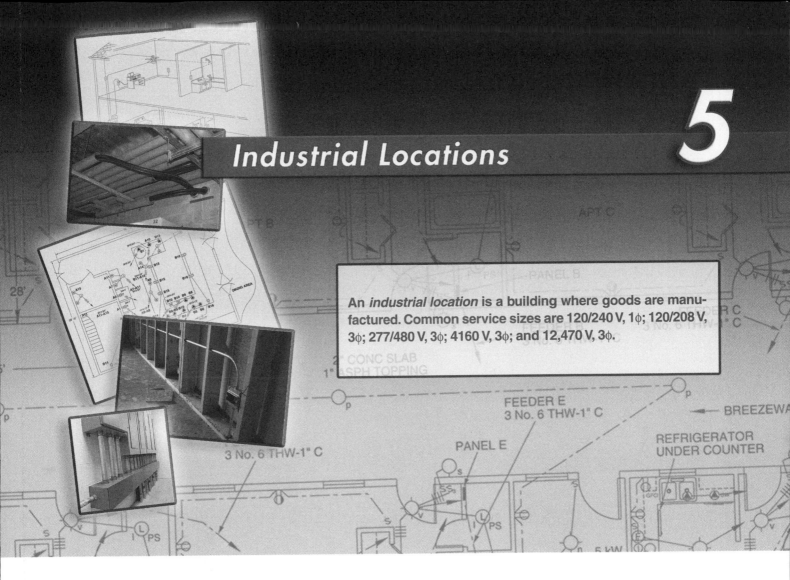

Industrial Locations

5

An *industrial location* is a building where goods are manufactured. Common service sizes are 120/240 V, 1φ; 120/208 V, 3φ; 277/480 V, 3φ; 4160 V, 3φ; and 12,470 V, 3φ.

FEEDER E
3 No. 6 THW-1" C

BREEZEWA

PANEL E

REFRIGERATOR
UNDER COUNTER

3 No. 6 THW-1" C

POWER INSTALLATION FOR INDUSTRIAL PLANT

The print shows only power outlets that are supplied by a 265/460 V, 3φ, 4-wire wye distribution system. Lighting and receptacle outlets are supplied by a separate 120/208 V, 3φ, 4-wire system and appear on a separate print. Specifications call for THW Cu conductors throughout the project.

MOTOR CIRCUITS

Article 430 explains when to use the tables at the end of 430 and when to use the motor nameplate rating to determine the ampacities and ratings of the various components of motor circuits. NEC® 430.6(A)(1) states that where the current rating of a motor is used to determine the ampacity of conductors or the ampere ratings of switches, controllers, branch-circuit overcurrent protection devices, etc., the values given in Tables 430.247,

430.248, 430.249, and 430.250, including notes, shall be used instead of the motor nameplate rating.

If only the full-load current is marked on the nameplate, the horsepower rating shall be determined by the corresponding value given in Tables 430.247 through 430.250. In the selection of separate motor-running overcurrent protection which will protect motors, motor-control apparatus, and motor branch-circuit conductors against excessive heating due to overloads and failure to start, the full-load current rating that appears on the motor nameplate shall be used.

NEC® 430.22 states that branch-circuit conductors supplying a single motor shall have an ampacity not less than 125% of the full-load current of the motor. NEC® 430.32(A)(1) specifies that continuous-duty motors rated more than 1 HP shall be protected against overload, in general, by a device that trips at not more than 125% of the full-load current of the motor.

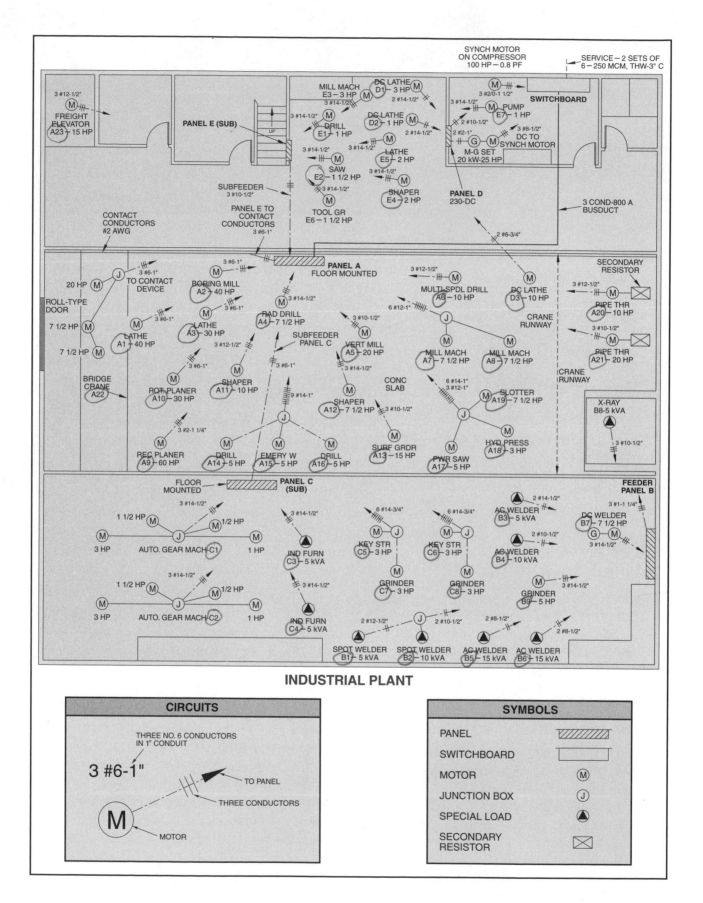

INDUSTRIAL PLANT

CIRCUITS		SYMBOLS	
THREE NO. 6 CONDUCTORS IN 1" CONDUIT		PANEL	▨
3 #6-1"		SWITCHBOARD	▭
M — TO PANEL, THREE CONDUCTORS, MOTOR		MOTOR	Ⓜ
		JUNCTION BOX	Ⓙ
		SPECIAL LOAD	▲
		SECONDARY RESISTOR	⊠

NEC® 430.52(B) requires that overcurrent protection devices protecting branch-circuit conductors shall be capable of carrying motor-starting current. Motor-starting current usually varies between 150% and 300% of the full-load current, depending on the type of motor. Table 430.52 lists information necessary to calculate the maximum rating or setting of motor branch-circuit, short-circuit, and ground-fault protective devices. In rare cases, starting current runs somewhat higher than values given there. NEC® 430.52 provides for this by allowing larger overcurrent protection devices, which permit the motor to start.

Application

A1 on the print shows a 40 HP, 460 V, 3φ squirrel-cage motor which has no Code letter marked on the nameplate, a marked service factor of 1.15, a full-load current of 49 A, and which is started by means of an autotransformer compensator. Table 430.150 gives the full-load current as 52 A. Under 430.22, the branch-circuit conductors shall be able to carry 65 A (52 A × 1.25 = 65 A). Table 310.15(B)(16) lists the nearest size of THW Cu wire as No. 6. Informative Annex C, Table C8 permits up to five No. 6 THW conductors in 1″ RMC.

Table 430.52 lists all types of motors including squirrel-cage and synchronous motors. Row 3 of Table 430.52, "squirrel-cage, other than Design B," applies here. Column 2 shows the branch-circuit nontime-delay fuse rating as 300% of the full-load current. For a motor current of 52 A, circuit protection should be 156 A (52 A × 3 = 156 A). The nearest standard fuse or non-adjustable CB rating that does not exceed the 300% is 150 A per 240.6(A).

Motor running overcurrent protection, per 430.32(A)(1), should be not greater than 61.25 A (49 A × 1.25 = 61.25 A). This protection may be obtained with an adjustable device set at 61.25 A or by the nearest size fuse or non-adjustable CB, which in this case is 60 A.

Vertical Milling Machine A5 is driven by a 20 HP, Code letter motor C, with a marked service factor of 1.15 and a full-load current of 25 A, which is protected with an instantaneous-trip circuit breaker (ITCB). Table 430.250 gives the full load current as 27 A. NEC® 430.22 calls for a No. 10 conductor (27 A × 1.25 = 33.75 A).

Informative Annex C, Table C8 specifies a ½″ RMC. Motor-running overcurrent protection per 430.32(A)(1) should be not greater than 31.25 A (25 A × 1.25 = 31.25 A). The nearest standard fuse or non-adjustable CB is rated at 30 A per 240.6(A). Table 430.52 gives 250% of the full-load current for branch-circuit protection, or 67.5 A (27 × 2.5 = 67.5 A). This allows a 70 A ITCB for branch-circuit protection.

The 7½ HP, Code letter F, full-voltage starting motor on Radial Drill A4 has a nameplate current of 10 A and a marked service factor of 1.15. Motor-running protection is 12.5 A (10 A × 1.25 = 12.5 A), for an adjustable protector per 430.32(A)(1). Branch-circuit protection is 300% of 11 A, or 33 A (11 A × 3 = 33 A), which permits a 35 A NTDF per Table 430.52.

Pipe-Threading Machine A21 uses a 20 HP wound-rotor motor. The size of circuit conductors, conduit, and motor-running overcurrent protection are the same as for the 20 HP squirrel-cage motor of Vertical Milling Machine A5, but the branch-circuit protection is different. Referring to Table 430.52, select the wound-rotor motor which allows 150% of the full-load current, or 40.5 A (27 A × 1.5 = 40.5 A). A 45 A NTDF or a 45 A time-limit CB may be used.

The motor nameplate lists a secondary current of 50 A. NEC® 430.23(A) states that for continuous-duty motors, the current-carrying capacity of wires between the motor secondary (rotor) and the controller shall not be less than 125% of the full-load secondary current. This value is 62.5 A (50 A × 1.25 = 62.5 A), the nearest conductor being No. 6. Table 430.23(C) specifies ampacities of conductors between the controller and resistors. In most cases, high-temperature wire is necessary because of the heat that is generated in resistor grids.

Lathe D3 is driven by a 10 HP, 230 V, DC motor with a full-load current of 37 A and a marked service factor of 1.15. Table 430.247 gives the full-load current as 38 A, requiring two No. 8 conductors in ¾″ conduit. The motor-running overcurrent protection is 46.25 A (37 A × 1.25 = 46.25 A) and the branch-circuit protection is 57 A (38 A × 1.5 = 57 A) per Table 430.52, which permits a 60 A NTDF.

Elevator Motor

A23 is a 15 HP, high-reactance, 30 and 60 minute rated motor with full-voltage starting. Table 430.22(E), Note states that the duty cycle on elevator motors shall be classed as intermittent duty. An *intermittent duty motor* is a motor that operates for alternate intervals of load and no load; load and rest; or load, no load, and rest. NEC® 430.33 provides that the ampacity of the branch-circuit conductors for intermittent duty motors shall be considered as protected against running overcurrent by a branch-circuit device which does not exceed the values specified in Table 430.52.

NEC® 430.22(E) states that the ampacity of the branch-circuit conductors for intermittent duty motors shall be not less than the percentage of the nameplate current rating of the motor as shown in Table 430.22(E) unless the AHJ grants special permission for conductors of a smaller size. In this case, the ampacity of the branch-circuit conductors for elevator motor A23 need be only 90% of the nameplate full-load current of the motor.

The full-load current rating of the elevator motor on A23 is listed as 21 A. The ampacity of the conductors is rated on the basis of 18.9 A (21 A × .9 = 18.9 A). A No. 12 conductor is needed. Table 430.52 shows that the branch-circuit protection may be 150% of the full-load current, or 31.5 A (21 A × 1.5 = 31.5 A) which permits a 35 A nontime-delay fuse or ITCB.

Feeder Supplying Two or More Motors

NEC® 430.24(1) states that conductors supplying two or more motors shall have an ampacity of not less than 125% of the full-load current rating of the highest rated motor in the group, plus the sum of the full-load current of the remaining motors. Thus, for two motors, one having a full-load current rating of 20 A and the other 14 A, the feeder ampacity shall be not less than 39 A (20 A × 1.25 = 25 A + 14 A = 39 A). A No. 8 THW Cu wire is required.

NEC® 430.62(A) provides that overcurrent protection for conductors supplying two or more motors shall be not greater than the branch-circuit overcurrent for the largest rating or setting plus the full-load currents of the other motors. With a 250% protective requirement for the larger motor in the preceding example, the feeder protective device should be set at not more than 64 A (20 A × 2.5 = 50 A + 14 A = 64 A). The nearest standard fuse or CB that does not exceed this value is 60 A per 240.6(A).

Crane Motors

Bridge Crane A22 has three motors, a 20 HP and two 7½ HP units. The nameplates specify 15-minute ratings. Noncontinuous-duty motors are commonly used on this type of service, the ratings being stated according to the nature of the application as 15 minutes, 30 minutes, or 60 minutes.

Table 610.14(A) has ampacities for conductors supplying this type of motor, the permissible currents being larger than those allowed by Table 310.15(B)(16). Columns of Table 610.14(A) list values according to types of insulation and ambient operating temperatures, the first of each set of two columns dealing with 60-minute ratings, the second with 30-minute ratings. A footnote states that allowable ampacities of conductors for 15-minute motors shall be equal to the 30-minute ratings increased by 12%.

In the present instance, the 75°C columns are applicable. Since the motors are 15-minute units, the sizes of the conductors are determined on the basis of the 30-minute rating increased by 12%. The full-load current of the largest motor is 27 A, and the others are 11 A each. According to the table, a No. 14 THW Cu conductor has a 30-minute current-carrying capacity of 26 A. A 12% increase gives this conductor a capacity of 29 A (26 A × 1.12 = 29.12 = 29 A), making it suitable for use with the larger motor.

NEC® 610.14(C) states that a No. 16 conductor may be used with crane motors, if otherwise suitable. The 30-minute rating of No. 16 is given as 12 A, and the allowable 12% increase raises the 15-minute rating to 13.4 A (12 A × 1.12 = 13.44 = 13.4 A), a value satisfactory for the two small motors.

NEC® 610.14(E)(2) states that the ampacity of the power supply conductors on the crane shall be not less than the combined short-time, full-load ampere rating of the largest motor or group for any single crane motor, plus 50% of the short-time full-load ampere rating of the next largest motor or group. Adding the 27 A of the largest motor to 50% of 22 A (full-load current of the other motors) equals 38 A (27 A + 11 A = 38 A). No. 8 THW conductor may be used per Table 310.15(B)(16). However, the specifications call for a No. 6 THW conductor here. NEC®

Table 610.14(D) lists the sizes of bare contact conductors which, in the present case where the runway is longer than 60', is No. 2.

Metalworking Machine Tools

NEC® 670.3(A) states that the nameplate shall give necessary data, including the full-load current and the ampere rating of the largest motor. Per 670.4(A), the ampacity of supply conductors shall be not less than the full-load current marked on the nameplate, plus 25% of the full-load current rating of the highest rated motor as indicated on the nameplate. This requirement is equivalent to the general rule stated in 430.24.

Automatic Gear Machines C1 and C2 fall within the metalworking/machine tool classification. The total nameplate rating is 10.2 A and the highest rated motor is 4.8 A. Per 670.4(A), the supply conductors shall have a current-carrying capacity of not less than 11.4 A (4.8 A × .25 = 1.2 A + 10.2 A = 11.4 A). The feeder to Panel C consists of three No. 14 conductors in ½" conduit. Short-circuit protection for each feeder per 430.62(A) is 19.8 A (4.8 A × 3 = 14.4 A; 2.6 A + 1.8 A + 1 A = 5.4 A; 14.4 A + 5.4 A = 19.8 A). A 15 A NTDF is acceptable.

Synchronous Motor

An air compressor in the switchboard room is driven by a 100 HP synchronous motor operating at a leading power factor of 80%. Table 430.250 gives the full-load current at unity power factor as 101 A. A footnote states that this figure shall be multiplied by 1.25 when the motor operates at 80% power factor, the current here becoming 126 A (101 A × 1.25 = 126.25 = 126 A). Circuit wires shall have an ampacity of at least 158 A (126 A × 1.25 = 157.5 = 158 A) per 430.22(A). Table 310.15(B)(16) lists No. 2/0 THW Cu wire at 175 A. Informative Annex C, Table C8 shows that 1½" RMC is required.

A footnote to Table 430.52 states that for synchronous motors used in applications of this nature, branch-circuit protection need not be greater than 200% of the full-load current. For a current of 126 A, the overcurrent protection device could be set as low as 252 A (126 A × 2 = 252 A). By using an ITCB, a 300 A CB is needed. Two

No. 10 conductors in ½" conduit, originating at Panel D, supply direct current excitation for the revolving field of the synchronous motor.

X-Ray Unit

The No. 10 conductors specified for this device are larger than might seem to be required. Their size is dictated by a desire to hold the voltage drop to a minimum.

Resistance Welders

NEC® 630.31(A)(1) states that the ampacity of conductors supplying a resistance welder shall be not less than 70% of the rated primary current for automatically fed welders and 50% for manually operated ones. NEC® 630.31(B) requires that rated ampacity of conductors which supply two or more such welders shall be not less than the sum of the above value for the largest welder plus 60% of such values for the smaller welders.

NEC® 630.32(A)(B) demands that each welder be protected by an overcurrent protection device set at not more than 300% of the rated primary current, and that a feeder supplying one or more units shall be protected by a device set at not more than 300% of the conductor rating.

Unit B1 is a 5 kVA spot welder rated at 11 A. Unit B2 is a 10 kVA spot welder rated at 22 A. Both are manually operated. Conductors supplying B1 may be No. 14. Those for B2 may also be No. 14. B1 may be protected by a 30 A fuse or a non-adjustable CB and B2 by a 60 A fuse or non-adjustable CB. The feeder should not be smaller than 14.3 A (22 A × 5 = 11 A; 11 A × 0.5 × 0.6 = 3.3 A; 11 A + 3.3 A = 14.3 A). The question of voltage drop must be considered if the units are to operate efficiently. Two No. 10 conductors are run in ½" conduit from the junction box to Panel B. Two No. 12 conductors extend from the small welder to the junction box.

Transformer Arc Welders

NEC® 630.11(B) provides that the conductor rating for a group of welders shall be based on the nature of the operation. The maximum demand for a similar group of welders used on an identical application is 50% of the total capacity of the units. The total capacity is 45 kVA

(5 kVA + 10 kVA + 15 kVA + 15 kVA = 45 kVA). One half of this amount is 22.5 kVA (45 kVA ÷ 2 = 22.5 kVA), and the current taken when they are balanced across a 460 V, 3φ supply line is equal to about 30 A per leg. By applying the formula $I = \dfrac{VA}{V \times \sqrt{3}}$, the current is found to be 28 A (I = 22,500 VA ÷ 797 = 28.23 = 28 A).

NEC® 630.12(A) requires that each welder have overcurrent protection not greater than 200% of the rated primary current. NEC® 630.12(B) requires that feeder protection be not greater than 200% of the conductor rating.

Motor-Generator Arc Welder

B7, a single-operator, motor-generator arc welder, has a duty cycle of 80%. Per 630.11(A), conductors supplying such a unit shall have an ampacity of 10 A, which is equal to 91% of the nameplate rating (11 A × 0.91 = 10.01 = 10 A). The smallest permissible conductor per Table 310.15(B)(16) is No. 14 Cu.

Feeder Calculations

The load on the sub-feeder from Panel C to Panel A totals about 53 A. Three No. 6 THW Cu conductors are used in 1″ RMC. The load on the sub-feeder from Panel E to Panel A is approximately 21.6 A. Specifications call for a No. 10 Cu feeder. Panel A by its own units, including 25% of the largest motor-running current, equals approximately 544 A. The total load on Panel A equals approximately 544 A plus the full-load current of the motors connected to Panels C and E, for a total feeder load to Panel A of 618 A (544 A + 52 A + 22 A = 618 A). The print shows an 800 A busway from the main switchboard to Panel A.

Feeder conductors for Panel B supply the following loads: 14.3 A, 1φ to resistance welders B1 and B2; 28 A, 3φ to transformer arc welders B3, B4, B5, and B6; 10 A, 3φ to motor-generator arc welder B7; 11 A, 1φ to the X-ray unit B8; and 7.6 A, 3φ to the grinder B9.

When this load is balanced across the three phases as equally as possible, the most heavily loaded conductor carries about 70 A. No. 4 THW Cu conductors would normally be used, but because of the nature of the equipment, especially the X-ray unit, three No. 1 conductors

are specified. Chapter 9, Table C8 shows that 1¼″ RMC is needed for these conductors.

Service Calculations

The service load is equal to the full-load motor currents of all of the motors plus all other equipment on Panels A and B plus the full-load current of the 25 HP motor-generator, plus that of the 100 HP synchronous motor, plus 25% of the full-load current of the synchronous motor. The total is:

Panel A =	525 A
Panel B =	70 A
25 HP MG Set =	34 A
100 HP Synchronous Motor =	126 A
+ 25% of Largest Motor =	32 A
	787 A

This value is larger than that for the conductors listed in Table 310.15(B)(16), so two or more conductors in parallel are required to be used.

One half of 787 A equals 393.5 A (787 A ÷ 2 = 393.5 A). Per Table 310.15(B)(16), the nearest listed conductor is 600 kcmil THW Cu. The service may consist of two sets of three 600 kcmil THW Cu conductors in two runs of 3″ RMC per Annex C, Table C8. For ease of handling, smaller conductors can be used. Table 310.15(B)(16) gives the ampacity of 250 kcmil THW Cu as 255 A. With six conductors in one conduit, the ampacity of each is reduced to 80% of 255 A, or 204 A (255 A × 0.80 = 204 A). Two such conductors on each leg furnish a capacity equal to 408 A (204 A + 204 A = 408 A). Annex C, Table C8 permits six 250 kcmil THW Cu conductors in 3″ RMC. The service then consists of two sets of six 250 kcmil THW Cu conductors in 3″ conduits.

Note: This service configuration is one of many possible alternatives. Supply conductors could be run in three or more sets of conduit, to avoid penalty with respect to ampacity, or even in the form of busduct if conditions warranted its use.

POWER INSTALLATION FOR RESTAURANT

The power installation is to be confined to the kitchen area, with the exception of a 5 HP exhaust fan on the roof. The service is 4-wire delta with 3φ power at 240 V between phase wires and 1ϕ power at 120/240 V between two phase wires and a neutral conductor, as indicated on the small diagram to the right of the print.

The print covers only the power installation. Lighting outlets appear on a separate drawing. This procedure is often used where either the power or the lighting is involved to avoid any confusion which may result from crowding both power and lighting onto the same sheet.

The panelboard is divided into two sections, A and B. Section A supplies 3ϕ loads and large 1ϕ loads which do not require a neutral conductor. Section B supplies any 1ϕ loads which require a neutral conductor. THW Cu conductors are used.

Circuit Analysis

Circuits A1 and A2 are connected to 21 kVA, 3ϕ ranges, with the nameplate current rating being 53 A. These loads, in normal operation, are likely to continue for long periods of time. NEC® 210.20(A) requires that conductors serving such continuous loads shall carry not more than 80% of the listed current. The conductors supplying each range are rated at a minimum ampacity of 67 A (53 A ÷ 0.8 = 66.25 = 67 A). Table 310.15(B)(16) rates No. 4 THW Cu conductors at 85 A. Informative Annex C, Table C8 indicates that 1″ RMC is needed for three No. 4 Cu conductors.

Deep fat fryers A3 and A4 are rated at 10 kVA, 3ϕ, with a current rating of 44 A. These loads are similar to those of the ranges in that they continue for long periods of time so that the 80% factor is used. The conductors have a normal ampacity of 55 A (44 A ÷ .8 = 55 A). Two No. 6 Cu conductors in ¾″ conduit will meet the requirements.

Circuit A5 supplies a 5 HP, 3ϕ blower motor on the roof with a nameplate current of 14.8 A and a service factor of 1.15. A riser leading to this unit may be seen at the left side of the panelboard. NEC® 430.6(A)(1) states that, except for hermetic motors, the ampacity of conductors, branch-circuit overcurrent devices, etc., shall be determined from the motor current tables at the end of Article 430 instead of the current ratings on motor nameplates. Table 430.250 lists the current of the 5 HP, 240 V, 3ϕ motor as 15.2 A.

NEC® 430.22 requires that branch-circuit conductors supplying a single motor shall have an ampacity of 125%

of 15.2 A, or 19 A (15.2 A × 1.25 = 19 A). No. 12 THW Cu conductors are rated at 20 A per Table 310.15(B)(16). Informative Annex C, Table C8 shows that three No. 12 Cu conductors may be run in ½″ RMC.

NEC® 430.52 and Table 430.52 require that the initial rating of a dual-element fuse, used for branch-circuit protection, shall not exceed 175% of the full-load current of the motor listed in Table 430.250. The rating of the dual-element fuse for the 5 HP motor should not exceed 26.6 A (15.2 A × 1.75 = 26.6 A).

NEC® 430.52(C)(1) requires the use of a 25 A fuse. NEC® 430.83(A)(1) states that a controller shall have a horsepower rating not less than that of the motor. NEC® 430.109 states a similar requirement for the disconnecting means. A 25 A CB satisfies the needs of a controller per 430.83(C) and of a disconnect per 430.110(A). The rating of a manually-operable switch on the roof of the building is 5 HP.

Per 430.32(A)(1), each continuous-duty motor rated at more than 1 HP shall be protected against running overcurrent by a separate overcurrent protection device rated at not more than 125% of the full-load current of the motor, or else contain an integral thermal protector. If the blower motor does not have built-in thermal protection, an overcurrent protection device shall be supplied (perhaps in a magnetic switch used to start and stop the motor). The rating of the device should be 18.5 A (14.8 A × 1.25 = 18.5 A).

NEC® 430.52 states that the maximum value of branch-circuit protection shall be limited to the value given in Table 430.52 (300% for NTDFs). NEC® 430.58 states that a circuit breaker is permitted to be used for motor branch-circuit protection if it conforms to 430.110(A), which in turn requires it to have an ampacity of 115% of the full-load current of the motor. The 5 HP motor has a full-load current of 17.5 A (15.2 A × 1.15 = 17.48 = 17.5 A). If a standard 20 A CB at the panel carries starting current of this motor without tripping, it satisfies the need for motor overcurrent protection per 430.52 and 430.58.

Circuit A6 supplies a 1½ HP, 3ϕ dishwasher motor with a built-in thermal device. Table 430.250 gives the current as 5.2 A, so that No. 14 Cu conductors are large enough. Per 430.83(C)(1), a controller located at the operating location may be a general-use switch, which has an ampere rating at least twice that of the full-load current of the motor.

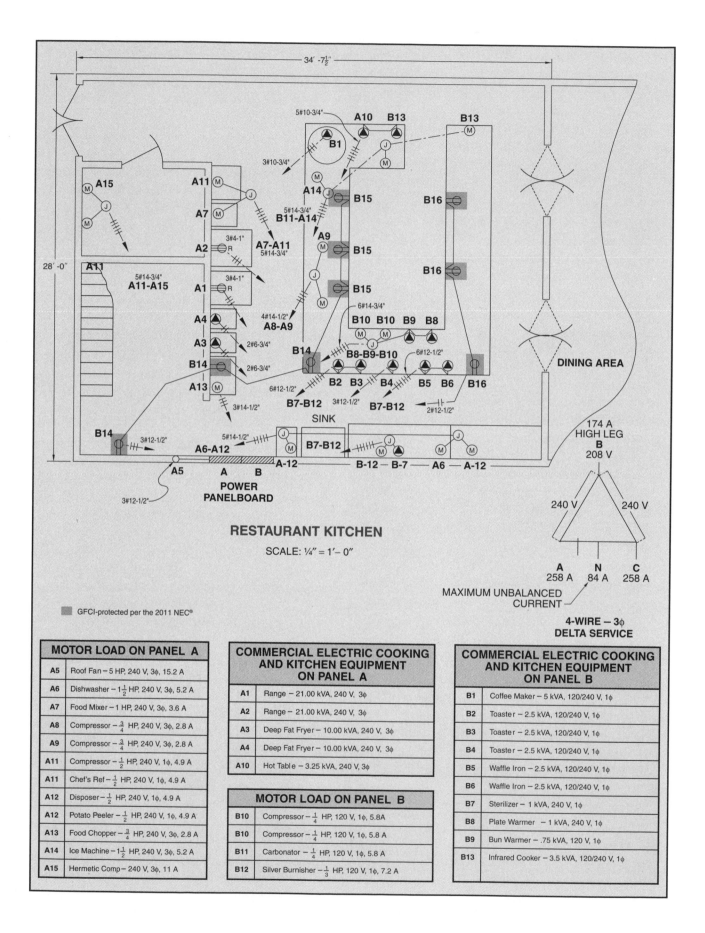

RESTAURANT KITCHEN

SCALE: ¼″ = 1′– 0″

■ GFCI-protected per the 2011 NEC®

MAXIMUM UNBALANCED CURRENT

174 A
HIGH LEG
B
208 V

240 V   240 V

A   N   C
258 A   84 A   258 A

4-WIRE — 3φ DELTA SERVICE

MOTOR LOAD ON PANEL A	
A5	Roof Fan – 5 HP, 240 V, 3φ, 15.2 A
A6	Dishwasher – 1½ HP, 240 V, 3φ, 5.2 A
A7	Food Mixer – 1 HP, 240 V, 3φ, 3.6 A
A8	Compressor – ¾ HP, 240 V, 3φ, 2.8 A
A9	Compressor – ¾ HP, 240 V, 3φ, 2.8 A
A11	Compressor – ½ HP, 240 V, 1φ, 4.9 A
A11	Chef's Ref – ½ HP, 240 V, 1φ, 4.9 A
A12	Disposer – ½ HP, 240 V, 1φ, 4.9 A
A12	Potato Peeler – ½ HP, 240 V, 1φ, 4.9 A
A13	Food Chopper – ¾ HP, 240 V, 3φ, 2.8 A
A14	Ice Machine – 1½ HP, 240 V, 3φ, 5.2 A
A15	Hermetic Comp – 240 V, 3φ, 11 A

COMMERCIAL ELECTRIC COOKING AND KITCHEN EQUIPMENT ON PANEL A	
A1	Range – 21.00 kVA, 240 V, 3φ
A2	Range – 21.00 kVA, 240 V, 3φ
A3	Deep Fat Fryer – 10.00 kVA, 240 V, 3φ
A4	Deep Fat Fryer – 10.00 kVA, 240 V, 3φ
A10	Hot Table – 3.25 kVA, 240 V, 3φ

MOTOR LOAD ON PANEL B	
B10	Compressor – ¼ HP, 120 V, 1φ, 5.8A
B10	Compressor – ¼ HP, 120 V, 1φ, 5.8 A
B11	Carbonator – ¼ HP, 120 V, 1φ, 5.8 A
B12	Silver Burnisher – ⅓ HP, 120 V, 1φ, 7.2 A

COMMERCIAL ELECTRIC COOKING AND KITCHEN EQUIPMENT ON PANEL B	
B1	Coffee Maker – 5 kVA, 120/240 V, 1φ
B2	Toaster – 2.5 kVA, 120/240 V, 1φ
B3	Toaster – 2.5 kVA, 120/240 V, 1φ
B4	Toaster – 2.5 kVA, 120/240 V, 1φ
B5	Waffle Iron – 2.5 kVA, 120/240 V, 1φ
B6	Waffle Iron – 2.5 kVA, 120/240 V, 1φ
B7	Sterilizer – 1 kVA, 240 V, 1φ
B8	Plate Warmer – 1 kVA, 240 V, 1φ
B9	Bun Warmer – .75 kVA, 120 V, 1φ
B13	Infrared Cooker – 3.5 kVA, 120/240 V, 1φ

Three No. 14 Cu conductors in ½″ conduit extend from the motor location to a junction box where they are joined by two No. 14 Cu conductors from a garbage disposer connected to A12. A ½″ conduit carries the five wires to a second junction box where two conductors from a potato peeler motor are tapped to Circuit A12 before the five wires continue to the panelboard.

Two motors are connected to Circuit A12 per 430.53(A), which states that motors not exceeding 1 HP each and having a full-load current not in excess of 6 A are permitted to be connected to a 20 A, 120 V, or less, circuit. Individual running overcurrent protection is required as these motors are permanently installed. See 430.32(B)(2). Usually, motors of this type are provided with a thermal protector integral with the motor which gives required protection.

Other motors are combined in this manner on Circuits B10, B11, and A11. Those on A11 are connected at the panelboard because they are run in separate conduits. The ¾ HP compressor motors on Circuits A8 and A9 may be combined on a single circuit but would require running protection, as they are automatically started per 430.32(B).

Circuit A10, for the hot table, and B13, for the infrared cooker, are taken to the panelboard as five No. 10 Cu conductors in ¾″ conduit. Circuit A10 is rated at 3250 VA or 13.5 A (3250 VA ÷ 240 V = 13.54 = 13.5 A) and Circuit B13 at 3500 VA or 14.6 A (3500 VA ÷ 240 V = 14.58 = 14.6 A). Since these are continuous loads, 210.20(A) requires that the total load on any overcurrent protection device not exceed 80% of its rating.

When there are from four to six current-carrying conductors in the same conduit supplying a continuous load, they are derated once as per 310.15(C)(2). No. 12 conductors are permissible for A10 and B13, but in this case a No. 10 THW Cu conductor was selected. Annex C, Table C8 permits six No. 10 THW Cu conductors in ¾″ RMC. Derating considerations also enter into calculations for toasters and waffle irons on Circuits B2, B3, B5, and B6.

Circuit A15, whose conduit is run exposed, supplies an 11 A sealed hermetic compressor motor. NEC® 440.6(A) states that the full-load current marked on the compressor nameplate shall be used to determine the ampacity of the branch-circuit conductors.

NEC® 440.4(A) requires that the locked rotor current for such motors shall be marked on the nameplate. Locked rotor current is marked as 53 A. NEC® 440.41(A) provides that controllers for sealed hermetic motors shall be selected on the basis of full-load current and locked rotor current.

Per Table 430.250, Column 3, the nearest current greater than 11 A (the full-load current on the motor nameplate) is 15.2 A, which applies to a 5 HP unit. In Table 430.251, the nearest 3ϕ locked rotor current to the 53 A value on the nameplate of the motor is 58 A, applying to a 3 HP motor. NEC® 440.41(A) states that where two different horsepower ratings are obtained from different tables, the higher of the two values shall be used. Here, the larger of them is used, necessitating a 5 HP controller. NEC® 440.12(A) imposes a similar demand with respect to the disconnecting means.

Service Calculation

The load on Panel A is obtained by converting motor loads to VA and adding them to the VA of the cooking and heating appliances. A value equal to 25% of the largest motor rating is included per 430.24(1) as well as a demand factor permitted by Table 220.56 for commercial electric cooking and other kitchen equipment. The total load on Panel A is 72,178 VA, or 174 A.

Note: When total wattage or VA is to be divided by phase-to-phase (3ϕ) voltage times 1.732, the following values may be substituted:

- for 208 V × 1.732, use 360
- for 230 V × 1.732, use 398
- for 240 V × 1.732, use 416
- for 440 V × 1.732, use 762
- for 460 V × 1.732, use 797
- for 480 V × 1.732, use 831

MOTOR LOAD ON PANEL A

A5 Roof Fan—5 HP, 240 V, 3ϕ, 15.2 A
VA = 15.2 × 416 = 6323 VA
A6 Dishwasher—1½ HP, 240 V, 3ϕ, 5.2 A
VA = 5.2 × 416 = 2163 VA
A7 Food Mixer—1 HP, 240 V, 3ϕ, 3.6 A
VA = 3.6 × 416 = 1498 VA
A8 Compressor—¾ HP, 240 V, 3ϕ, 2.8 A
VA = 2.8 × 416 = 1165 VA

A9 Compressor—¾ HP, 240 V, 3φ, 2.8 A

VA = 2.8 × 416 =   1165 VA

A11 Compressor— ½ HP, 240 V, 1φ, 4.9 A

VA = 4.9 × 240 =   1176 VA

A11 Chef's Ref—½ HP, 240 V, 1φ, 4.9 A

VA = 4.9 × 240 =   1176 VA

A12 Disposer—½ HP, 240 V, 1φ, 4.9 A

VA = 4.9 ×240 =   1176 VA

A12 Potato Peeler—½ HP, 240 V, 1φ, 4.9 A

VA = 4.9 × 240 =   1176 VA

A13 Food Chopper—¾ HP, 240 V, 3φ, 2.8 A

VA = 2.8 × 416 =   1165 VA

A14 Ice Machine—½ HP, 240 V, 3φ, 5.2 A

VA = 5.2 × 416 =   2163 VA

A15 Hermetic Comp—240 V, 3φ, 11 A

VA = 11 × 416 =   4576 VA

Total Motor VA =   24,922 VA

220.14(C): + 25% of largest motor

6323 VA × 0.25 =   1581 VA

Total Motor Load on Panel A =   **26,503 VA**

COMMERCIAL ELECTRIC COOKING AND KITCHEN EQUIPMENT LOAD ON PANEL A

A1 Range—21 kVA, 240 V, 3φ   21,000 VA

A2 Range—21 kVA, 240 V, 3φ   21,000 VA

A3 Deep Fat Fryer—10 kVA, 240 V, 3φ   10,000 VA

A4 Deep Fat Fryer—10 kVA, 240 V, 3φ   10,000 VA

A10 Hot Table—3.25 kVA, 240 V, 3φ   3250 VA

Total =   65,250 VA

Table 220.56: 65,250 VA × 0.70 = 45,675 VA

Total Commercial Electric Cooking and Kitchen Equipment Load on Panel A =   **45,675 VA**

Total VA on Panel A =   **72,178 VA**

$$I = \frac{VA}{V \times \sqrt{3}}$$

$$I = \frac{72,178}{240 \times 1.73}$$

$$I = \frac{72.178}{416} = 173.5 = 174 \text{ A}$$

Panel A = 174 A

The load on Panel B, found in the same manner, is 20,226 VA or 84 A.

MOTOR LOAD ON PANEL B

B10 Compressor—¼ HP, 120 V, 1φ, 5.8 A

VA = 5.8 × 120 =   696 VA

B10 Compressor—¼ HP, 120 V, 1φ, 5.8 A

VA = 5.8 ×120 =   696 VA

B11 Carbonator—¼ HP, 120 V, 1φ, 5.8 A

VA = 5.8 × 120 =   696 VA

B12 Silver Burn.—⅓ HP, 120 V, 1φ, 7.2 A

VA = 7.2 × 120 =   864 VA

Total Motor VA =   2952 VA

220.14(C): + 25% of largest motor

864 VA × 0.25 =   216 VA

Total Motor Load on Panel B   **3168 VA**

COMMERCIAL ELECTRIC COOKING AND KITCHEN EQUIPMENT LOAD ON PANEL B

B1 Coffee Maker—5 kVA, 120/240 V, 1φ   5000 VA

B2 Toaster—2.5 kVA, 120/240 V, 1φ   2500 VA

B3 Toaster—2.5 kVA, 120/240 V, 1φ   2500 VA

B4 Toaster—2.5 kVA, 120/240 V, 1φ,   2500 VA

B5 Waffle Iron—2.5 kVA, 120/240 V, 1φ   2500 VA

B6 Waffle Iron—2.5 kVA, 120/240 V, 1φ   2500 VA

B7 Sterilizer—1 kVA, 240 V, 1φ   1000 VA

B8 Plate Warmer—1 kVA, 240 V, 1φ   1000 VA

B9 Bun Warmer—0.75 kVA, 120 V, 1φ   750 VA

B13 Infrared Cooker—3.5 kVA, 120/240 V, 1φ   3500 VA

Total =   **23,750 VA**

Table 220.56: 23,750 VA × 0.65 =   15,438 VA

Total Commercial Electric Cooking and Kitchen Equipment on Panel B =   15,438 VA

Receptacles, 120 V: 180 VA × 9 =   1620 VA

Total VA on Panel B =   **20,226 VA**

$$I = \frac{VA}{V}$$

$$I = \frac{20,226}{240} = 84.275 = 84 \text{ A}$$

Panel B = 84 A

These values have been marked on the diagram at the right of the drawing, the load on phase B being 174 A, that on A or C is 258 A, and the current in N (neutral) is 84 A.

This method of adding 1φ current directly to 3φ current in order to determine the total current on a particular supply wire is the one customarily used and gives results sufficiently accurate for practical needs.

Table 310.15(B)(16) gives the nearest ampacities for THW Cu conductors as No. 2/0 for Phase *B*, 300 kcmil for Phase *A* and Phase *C*, and No. 4 for *N*. Chapter 9, Table 5 lists cross-sectional areas as 0.2624 sq in. for No. 2/0, 0.5281 sq in. for 300 kcmil, and 0.0973 sq in. for No. 4.

The cross-sectional areas of conductors are given in Chapter 9, Table 5. These areas are multiplied by the number of conductors to determine the total cross-sectional conductor area.

The total cross-sectional conductor area equals:

No. 2/0 THW =	$0.2624 \times 1 =$	0.2624
300 kcmil THW =	$0.5281 \times 2 =$	1.0562
No. 4 THW =	$0.0973 \times 1 =$	0.0973
	Total	**1.4159 sq in.**

For 40% fill, the cross-sectional area of conduit must be at least 3.5398 sq in. (1.4159 sq in. ÷ 0.4 = 3.5398 sq in.). Chapter 9, Table 4 lists the nearest size of conduit as 2½″.

Per 250.24(C)(1) the size of a copper grounded conductor is No. 2. The increased size provides an adequate return path for fault current. The conduit size remains the same in this case.

As in the case of the store building, parallel service conduits could be used in order to reduce the size of the conductors. The CB may be a 400 A frame, with current settings suited to protect each of the 3ϕ conductors. See 240.6(C) for the conditions that shall be adhered to with adjustable trip CBs.

Table 250.66 states that a No. 2 Cu GEC is required for this application.

MOTOR CONTROLLERS — 430, PART VII

430.2 Controller— A switch or device normally used to start or stop a motor by making or breaking the motor circuit current.

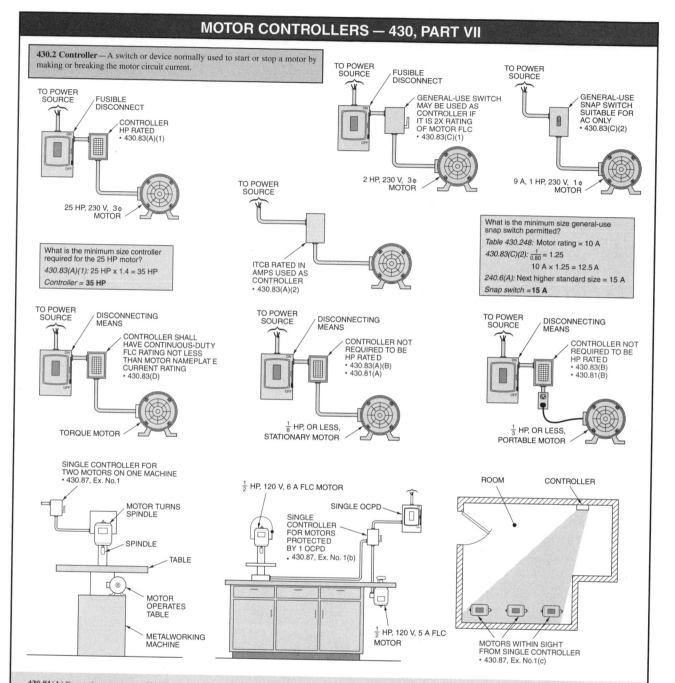

TO POWER SOURCE
FUSIBLE DISCONNECT
CONTROLLER HP RATED
• 430.83(A)(1)
25 HP, 230 V, 3ϕ MOTOR

What is the minimum size controller required for the 25 HP motor?
430.83(A)(1): 25 HP x 1.4 = 35 HP
Controller = **35 HP**

TO POWER SOURCE
FUSIBLE DISCONNECT
GENERAL-USE SWITCH MAY BE USED AS CONTROLLER IF IT IS 2X RATING OF MOTOR FLC
• 430.83(C)(1)
2 HP, 230 V, 3ϕ MOTOR

TO POWER SOURCE
GENERAL-USE SNAP SWITCH SUITABLE FOR AC ONLY
• 430.83(C)(2)
9 A, 1 HP, 230 V, 1ϕ MOTOR

What is the minimum size general-use snap switch permitted?
Table 430.248: Motor rating = 10 A
430.83(C)(2): $\frac{1}{0.80}$ = 1.25
10 A x 1.25 = 12.5 A
240.6(A): Next higher standard size = 15 A
Snap switch = **15 A**

TO POWER SOURCE
ITCB RATED IN AMPS USED AS CONTROLLER
• 430.83(A)(2)

TO POWER SOURCE
DISCONNECTING MEANS
CONTROLLER SHALL HAVE CONTINUOUS-DUTY FLC RATING NOT LESS THAN MOTOR NAMEPLATE CURRENT RATING
• 430.83(D)
TORQUE MOTOR

TO POWER SOURCE
DISCONNECTING MEANS
CONTROLLER NOT REQUIRED TO BE HP RATED
• 430.83(A)(B)
• 430.81(A)
$\frac{1}{8}$ HP, OR LESS, STATIONARY MOTOR

TO POWER SOURCE
DISCONNECTING MEANS
CONTROLLER NOT REQUIRED TO BE HP RATED
• 430.83(B)
• 430.81(B)
$\frac{1}{3}$ HP, OR LESS, PORTABLE MOTOR

SINGLE CONTROLLER FOR TWO MOTORS ON ONE MACHINE
• 430.87, Ex. No.1
MOTOR TURNS SPINDLE
SPINDLE
TABLE
MOTOR OPERATES TABLE
METALWORKING MACHINE

$\frac{1}{2}$ HP, 120 V, 6 A FLC MOTOR
SINGLE OCPD
SINGLE CONTROLLER FOR MOTORS PROTECTED BY 1 OCPD
• 430.87, Ex. No. 1(b)
$\frac{1}{2}$ HP, 120 V, 5 A FLC MOTOR

ROOM   CONTROLLER
MOTORS WITHIN SIGHT FROM SINGLE CONTROLLER
• 430.87, Ex. No.1(c)

430.81(A) For stationary motors of ⅛ HP or less that are normally left running, the branch-circuit disconnecting means is permitted to serve as the controller.

430.81(B) For portable motors of ⅓ HP or less, the controller is permitted to be an attachment plug and receptacle or cord connector.

430.83(A)(1) The horsepower rating of the controller shall not be less than the horsepower rating of the motor.

430.83(A)(2) A branch-circuit inverse-time circuit breaker (ITCB) rated in amperes only is permitted to be used as a controller.

430.83(B) Devices permitted by 430.81(A – B) are not required to be horsepower-rated to serve as a controller.

430.83(C) For stationary motors of 2 HP or less and 300 V or less, the controller is permitted to be a general-use switch having an ampere rating at least

twice the FLC rating of the motor. For AC circuits, a general-use snap switch suitable for use on an AC circuit shall be used to control a 2 HP or less and 300 V or less motor with a current rating not more than 80% of the ampere rating of the switch.

430.83(D) Motor controllers for torque motors shall have a continuous-duty, full-load current rating not less than the motor's nameplate current rating.

430.84 Controllers are not required to open all conductors unless they also serve as the disconnecting means per 430.111. See Exception.

430.87 In general, each motor of 600 V or less shall have its own controller.

430.87, Ex. No. 1 A single controller rated at no less than the total horsepower ratings of all the motors in a group is permitted, provided (a) the motors drive parts of the same machine; (b) the motors have one overcurrent device per 430.53(A); or (c) the group of motors is in a single room within sight from the controller.

SWITCHBOARDS — 408, PART II

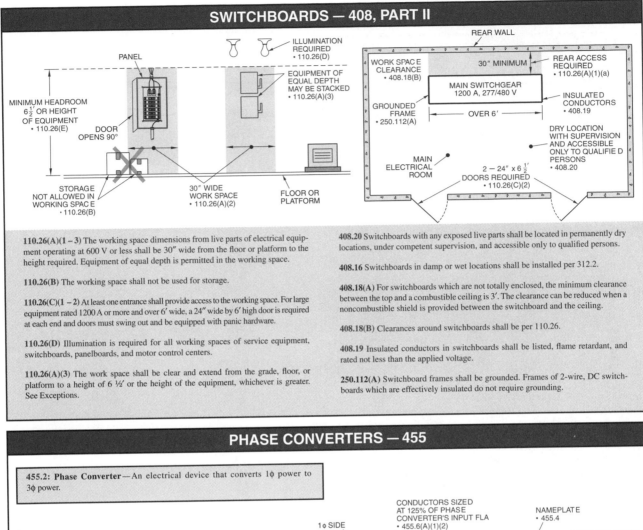

110.26(A)(1 – 3) The working space dimensions from live parts of electrical equipment operating at 600 V or less shall be 30″ wide from the floor or platform to the height required. Equipment of equal depth is permitted in the working space.

110.26(B) The working space shall not be used for storage.

110.26(C)(1 – 2) At least one entrance shall provide access to the working space. For large equipment rated 1200 A or more and over 6′ wide, a 24″ wide by 6′ high door is required at each end and doors must swing out and be equipped with panic hardware.

110.26(D) Illumination is required for all working spaces of service equipment, switchboards, panelboards, and motor control centers.

110.26(A)(3) The work space shall be clear and extend from the grade, floor, or platform to a height of 6 ½′ or the height of the equipment, whichever is greater. See Exceptions.

408.20 Switchboards with any exposed live parts shall be located in permanently dry locations, under competent supervision, and accessible only to qualified persons.

408.16 Switchboards in damp or wet locations shall be installed per 312.2.

408.18(A) For switchboards which are not totally enclosed, the minimum clearance between the top and a combustible ceiling is 3′. The clearance can be reduced when a noncombustible shield is provided between the switchboard and the ceiling.

408.18(B) Clearances around switchboards shall be per 110.26.

408.19 Insulated conductors in switchboards shall be listed, flame retardant, and rated not less than the applied voltage.

250.112(A) Switchboard frames shall be grounded. Frames of 2-wire, DC switchboards which are effectively insulated do not require grounding.

PHASE CONVERTERS — 455

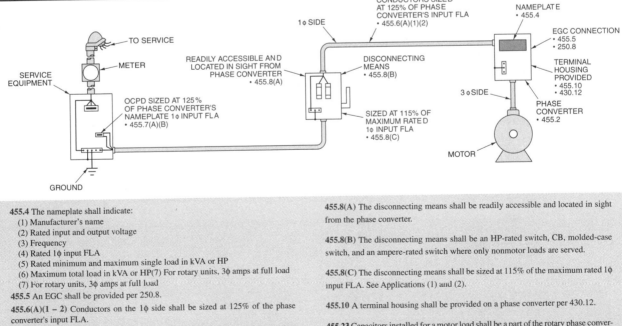

455.2: Phase Converter—An electrical device that converts 1ϕ power to 3ϕ power.

455.4 The nameplate shall indicate:
(1) Manufacturer's name
(2) Rated input and output voltage
(3) Frequency
(4) Rated 1ϕ input FLA
(5) Rated minimum and maximum single load in kVA or HP
(6) Maximum total load in kVA or HP(7) For rotary units, 3ϕ amps at full load
(7) For rotary units, 3ϕ amps at full load

455.5 An EGC shall be provided per 250.8.

455.6(A)(1 – 2) Conductors on the 1ϕ side shall be sized at 125% of the phase converter's input FLA.

455.7(A – B) The OCPD shall be sized at 125% of the phase converter's nameplate 1ϕ input FLA.

455.8(A) The disconnecting means shall be readily accessible and located in sight from the phase converter.

455.8(B) The disconnecting means shall be an HP-rated switch, CB, molded-case switch, and an ampere-rated switch where only nonmotor loads are served.

455.8(C) The disconnecting means shall be sized at 115% of the maximum rated 1ϕ input FLA. See Applications (1) and (2).

455.10 A terminal housing shall be provided on a phase converter per 430.12.

455.23 Capacitors installed for a motor load shall be a part of the rotary phase conversion system and shall be installed on the line side of the motor overload device.

BUSWAYS — 368

Busway—A grounded metal enclosure with factory-mounted, bare or insulated conductors, which are usually copper or aluminum bars, rods, or tubes.

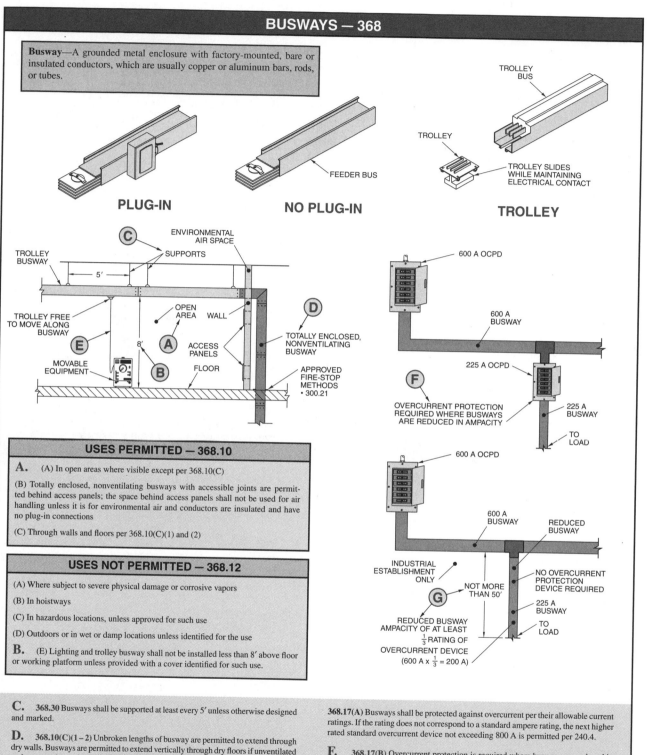

PLUG-IN

NO PLUG-IN

TROLLEY

USES PERMITTED — 368.10

A. (A) In open areas where visible except per 368.10(C)

(B) Totally enclosed, nonventilating busways with accessible joints are permitted behind access panels; the space behind access panels shall not be used for air handling unless it is for environmental air and conductors are insulated and have no plug-in connections

(C) Through walls and floors per 368.10(C)(1) and (2)

USES NOT PERMITTED — 368.12

(A) Where subject to severe physical damage or corrosive vapors

(B) In hoistways

(C) In hazardous locations, unless approved for such use

(D) Outdoors or in wet or damp locations unless identified for the use

B. (E) Lighting and trolley busway shall not be installed less than 8' above floor or working platform unless provided with a cover identified for such use.

C. 368.30 Busways shall be supported at least every 5' unless otherwise designed and marked.

D. 368.10(C)(1 – 2) Unbroken lengths of busway are permitted to extend through dry walls. Busways are permitted to extend vertically through dry floors if unventilated and protected to at least 6' above the floor.

368.58 All dead ends of busways shall be closed.

E. 368.56 Branches from busways shall be per Articles listed in 368.56(A–C) (1 – 16). Cord and cable assemblies approved for hard usage and listed bus drop cable are also permitted. Where a separate EGC is installed, the connection shall be per 250.8 and 250.12.

368.17 Overcurrent protection for busways shall be per 368.17(A – D).

368.17(A) Busways shall be protected against overcurrent per their allowable current ratings. If the rating does not correspond to a standard ampere rating, the next higher rated standard overcurrent device not exceeding 800 A is permitted per 240.4.

F. 368.17(B) Overcurrent protection is required where busways are reduced in ampacity. See Exception.

G. 368.17(B), Ex. For industrial establishments only, overcurrent devices may be omitted where the size is reduced if a smaller busway is not over 50' long and at least the rating of the next overcurrent device back on the line, provided it is not in contact with combustible material.

368.17(C), Ex. 2 When used to supply lighting units, the branch-circuit overcurrent device may be part of the fixture cord cap or may be made part of the fixture.

CABLEBUS — 370

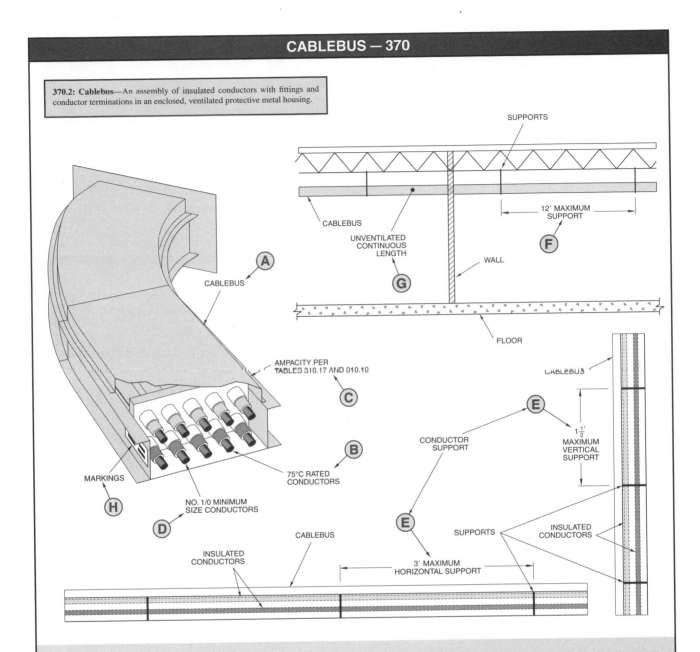

370.2: Cablebus—An assembly of insulated conductors with fittings and conductor terminations in an enclosed, ventilated protective metal housing.

SUPPORTS

CABLEBUS

UNVENTILATED CONTINUOUS LENGTH

12' MAXIMUM SUPPORT

WALL

FLOOR

CABLEBUS

AMPACITY PER TABLES 310.17 AND 010.10

CABLEBUS

1½' MAXIMUM VERTICAL SUPPORT

CONDUCTOR SUPPORT

75°C RATED CONDUCTORS

INSULATED CONDUCTORS

MARKINGS

NO. 1/0 MINIMUM SIZE CONDUCTORS

CABLEBUS

SUPPORTS

3' MAXIMUM HORIZONTAL SUPPORT

INSULATED CONDUCTORS

INSULATED CONDUCTORS

A. **370.3** Approved cablebus is permitted at any voltage or current for which the conductors are rated. It shall be exposed. Cablebus shall be identified for the use when installed outdoors or in corrosive, wet, or damp locations. It shall not be installed in hoistways or hazardous locations unless approved for the use. It may be used for branch circuits, feeders, and services.

B. **370.4(A)** Current-carrying conductors in cablebus shall have an insulation rating of 75°C (167°F) or higher and an approved type suitable for the application.

C. **370.4(B)** The ampacity of conductors shall be per Tables 310.15(B)(17) and 310.15(B)(19) or Tables 310.60(C)(69) and 310.60(C)(70) for over 600 V.

D. **370.4(C)** Conductor size shall be per the cablebus design and not smaller than No. 1/0.

E. **370.4(D)** Insulated conductors shall be supported at least every 3' horizontally and 1½' vertically.

F. **370.6(A)** Cablebus shall be supported at least every 12' unless the structure is designed for greater lengths.

G. **370.6(B)** Cablebus in continuous sections is permitted to pass through partitions or walls when protected against physical damage and unventilated.

370.6(C) Cablebus is permitted to pass through dry floors and platforms except where firestops are required. It shall be totally enclosed and protected for up to 6'.

370.6(D) Cablebus is permitted to pass through wet floors and platforms except where firestops are required if curbs are present to prevent water flow and it is totally enclosed and protected for up to 6'.

370.7 Cablebus shall have approved fittings for: (1) changes in directions; (2) dead ends; (3) terminations for connected equipment; and (4) additional physical protection where subject to severe physical damage.

370.9 Cablebus shall be grounded per 250, excluding 250.86, Ex. 2.

H. **370.10** Each section of cablebus shall be marked to include the manufacturer's name and the maximum diameter, number, voltage rating, and ampacity of conductors. Markings shall be visible after installation.

MOTORS — 430

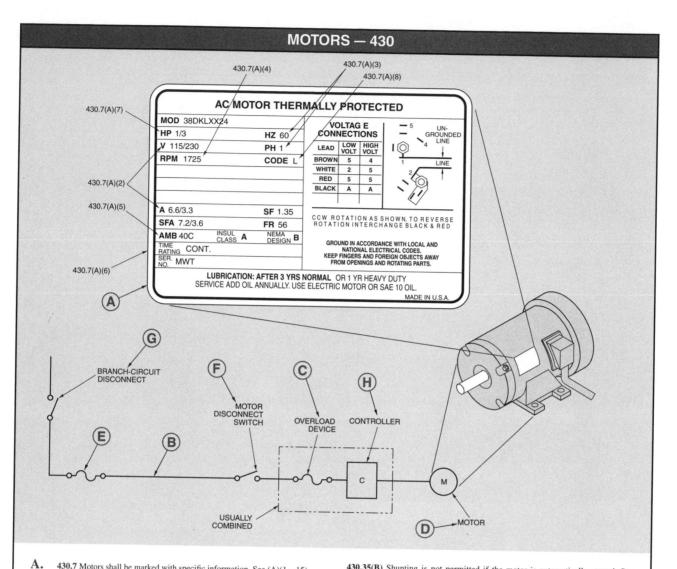

A. **430.7** Motors shall be marked with specific information. See (A)(1 – 15).

B. **430.22** The ampacity of branch-circuit conductors supplying a single motor shall be not less than 125% of the motor current rating.

C. **430.32(A)(1)** Continuous-duty motors over 1 HP shall be protected against overload by a separate overload device rated at not more than 125% of the full-load current rating for motors with a service factor not less than 1.15 or a temperature rise not over 40°C, and at not over 115% for all other motors.

430.32(B)(1) Continuous-duty motors of 1 HP or less which are not permanently installed and are automatically started and within sight from the controller are permitted to be protected by the branch-circuit, short-circuit, and ground-fault protective device.

430.32(C) Where the overcurrent relay selected does not allow the motor to start, the next higher size overload relay is selected provided the trip current does not exceed 140% of the motor's full-load current rating for motors with a service factor of not less than 1.15 or a temperature rise not over 40°C, and 130% for all other motors.

430.32(D)(1) Continuous-duty motors of 1 HP or less which are nonautomatically started shall be protected against overload by a separate overload device rated at not more than 125% of the full-load current rating for motors with a service factor not less than 1.15 or a temperature rise not over 40°C, and at not over 115% for all other motors.

D. **430.33** Any motor applications shall be considered as continuous duty unless the driven apparatus is such that the motor cannot operate continuously.

430.35(A) A running overcurrent device may be shunted at starting of a manual-start motor if no hazard is introduced and the branch-circuit device of not over 400% is operative in the circuit during the starting period.

430.35(B) Shunting is not permitted if the motor is automatically started. See Exception.

430.40 Thermal cutouts and overload relays for motor-running protection not capable of opening short circuits shall be protected per 430.52 unless approved for group installation and marked with the maximum size of required protection.

E. **430.52** The motor branch-circuit overcurrent device shall be able to carry the starting current (150%–300% per Table 430.52; absolute maximum 400% with NTDFs and Class CC fuses; and 225% with TDFs).

430.102(A) A disconnecting means shall be in sight from the controller location and shall disconnect the controller. See Ex. 1, 2, and 3.

F. **430.102(B)** A disconnecting means shall be in sight from the motor location and the driven machinery location. See Exception.

430.107 One of the disconnecting means shall be readily accessible.

430.108 All disconnecting means shall comply with 430.109 and 430.110.

G. **430.109** The disconnecting means shall be a (1) motor circuit switch, (2) molded case circuit breaker, (3) molded case switch, (4) instantaneous trip circuit breaker, (5) self-protected combination controller, (6) manual motor controller, or (7) system isolation equipment.

430.110(A) All disconnecting means shall have an ampere rating of at least 115% of the motor's FLC, taken from FLC tables per 430.6(A).

H. **430.111** A suitable switch or CB may serve as both the disconnecting means and the controller.

MOTOR DISCONNECTING MEANS — 430.109

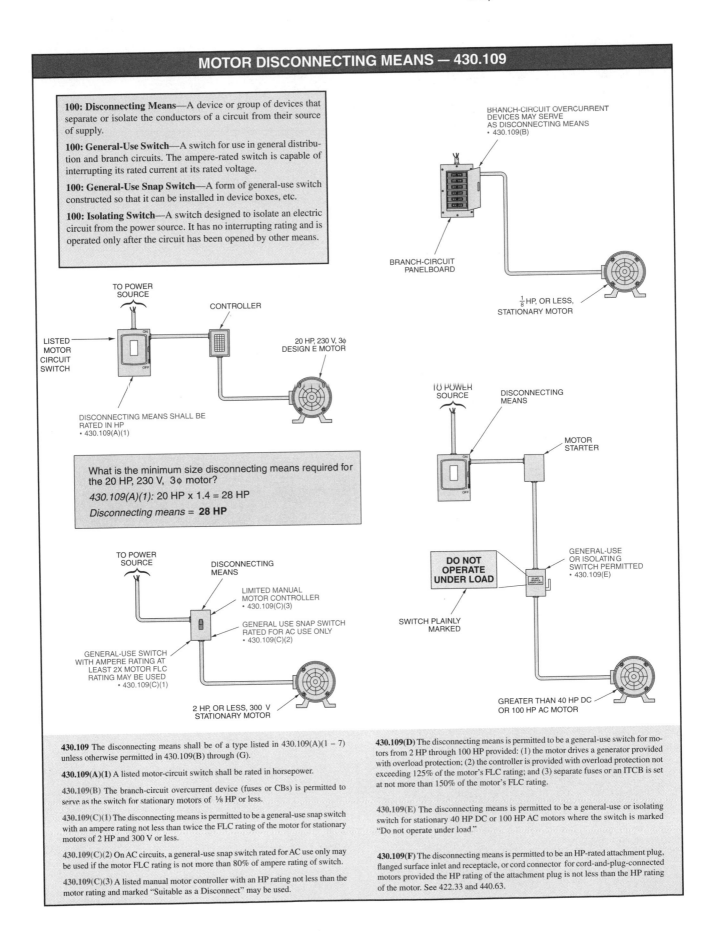

100: Disconnecting Means—A device or group of devices that separate or isolate the conductors of a circuit from their source of supply.

100: General-Use Switch—A switch for use in general distribution and branch circuits. The ampere-rated switch is capable of interrupting its rated current at its rated voltage.

100: General-Use Snap Switch—A form of general-use switch constructed so that it can be installed in device boxes, etc.

100: Isolating Switch—A switch designed to isolate an electric circuit from the power source. It has no interrupting rating and is operated only after the circuit has been opened by other means.

BRANCH-CIRCUIT OVERCURRENT DEVICES MAY SERVE AS DISCONNECTING MEANS
• 430.109(B)

BRANCH-CIRCUIT PANELBOARD

$\frac{1}{8}$ HP, OR LESS, STATIONARY MOTOR

TO POWER SOURCE

CONTROLLER

LISTED MOTOR CIRCUIT SWITCH

20 HP, 230 V, 3φ DESIGN E MOTOR

DISCONNECTING MEANS SHALL BE RATED IN HP
• 430.109(A)(1)

What is the minimum size disconnecting means required for the 20 HP, 230 V, 3φ motor?

430.109(A)(1): 20 HP x 1.4 = 28 HP

Disconnecting means = **28 HP**

TO POWER SOURCE

DISCONNECTING MEANS

MOTOR STARTER

DO NOT OPERATE UNDER LOAD

GENERAL-USE OR ISOLATING SWITCH PERMITTED
• 430.109(E)

SWITCH PLAINLY MARKED

GREATER THAN 40 HP DC OR 100 HP AC MOTOR

TO POWER SOURCE

DISCONNECTING MEANS

LIMITED MANUAL MOTOR CONTROLLER
• 430.109(C)(3)

GENERAL USE SNAP SWITCH RATED FOR AC USE ONLY
• 430.109(C)(2)

GENERAL-USE SWITCH WITH AMPERE RATING AT LEAST 2X MOTOR FLC RATING MAY BE USED
• 430.109(C)(1)

2 HP, OR LESS, 300 V STATIONARY MOTOR

430.109 The disconnecting means shall be of a type listed in 430.109(A)(1 – 7) unless otherwise permitted in 430.109(B) through (G).

430.109(A)(1) A listed motor-circuit switch shall be rated in horsepower.

430.109(B) The branch-circuit overcurrent device (fuses or CBs) is permitted to serve as the switch for stationary motors of ⅛ HP or less.

430.109(C)(1) The disconnecting means is permitted to be a general-use snap switch with an ampere rating not less than twice the FLC rating of the motor for stationary motors of 2 HP and 300 V or less.

430.109(C)(2) On AC circuits, a general-use snap switch rated for AC use only may be used if the motor FLC rating is not more than 80% of ampere rating of switch.

430.109(C)(3) A listed manual motor controller with an HP rating not less than the motor rating and marked "Suitable as a Disconnect" may be used.

430.109(D) The disconnecting means is permitted to be a general-use switch for motors from 2 HP through 100 HP provided: (1) the motor drives a generator provided with overload protection; (2) the controller is provided with overload protection not exceeding 125% of the motor's FLC rating; and (3) separate fuses or an ITCB is set at not more than 150% of the motor's FLC rating.

430.109(E) The disconnecting means is permitted to be a general-use or isolating switch for stationary 40 HP DC or 100 HP AC motors where the switch is marked "Do not operate under load."

430.109(F) The disconnecting means is permitted to be an HP-rated attachment plug, flanged surface inlet and receptacle, or cord connector for cord-and-plug-connected motors provided the HP rating of the attachment plug is not less than the HP rating of the motor. See 422.33 and 440.63.

MOTOR CONTROL CIRCUITS — 430, PART VI

430.2: Motor Control Circuit—The circuit of a control apparatus that carries electrical signals directing the controller's performance, but does not carry the main power current.

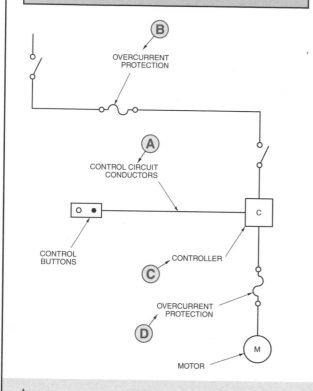

A.   300.3(C)(2)c Excitation, control, relay, and ammeter conductors used in connection with the motor or starter may occupy the same enclosures as the motor-circuit conductors.

B.   430.72(A) A motor control circuit tapped from the load side of a branch-circuit, short-circuit, and ground-fault protective device shall be protected against overcurrent.

430.72(B)(1) (General Rule) Motor control conductor overcurrent protection shall not exceed the values listed in column A of Table 430.72(B).

C.   430.72(B)(2) Control conductors that do not leave the controller are permitted to be protected by the motor branch-circuit protective device not exceeding the values listed in Table 430.72(B), Col B. Control conductors that extend beyond the controller are permitted to be protected by the motor branch-circuit protective device not exceeding the values listed in Table 430.72(B), Col C.

D.   430.72(B), Ex. 1 Motor branch-circuit overcurrent protective devices are permitted where the opening of a control circuit creates a hazard. (Example: The control circuit of a fire pump motor.)

430.72(B), Ex. 2 Transformer-fed, 2-wire primary, 2-wire secondary control conductors are considered protected if the motor branch-circuit overcurrent protective device protects secondary wires per Table 430.72(B). See 240.4(F).

430.72(C) Control circuit transformers are required to be protected per 430.72(C)(1–5). See Special Applications (1–5).

430.73 Remote control circuits shall be installed in a raceway or have physical protection and be so arranged that an accidental ground will not start the motor or bypass the shutdown devices.

430.75(A) Control circuit shall be disconnected from the supply when the motor circuit is opened. If two separate devices are used, they shall be immediately adjacent. See Ex. 1 and 2.

430.75(B) Control transformers located inside the controller shall be on the load side of the disconnect for the control circuit.

WOUND-ROTOR MOTORS

Wound-Rotor Motor—A 3φ motor in which the rotor windings are connected to slip rings.

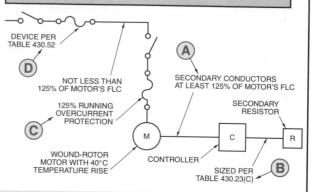

A.   430.23(A) For continuous duty, the conductors for the secondary of a wound-rotor motor to its controller shall be not less than 125% of the motor's FLC rating.

430.23(B) For other than continuous duty, the conductors for the secondary of a wound-rotor motor to its controller shall be not less than the values in Table 430.22(E).

B.   430.23(C) If the secondary resistor is separate from the controller, the ampacity of the conductors between the controller and resistor shall be not less than the values in Table 430.23(C).

C.   430.32(A)(1) The running overcurrent protection shall be not more than 125% for motors with service factors not less than 1.15 and motors with a temperature rise not over 40°C. For all other motors, the running overcurrent protection shall be not more than 115%.

D.   430.52 The motor branch-circuit, short-circuit, and ground-fault protective device shall be selected from Table 430.52. For NTDFs and dual-element fuses, the rating shall not exceed 150% of the motor's FLC rating.

PART — WINDING MOTORS

Part-Winding Motor—A motor arranged for starting by first energizing part of its primary (armature) winding and then energizing the rest of the winding in one or more steps.

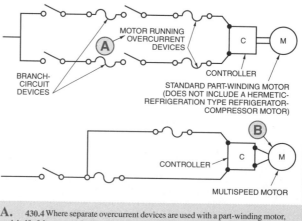

A.   430.4 Where separate overcurrent devices are used with a part-winding motor, each half of the motor winding shall be individually protected per 430.32 and 430.37 with one half of the trip current specified.

B.   430.22(B) The size of the branch-circuit conductors for multispeed motors is based on the highest of the FLC ratings on the motor nameplate.

DIRECT CURRENT MOTORS

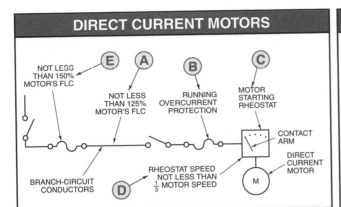

A. **430.22** Branch-circuit conductors shall be no less than 125% of the motor's FLC rating.

430.22(A) DC motors supplied from a rectified 1φ supply shall have conductor ampacity of: (A) 190% for 1φ half-wave rectifier; or (B) 150% for 1φ full-wave rectifier.

B. **430.32(A)(1)** The running overcurrent protection shall be not more than 125% for motors with service factors not less than 1.15 and motors with a temperature rise not over 40°C. For all other motors, the protection shall be not more than 115%.

C. **430.82(C)(1)** The rheostat shall be designed so the contact arm cannot be left on intermediate segments.

D. **430.82(C)(2)** An under-voltage release is required on rheostats to interrupt the power supply where the motor's speed falls to one-third of its normal rate.

430.88 The adjustable-speed control shall be arranged so the motor cannot start on a weakened field, except where so designed.

430.89 Separately excited DC motors, series motors, and motor generators and converters shall be provided with speed-limiting devices.

E. **430.52** The motor branch-circuit, short-circuit, and ground-fault protective device shall be selected from Table 430.52. For NTDFs and dual-element fuses, the rating shall not exceed 150% of the motor's FLC rating.

MOTOR FEEDER CALCULATIONS

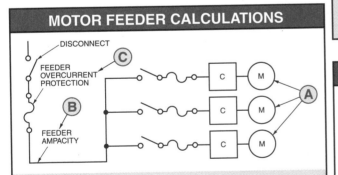

A. **430.6(A)(1)** For general motor applications, except for torque motors, Tables 430.247, 248, 249, and 250, including notes, shall be used to determine the conductor ampacity, ampere rating of switches, short-circuit and ground-fault protection, etc. The motor nameplate FLA is used to determine the overload protection. Where the motor is marked in amps but no HP, the HP rating is assumed to be the corresponding value given in Tables 430.247, 248, 249, and 250, interpolated if necessary.

B. **430.24** The minimum feeder ampacity is 125% of the FLC rating of the largest motor plus the full-load currents of all other motors. Where the motors are for types of service listed in 430.22(a), similar procedures are followed, taking into consideration values obtained from Table 430.22(E). Where circuitry prevents all motors from operating at the same time, the conductor size is determined on the basis of the maximum load at any given time.

C. **430.62(A)** The maximum feeder protection is determined by the largest branch-circuit protection plus the sum of the full-load currents of all other motors.

HERMETIC MOTORS

440.2: Hermetic Refrigerant Motor-Compressor—A combination compressor and motor enclosed in the same housing, without an external shaft or shaft seals, and the motor operating in the refrigerant.

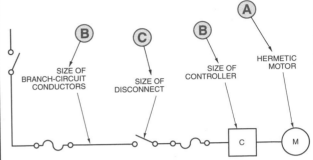

A. **440.4(A)** The LRC shall be marked on the nameplate of all polyphase motors and 1φ motors over 9 A and 115 V or 4.5 A and 230 V.

B. **440.6(A)** The data marked on the nameplate of the equipment shall be used to determine the rating of the controller. If no rated-load current is shown on the equipment nameplate, the rated-load current of the compressor shall be used. The full-load current marked on the equipment nameplate is used to determine the size of the branch-circuit conductors.

C. **440.12(A)** The disconnecting means shall be sized based on the nameplate rated-load current or branch-circuit selection current, whichever is greater, and the LRC of the motor compressor per 440.12(A)(1 – 2).

440.22(A) The motor is considered protected by the branch-circuit short circuit device rated for 175% of the rated-load current or the branch-circuit selection current, whichever is greater.

440.22(C) Where ratings shown on the manufacturer's heater table are less than those permitted by 440.22(A – B), then the lower values listed on the manufacturer's heater table shall be used.

440.52(A)(1 – 3) The overload setting shall not be over 140% of the motor's FLC for overload relays and 125% for other devices.

MOTOR CONTROLLERS

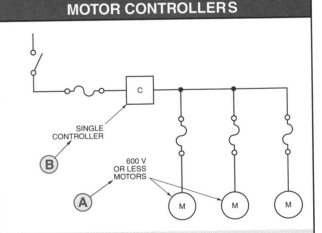

A. **430.87** (General Rule) Each motor of 600 V or less shall have its own controller.

B. **430.87, Ex. No. 1 (a – c)** A single controller is permissible if the motors are part of the same machine, or if in a group per 430.53(A), or if in a group in a single room in sight of the controller.

TWO OR MORE MOTORS ON ONE BRANCH CIRCUIT

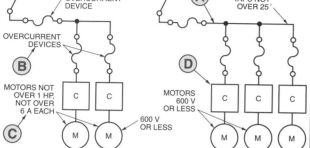

A. 430.24 The minimum feeder current rating is 125% of the largest motor current, the sum of the other motors, 100% of noncontinuous nonmotor loads, and 125% of continuous nonmotor loads.

B. 430.53(A) The maximum setting of the overcurrent device shall be 20 A, 120 V or less; 15 A, 600 V or less. Two or more motors not over 1 HP are permitted on the same 120 V, 20 A branch circuit or 600 V, 15 A branch circuit, if: 1. not over 6 A; 2. controller maximum branch circuit overcurrent device is not exceeded; 3. each overload is per 430.32.

C. 430.53(B) All motors have overload protection and the smallest motor is protected per 430.52, and where the branch-circuit device will not open under the most severe normal service conditions that might be encountered.

D. 430.53(C) Two or more motors of any size may be connected to one branch circuit if each has running overcurrent protection and all of the six requirements of 430.53(C) are met.

E. 430.53(D)(2) Ampacity of taps shall be not less than that of the branch circuit conductors, not over 25′ long, and protected from physical damage by an enclosure in an approved raceway or by other approved means.

430.62(A) The maximum feeder protection is determined by the largest branch circuit protection plus the sum of the full-load currents of all other motors. Use Table 430.52 or 440.22(A) for hermetic motors.

430.112 In general, individual disconnects are required for each motor. See Exception.

430.112, Ex. (a – c) Individual disconnects are required unless the group is part of a single machine or if the group is in a single room in sight of the disconnect.

METALWORKING MACHINE TOOLS

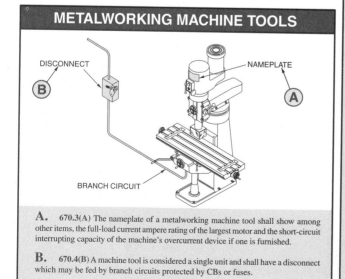

A. 670.3(A) The nameplate of a metalworking machine tool shall show among other items, the full-load current ampere rating of the largest motor and the short-circuit interrupting capacity of the machine's overcurrent device if one is furnished.

B. 670.4(B) A machine tool is considered a single unit and shall have a disconnect which may be fed by branch circuits protected by CBs or fuses.

MISCELLANEOUS MOTOR RULES

Fluke Corporation

Motors shall be located where maintenance can be accomplished.

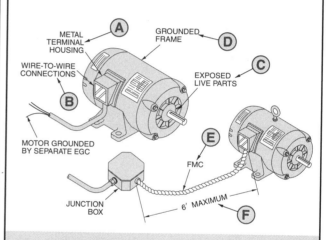

A. 430.12(A) In general, terminal housings shall be metal. Nonmetallic and non-combustible housings are permitted in other-than-hazardous areas. See Exception.

B. 430.12(B) Table 430.12(B) gives sizes for wire-to-wire connections.

430.14(A) The motor shall be located where adequate ventilation is provided and where maintenance can be accomplished.

430.16 The motor shall be protected from the accumulation of dust.

430.43 An overload which can automatically restart a motor shall be approved for the motor and if motor restarts could not result in injury.

C. 430.232 Exposed live parts shall be guarded if over 50 V or more between terminals. See Means (1 – 3) for grounding provisions.

D. 430.242 Frames of stationary motors shall be grounded if: (1) metal enclosed wiring; (2) in wet location; (3) in hazardous location; or (4) over 150 V to ground. If the motor frame is not grounded, it shall be insulated from ground.

E. 430.245(A) Where the motor terminals of fixed motors are in metal raceways or metal enclosed cable, junction boxes for motor terminals are required. Metal raceways and the armor of metal cables shall be connected to the terminal housings per Article 250.96(A) and 250.97.

F. 430.245(B) The junction box shall not be more than 6′ from the motor under conditions specified.

430.245(C) Instrument transformer secondaries and exposed metal parts of instruments or relays used in connection with motors shall be grounded per 250.170 through 250.178.

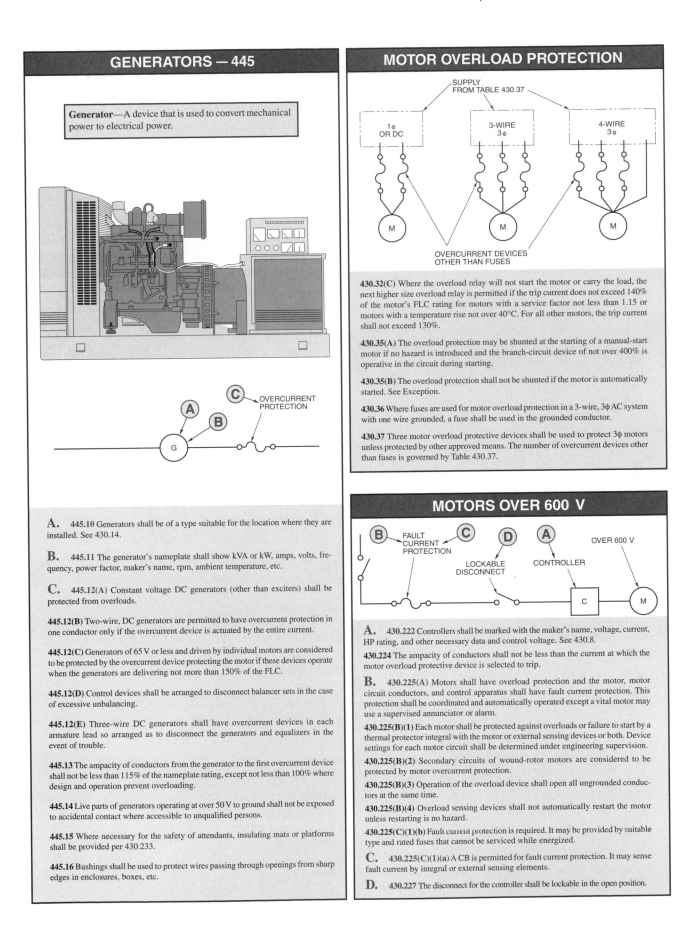

GENERATORS — 445

Generator—A device that is used to convert mechanical power to electrical power.

OVERCURRENT PROTECTION

A. **445.10** Generators shall be of a type suitable for the location where they are installed. See 430.14.

B. **445.11** The generator's nameplate shall show kVA or kW, amps, volts, frequency, power factor, maker's name, rpm, ambient temperature, etc.

C. **445.12(A)** Constant voltage DC generators (other than exciters) shall be protected from overloads.

445.12(B) Two-wire, DC generators are permitted to have overcurrent protection in one conductor only if the overcurrent device is actuated by the entire current.

445.12(C) Generators of 65 V or less and driven by individual motors are considered to be protected by the overcurrent device protecting the motor if these devices operate when the generators are delivering not more than 150% of the FLC.

445.12(D) Control devices shall be arranged to disconnect balancer sets in the case of excessive unbalancing.

445.12(E) Three-wire DC generators shall have overcurrent devices in each armature lead so arranged as to disconnect the generators and equalizers in the event of trouble.

445.13 The ampacity of conductors from the generator to the first overcurrent device shall not be less than 115% of the nameplate rating, except not less than 100% where design and operation prevent overloading.

445.14 Live parts of generators operating at over 50 V to ground shall not be exposed to accidental contact where accessible to unqualified persons.

445.15 Where necessary for the safety of attendants, insulating mats or platforms shall be provided per 430.233.

445.16 Bushings shall be used to protect wires passing through openings from sharp edges in enclosures, boxes, etc.

MOTOR OVERLOAD PROTECTION

SUPPLY FROM TABLE 430.37

1φ OR DC

3-WIRE 3φ

4-WIRE 3φ

OVERCURRENT DEVICES OTHER THAN FUSES

430.32(C) Where the overload relay will not start the motor or carry the load, the next higher size overload relay is permitted if the trip current does not exceed 140% of the motor's FLC rating for motors with a service factor not less than 1.15 or motors with a temperature rise not over 40°C. For all other motors, the trip current shall not exceed 130%.

430.35(A) The overload protection may be shunted at the starting of a manual-start motor if no hazard is introduced and the branch-circuit device of not over 400% is operative in the circuit during starting.

430.35(B) The overload protection shall not be shunted if the motor is automatically started. See Exception.

430.36 Where fuses are used for motor overload protection in a 3-wire, 3φ AC system with one wire grounded, a fuse shall be used in the grounded conductor.

430.37 Three motor overload protective devices shall be used to protect 3φ motors unless protected by other approved means. The number of overcurrent devices other than fuses is governed by Table 430.37.

MOTORS OVER 600 V

FAULT CURRENT PROTECTION

OVER 600 V

LOCKABLE DISCONNECT

CONTROLLER

A. **430.222** Controllers shall be marked with the maker's name, voltage, current, HP rating, and other necessary data and control voltage. See 430.8.

430.224 The ampacity of conductors shall not be less than the current at which the motor overload protective device is selected to trip.

B. **430.225(A)** Motors shall have overload protection and the motor, motor circuit conductors, and control apparatus shall have fault current protection. This protection shall be coordinated and automatically operated except a vital motor may use a supervised annunciator or alarm.

430.225(B)(1) Each motor shall be protected against overloads or failure to start by a thermal protector integral with the motor or external sensing devices or both. Device settings for each motor circuit shall be determined under engineering supervision.

430.225(B)(2) Secondary circuits of wound-rotor motors are considered to be protected by motor overcurrent protection.

430.225(B)(3) Operation of the overload device shall open all ungrounded conductors at the same time.

430.225(B)(4) Overload sensing devices shall not automatically restart the motor unless restarting is no hazard.

430.225(C)(1)(b) Fault current protection is required. It may be provided by suitable type and rated fuses that cannot be serviced while energized.

C. **430.225(C)(1)(a)** A CB is permitted for fault current protection. It may sense fault current by integral or external sensing elements.

D. **430.227** The disconnect for the controller shall be lockable in the open position.

CRANES AND HOISTS — 610

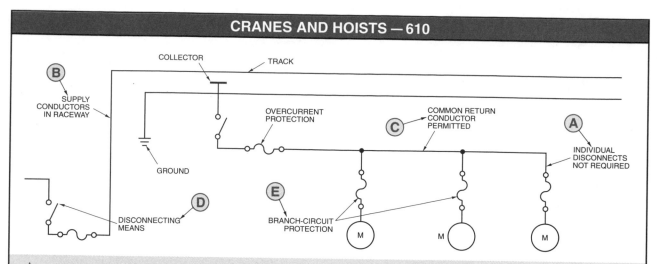

A. **430.112, Ex. a.** Individual disconnects are not required where a number of motors drive a single machine.

B. **610.11** Supply conductors shall be in raceways, AC cable with insulated EGC, MI, or MC cable, except: (a) bar contact conductors; (b) short lengths of exposed conductors at resistors, collectors, etc.; (c) flexible stranded wire installed in FMC, LFMC, LFNC, multiconductor cable, or approved PVC raceway; (d) multiconductor cable to pendant pushbutton station; or (e) flexibility for moving parts, listed suitable cable or cord is permitted.

610.14(A) Table 610.14(A) provides allowable ampacity for conductors.

610.14(D) Bare conductors No. 6 span 0'–30'; No. 4 span 30'–60'; No. 2 over 60'.

610.14(E)(2) The ampacity of supply conductors shall be not less than the sum of the short-time full-load current rating for the largest motor, or group, for a single crane motion plus 50% of that for the next larger motor or group.

C. **610.15** Where there is more than one motor, a common return conductor is permitted.

610.21(F) The grounded track may serve as a conductor if the supply is by way of an insulating transformer not over 300 V and other conductors are insulated. See Conditions (1 – 4).

D. **610.31** The disconnecting means shall be operable from the ground and in sight of the crane and runway contact conductors. The disconnect rating shall be per 610.14(E – F).

610.32 A lockable motor disconnect is required in leads from runway contact conductors or other power supply on all cranes and monorail hoists. Where the disconnect is not readily accessible from the operating station, means are required at the operating station to open the power circuit of the crane or monorail hoist.

610.33 The minimum rating of the disconnecting means shall be 50% of the combined short-time ampere ratings of all motors and 75% of the sum of the short-time ampere ratings of motors for single crane operation.

610.41(A) An overcurrent device is required for the main contact conductors.

E. **610.42(A)** Each motor shall have branch-circuit protection in accordance with Table 430.52.

610.42(B)(1 – 3) Two motors operating as a unit may have a single overload protector for taps to brake coils.

610.55 A limit switch or other device is required on a hoist unit.

610.61 Grounding of metal parts is required per Article 250.

ELEVATORS, DUMBWAITERS, ESCALATORS, AND MOVING WALKS — 620

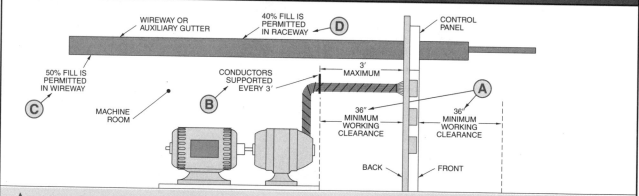

A. **620.5** There shall be sufficient, clear working space around control panels to permit safe and convenient access to all live parts for maintenance and adjustment. The minimum space shall be per 110.26(A). Where the minimum working space cannot be provided for escalator and moving-walk panels and the panel is in the same space as the drive machine, the clearances may be waived where the entire panel and disconnecting means are arranged to be readily removed from the machine space and has flexible leads to all external connections.

B. **620.21(A)(3)(d)** For elevator machine or control room wiring, conductors may be grouped and bound together without being installed in a raceway if supported every 3' and free from physical damage.

C. **620.32** Fifty percent fill is permitted in a wireway. Vertical runs require support at least every 15'.

D. **620.33** Forty percent fill is permitted in raceways.

620.35 The 30' length restriction of 366.12(2) and number of wires restriction of 366.22 do not apply to auxiliary gutters.

620.36 Conductor of power, control, lighting, and signal systems not over 600 V may be run in the same raceway or cable if insulated for the maximum voltage found in the cable or raceway system.

OPERATING REQUIREMENTS

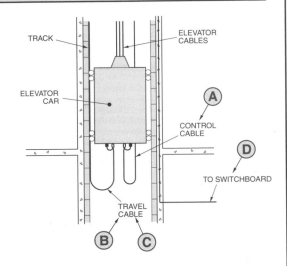

A. 620.13(A) The ampacity of conductors for a single motor shall conform to 430.22(A) and (E).

620.13(D) Conductors supplying several motors shall comply with the standard rule governing motor feeders unless a demand factor is allowed per 620.14.

620.21 In general, all elevator wiring shall be in RMC, IMC, EMT, RNC, wireways, or in Types MC, MI, or AC cable. See Applications (A – C).

B. 620.51 An identified disconnect is required for the power supply to each unit.

C. 620.51(A) The disconnect shall be an enclosed, externally operable, fused motor switch or a CB which can be locked in the open position.

620.51(C) The disconnect shall be readily accessible to qualified persons.

D. 620.52(B) For power from more than one source and multiple disconnects, a clearly legible warning sign shall read: "**WARNING: PARTS OF THE CONTROL-LER ARE NOT DE-ENERGIZED BY THIS SWITCH**." The sign shall be mounted on or adjacent to the disconnecting means.

620.61(B)(1) Elevators and dumbwaiters are protected on an intermittent-duty basis. Elevator motors shall be classified as intermittent-duty motors and protected against overload per 430.33.

620.61(B)(2) Escalator and moving-walk driving motors shall be classified as continuous-duty motors and shall be protected against overload per 430.32.

620.82 In general, operating equipment shall be bonded per Article 250.

620.91(C) Where emergency power is provided, a disconnect is required by 620.51. It shall disconnect both emergency and normal power.

VOLTAGE TRANSFORMERS

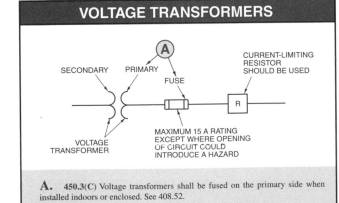

A. 450.3(C) Voltage transformers shall be fused on the primary side when installed indoors or enclosed. See 408.52.

TRAVELING CABLES AND CONTROL HOISTWAY

> **620.2: Control System**—The overall system governing the starting, stopping, motion direction, acceleration, speed, and retardation of the moving member.
>
> **620.2: Controller, Motor** —The operative units of the control system consisting of the starter device(s) and power conversion equipment used to drive an electric motor or the pump for hydraulic control equipment.
>
> **620.2: Controller, Operation**—The electric device(s) for the part of the control system that initiates starting, stopping, and direction of motion in response to a signal from an operating device.
>
> **620.2: Operating Device**—The car switch, pushbuttons, etc. used to activate the operation controller.

A. 620.3 Control and signal voltage limited to 300 V except for special limited-energy circuits. See Applications (A – C).

B. 620.11(B) Traveling cables used as flexible connections shall be of the types listed in Table 400.4 as elevator cable or be of other approved types.

C. 620.12(A)(1 – 2) Traveling cables: (1) for lighting circuits, minimum No. 14 or equivalent in parallel conductors No. 20 or larger; (2) for other circuits, minimum size shall be No. 20.

620.21(A)(1)(a) FMC, LFMC, and LFNC are permitted in the hoistway or in the car for limit switches, interlocks, doors, etc.

D. 620.37(A)(C) Only wiring directly related to elevator or related systems shall be permitted inside the hoistway and the machine room. Main power feeders supplying elevator or dumbwaiters shall be installed outside the hoistway unless necessary for elevators with driving machine motors located in the hoistway, on car, or on counterweight, or by special permission.

620.41 Traveling cables shall be suspended at the car and hoistway ends to reduce strain.

620.41(1 – 3) Traveling cables supported by: (1) steel supporting members; (2) looping cables around supports for unsupported lengths less than 100′; (3) means that automatically tighten around cable (for cables up to 200′) when tension is increased.

620.53 A single disconnecting means shall be provided for all ungrounded car light and accessory power supply conductors. See Exception.

620.81 Metal raceways and cable assemblies attached to elevators shall be bonded to the grounded metal frame of car that is bonded to the EGC.

SECONDARY TIES

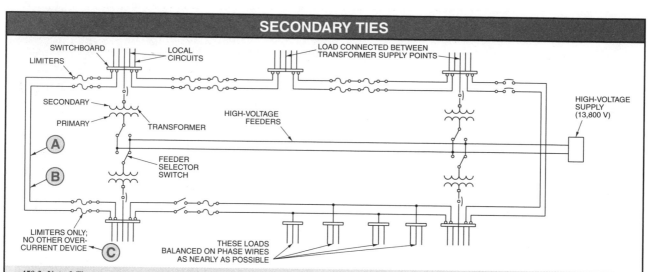

450.3, Note 1 The overcurrent protection of conductors is per 240.4, 240.21, 240.100, and 240.101.

A. **450.6** A secondary tie is a 600 V or less circuit between phases that connects two power sources or power supply points. As used in this section, the word "transformer" means one transformer or a bank of transformers operating as a unit.

450.6(A) The circuits shall be provided at each end with overcurrent protection per 240. See Applications (1), (2), (3), (4), and (5).

B. **450.6(A)(1)** The rated ampacity of the tie shall not be less than 67% of the rated secondary current of the largest transformer connected to the secondary tie system if overcurrent protection is not provided per 240.

450.6(A)(2) Where the load is connected to the tie between transformer supply points and overcurrent protection is not provided per 240, the rated ampacity of the tie shall not be less than 100% of the rated secondary current of the largest transformer connected to the secondary tie system.

C. **450.6(A)(3)** Limiters or CBs having comparable current characteristics may be used under conditions described in (A)(1) and (A)(2).

450.6(A)(4)(a) Ordinarily, corresponding phase wires shall be interconnected together at tap points.

450.6(A)(4)(b) Loads shall be connected to individual conductors and without limiters where the total ampacity of the ties is not less than 133% of the rated secondary current of the largest transformer, and the total load of the taps does not exceed this value.

450.6(A)(5) If the operating voltage is over 150 V to ground, a disconnect switch is required at each end.

450.6(B) For secondary ties, a CB set at not more than 250% of the rated secondary current of the transformers shall be provided. Additionally, an automatic CB shall be provided.

IMPEDANCE OF TRANSFORMERS

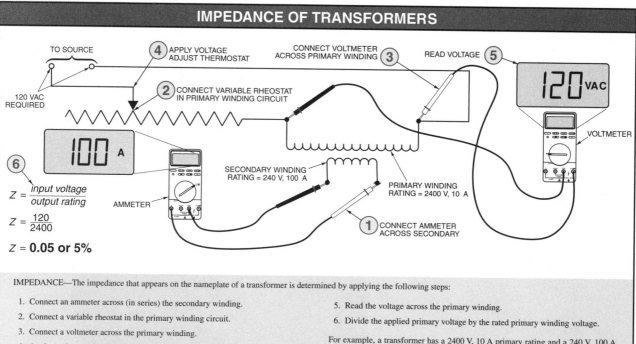

$$Z = \frac{input\ voltage}{output\ rating}$$

$$Z = \frac{120}{2400}$$

$$Z = \textbf{0.05 or 5\%}$$

IMPEDANCE—The impedance that appears on the nameplate of a transformer is determined by applying the following steps:

1. Connect an ammeter across (in series) the secondary winding.
2. Connect a variable rheostat in the primary winding circuit.
3. Connect a voltmeter across the primary winding.
4. Apply a voltage to the primary circuit. Adjust the rheostat to increase the voltage across the primary winding until the secondary winding current reaches its rated current.

5. Read the voltage across the primary winding.
6. Divide the applied primary voltage by the rated primary winding voltage.

For example, a transformer has a 2400 V, 10 A primary rating and a 240 V, 100 A secondary rating. A voltage of 120 V is needed across the primary winding for the ammeter in the secondary winding circuit to read 100 A. Dividing 120 by 2400 equals 0.05 or 5% impedance.

OVERCURRENT PROTECTION OF TRANSFORMERS (OVER 600 V) — 450.3(A)

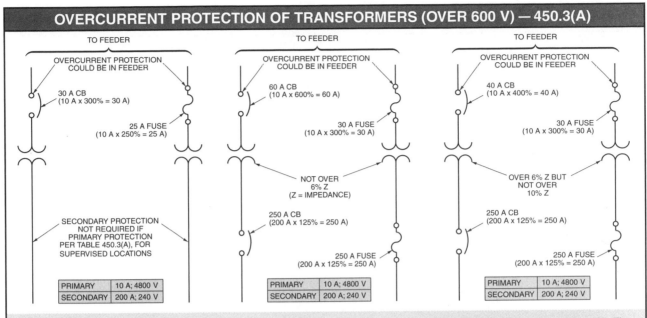

450.3(A) Transformers over 600 V shall have primary and secondary protection per Table 450.3(A). Where the fuse or CB rating exceeds a standard rating, the next higher rating is permitted. See Table 450.3(A), Note 1.

Table 450.3(A) Primary protection of 600% for CBs and 300% for fuses is permitted if not over 6% Z and 400% for CBs and 300% for fuses if over 6% but not over 10% Z. See Table 450.3(A) for secondary protection. See Table 450.3(A), Notes 1 – 5.

Supervised installations require individual protection on the primary side. The continuous rating of the primary fuse shall not be over 250% of the transformer and for CBs, the rating shall not be over 300%.

Table 450.3(A), Note 1 The next higher fuse or CB rating is permitted if the fuse or CB does not correspond to the next higher standard size.

OVERCURRENT PROTECTION OF TRANSFORMERS (600 V OR LESS) — 450.3(B)

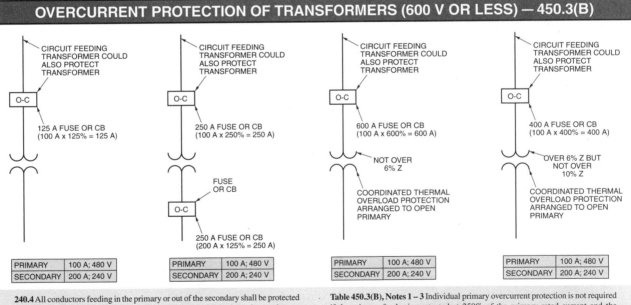

240.4 All conductors feeding in the primary or out of the secondary shall be protected at their rated ampacities. This protection shall be provided where the conductors receive their supply per 240.21.

Table 450.3(B) Each transformer of 600 V or less is required to have individual primary overcurrent protection. This overcurrent device is to be rated or set at not over 125% of the rated primary current except: (1) if the primary current is 9 A or more and 125% is not a standard size fuse or CB, then the next larger size device may be used. Where the primary is less than 2 A, the overcurrent device may be rated or set at not more than 300% of the rated primary amps; (2) the primary circuit protection may also protect transformer if of proper size; (3) if secondary is protected per Table 450.3(B).

Table 450.3(B), Notes 1 – 3 Individual primary overcurrent protection is not required if the primary feeder is protected at 250% of the primary rated current and the secondary is protected at not over 125% of the rated secondary current. Individual transformer primary protection is not required if the transformer has coordinated thermal overload protection by the manufacturer, is arranged to open the primary circuit, and if the primary overcurrent device is not over 6 times the rated primary current for transformers with not more than 6% Z and not more than 4 times the rated primary current for transformers having more than 6% but not more than 10% Z except: where the secondary current is 9 A or more and 125% is not a standard size fuse or CB, then the next larger device may be used. Where secondary is less than 9 A, the overcurrent device may be rated or set at not more than 167% of the rated primary current.

DRY-TYPE TRANSFORMERS — INSTALLED INDOORS

ABB Power T&D Company Inc.

Dry-type transformers may be installed indoors when they are rated at not more than 112 kVA and not more than 600 V, and the transformer is completely enclosed except for ventilating openings.

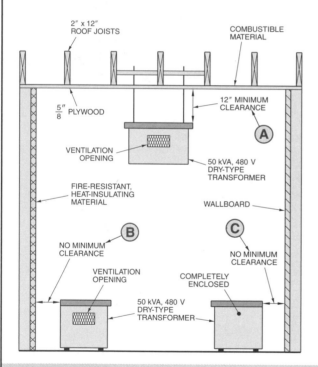

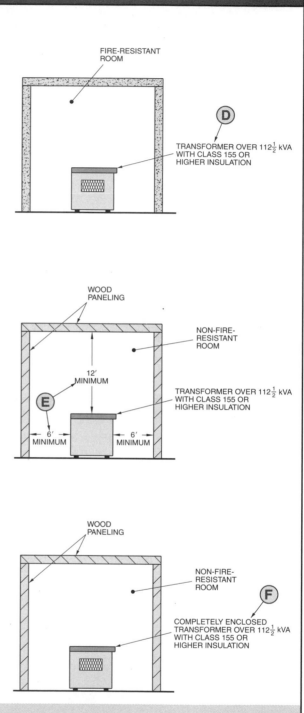

A. **450.21(A)** Dry-type transformers installed indoors and rated at 112½ kVA or less shall be a minimum of 12″ from combustible material.

B. **450.21(A)** The 12″ minimum is not required when transformers are separated from combustible material by fire-resistant, heat-insulating barriers.

C. **450.21(A), Ex.** The 12″ minimum is not required when the transformer is 600 V or less and completely enclosed, with or without ventilating openings.

D. **450.21(B)** Dry-type transformers over 112½ kVA shall be installed in a transformer room of fire-resistant construction.

E. **450.21(B), Ex. 1** Transformers with Class 155 or higher insulation are permitted to be installed in a non-fire-resistant room provided they are separated from combustible material by a minimum of 6′ horizontally and 12′ vertically.

F. **450.21(B), Ex. 2** Transformers with Class 155 or higher insulation and completely enclosed except for ventilation openings are permitted to be enclosed in a non-fire-resistant room.

TRANSFORMERS IN PARALLEL

OVERCURRENT PROTECTION

(A)

TRANSFORMERS CONNECTED IN PARALLEL

T1   T2

A. **450.7** Transformers may be connected in parallel and protected as a unit when their electrical characteristics are suitable, provided overcurrent protection for each transformer meets requirements of 450.3(A) or 450.3(B).

OIL-INSULATED TRANSFORMERS — INSULATED INDOORS AND OUTDOORS

OIL-INSULATED TRANSFORMER

450.26 (General Rule) A transformer vault is required for an oil-insulated transformer indoors.

450.26, Ex. 1 If the transformer is not over 112½ kVA, the thickness of the vault wall need be only 4″.

450.26, Ex. 2 A vault is not required if not over 600 V, not over 10 kVA in combustible area, or 75 kVA in fire-resistant area.

450.26, Ex. 3 Electric furnace transformers totaling not more than 75 kVA are permitted to be installed in a fire-resistive location if provisions are made to prevent the spread of an oil fire.

450.26, Ex. 5 A vault is not required if the transformer is in a suitable detached building accessible only to qualified persons.

450.27 Oil-insulated transformers installed outdoors shall be installed so combustible material, combustible buildings, parts of buildings, door and window openings, fire escapes, etc. are safeguarded from transformer oil fires when the transformer is installed, attached to, or adjacent to a building or combustible material, or on roofs.

AUTOTRANSFORMERS

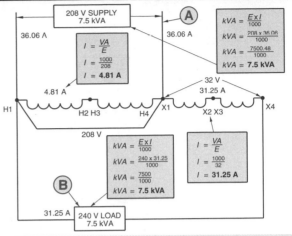

A. **210.9** Autotransformers are not permitted to supply branch circuits unless the grounded conductor is electrically connected to both supply and load.

B. **210.9, Ex. 1** Autotransformer is permitted to extend or add individual branch circuit to the existing installation for equipment load without an electrically continuous grounded conductor when transforming nominal 208 V to 240 V or 240 V to 208 V.

210.9, Ex. 2 Autotransformers in industrial occupancies are permitted to supply nominal 600 V loads from nominal 480 V systems and nominal 480 V loads from nominal 600 V systems, without connection to a similar grounded conductor, if installed by a qualified person.

ASKAREL OR NONFLAMMABLE TRANSFORMERS — INSTALLED INDOORS

ASKAREL-INSULATED TRANSFORMER INSTALLED INDOORS

PRESSURE-RELIEF VENT

(A)

450.24 Fluid-insulated transformers insulated with a dielectric fluid identified as nonflammable are permitted indoors and outdoors. When over 35,000 V and indoors, a vault is required. In this section a nonflammable dielectric fluid does not have a flash or fire point and is not flammable in air.

A. **450.25** Askarel-insulated transformers 25 kVA or over shall have pressure-relief vent. If more than 35,000 V, they shall be installed in an approved vault. Where askarel-insulated transformers are installed in a poorly ventilated area, means shall be provided to absorb any gas, or a chimney or flue shall be connected to carry such gases outside buildings.

HIGH-VOLTAGE INSTALLATION IN APPROVED TRANSFORMER VAULT

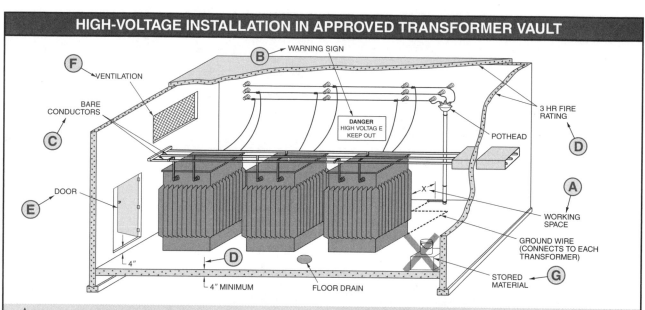

A. 110.34(E) Unguarded live parts are insulated by elevation; 9' from 601 volts to 7500 V. Minimum depth of working space to 9000 V is 4'. See Table 110.34(A).

230.202(B) For services exceeding 600 V, wiring methods shall be per 300.37 and 300.50.

B. 110.34(C) Warning signs that read "DANGER HIGH VOLTAGE KEEP OUT" shall be provided wherever unauthorized persons may come in contact with energized parts.

C. 300.37 Conductors above ground shall be in RMC, IMC, EMT, RTRC, PVC, cable trays, auxiliary gutters, busways, cablebus, MI cable, or other identified raceways. Bare conductors and busbars permitted where accessible only to qualified persons. Exposed runs of metal-clad cable suitable for the use and purpose are also permitted.

300.50 Underground conductors must be suitable for voltage and prevailing conditions and, if buried, per 310.10(F). Underground cables shall be installed per 300.50 (A)(1) or (A)(2). See Table 300.50 and Exceptions for other wiring methods and unusual conditions.

450.8(C) Energized parts shall be guarded per 110.27 and 110.34.

450.10 Exposed noncurrent-carrying metal parts of transformer installations shall be grounded and bonded per Article 250.

D. 450.42 Walls and roof of vaults shall be of fire-resistive construction with a minimum fire resistance of 3 hr. See Ex. for 1-hr rating. Concrete floors shall be at least 4" thick.

E. 450.43(A) Vault doors shall have a minimum fire rating of 3 hr. The door sill shall be at least 4" high. Vault doors shall be locked, accessible only to qualified persons, and out-swinging.

F. 450.45(C) Natural ventilation shall be no less than 3 sq in. per kVA. For under 50 kVA, the minimum shall be 1 sq ft.

450.47 Foreign pipes or ducts shall not pass through a transformer vault.

G. 450.48 Material shall not be stored in transformer vaults.

490.24 For minimum air separation between bare live parts, see Table 490.24. Equipment designed, manufactured, and tested in accordance with national standards is exempt.

SURGE ARRESTERS

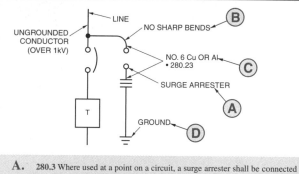

A. 280.3 Where used at a point on a circuit, a surge arrester shall be connected to each ungrounded conductor.

280.11 Surge arresters are permitted indoors or outdoors and accessible only to qualified persons unless listed for accessible locations.

B. 280.12 The conductor from the surge arrester to the line and ground shall be as short as possible and contain no unnecessary bends.

C. 280.21 The arrester shall be connected to one of the following: (1) grounded service conductor; (2) grounding electrode conductor; (3) service grounding electrode; or (4) equipment grounding bus in service equipment.

D. 280.25 Grounding electrode connections shall be per Article 250 except as given in Article 280.

PROTECTION REQUIREMENTS

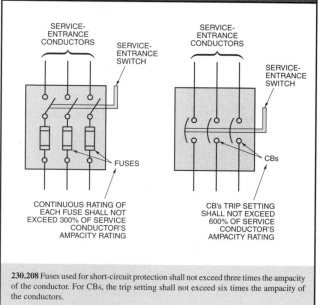

230.208 Fuses used for short-circuit protection shall not exceed three times the ampacity of the conductor. For CBs, the trip setting shall not exceed six times the ampacity of the conductors.

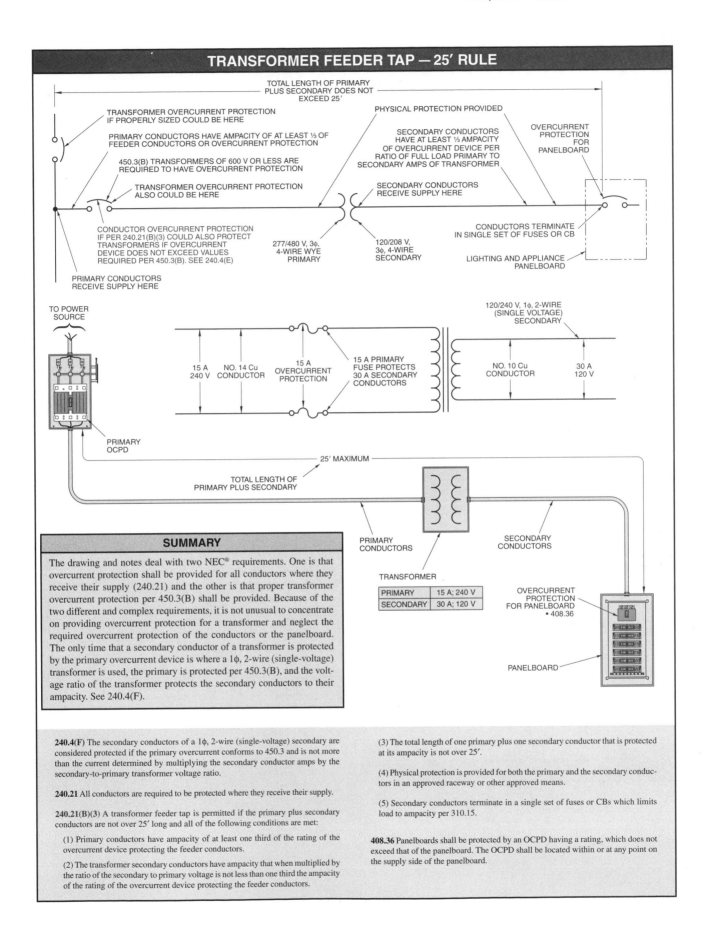

TRANSFORMER FEEDER TAP — 25' RULE

TOTAL LENGTH OF PRIMARY PLUS SECONDARY DOES NOT EXCEED 25'

TRANSFORMER OVERCURRENT PROTECTION IF PROPERLY SIZED COULD BE HERE

PHYSICAL PROTECTION PROVIDED

OVERCURRENT PROTECTION FOR PANELBOARD

PRIMARY CONDUCTORS HAVE AMPACITY OF AT LEAST ⅓ OF FEEDER CONDUCTORS OR OVERCURRENT PROTECTION

SECONDARY CONDUCTORS HAVE AT LEAST ⅓ AMPACITY OF OVERCURRENT DEVICE PER RATIO OF FULL LOAD PRIMARY TO SECONDARY AMPS OF TRANSFORMER

450.3(B) TRANSFORMERS OF 600 V OR LESS ARE REQUIRED TO HAVE OVERCURRENT PROTECTION

SECONDARY CONDUCTORS RECEIVE SUPPLY HERE

TRANSFORMER OVERCURRENT PROTECTION ALSO COULD BE HERE

CONDUCTOR OVERCURRENT PROTECTION IF PER 240.21(B)(3) COULD ALSO PROTECT TRANSFORMERS IF OVERCURRENT DEVICE DOES NOT EXCEED VALUES REQUIRED PER 450.3(B). SEE 240.4(E)

277/480 V, 3ɸ, 4-WIRE WYE PRIMARY

120/208 V, 3ɸ, 4-WIRE SECONDARY

CONDUCTORS TERMINATE IN SINGLE SET OF FUSES OR CB

LIGHTING AND APPLIANCE PANELBOARD

PRIMARY CONDUCTORS RECEIVE SUPPLY HERE

TO POWER SOURCE

120/240 V, 1ɸ, 2-WIRE (SINGLE VOLTAGE) SECONDARY

15 A 240 V

NO. 14 Cu CONDUCTOR

15 A OVERCURRENT PROTECTION

15 A PRIMARY FUSE PROTECTS 30 A SECONDARY CONDUCTORS

NO. 10 Cu CONDUCTOR

30 A 120 V

PRIMARY OCPD

25' MAXIMUM

TOTAL LENGTH OF PRIMARY PLUS SECONDARY

PRIMARY CONDUCTORS

SECONDARY CONDUCTORS

TRANSFORMER

PRIMARY	15 A; 240 V
SECONDARY	30 A; 120 V

OVERCURRENT PROTECTION FOR PANELBOARD
• 408.36

PANELBOARD

SUMMARY

The drawing and notes deal with two NEC® requirements. One is that overcurrent protection shall be provided for all conductors where they receive their supply (240.21) and the other is that proper transformer overcurrent protection per 450.3(B) shall be provided. Because of the two different and complex requirements, it is not unusual to concentrate on providing overcurrent protection for a transformer and neglect the required overcurrent protection of the conductors or the panelboard. The only time that a secondary conductor of a transformer is protected by the primary overcurrent device is where a 1ɸ, 2-wire (single-voltage) transformer is used, the primary is protected per 450.3(B), and the voltage ratio of the transformer protects the secondary conductors to their ampacity. See 240.4(F).

240.4(F) The secondary conductors of a 1ɸ, 2-wire (single-voltage) secondary are considered protected if the primary overcurrent conforms to 450.3 and is not more than the current determined by multiplying the secondary conductor amps by the secondary-to-primary transformer voltage ratio.

240.21 All conductors are required to be protected where they receive their supply.

240.21(B)(3) A transformer feeder tap is permitted if the primary plus secondary conductors are not over 25' long and all of the following conditions are met:

(1) Primary conductors have ampacity of at least one third of the rating of the overcurrent device protecting the feeder conductors.

(2) The transformer secondary conductors have ampacity that when multiplied by the ratio of the secondary to primary voltage is not less than one third the ampacity of the rating of the overcurrent device protecting the feeder conductors.

(3) The total length of one primary plus one secondary conductor that is protected at its ampacity is not over 25'.

(4) Physical protection is provided for both the primary and the secondary conductors in an approved raceway or other approved means.

(5) Secondary conductors terminate in a single set of fuses or CBs which limits load to ampacity per 310.15.

408.36 Panelboards shall be protected by an OCPD having a rating, which does not exceed that of the panelboard. The OCPD shall be located within or at any point on the supply side of the panelboard.

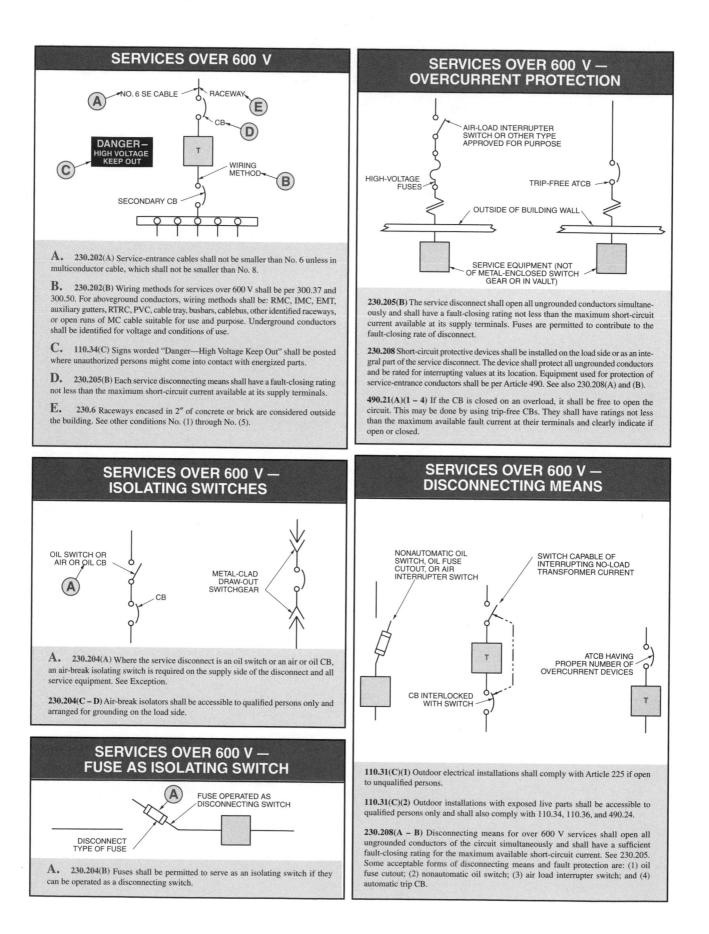

SERVICES OVER 600 V

A. 230.202(A) Service-entrance cables shall not be smaller than No. 6 unless in multiconductor cable, which shall not be smaller than No. 8.

B. 230.202(B) Wiring methods for services over 600 V shall be per 300.37 and 300.50. For aboveground conductors, wiring methods shall be: RMC, IMC, EMT, auxiliary gutters, RTRC, PVC, cable tray, busbars, cablebus, other identified raceways, or open runs of MC cable suitable for use and purpose. Underground conductors shall be identified for voltage and conditions of use.

C. 110.34(C) Signs worded "Danger—High Voltage Keep Out" shall be posted where unauthorized persons might come into contact with energized parts.

D. 230.205(B) Each service disconnecting means shall have a fault-closing rating not less than the maximum short-circuit current available at its supply terminals.

E. 230.6 Raceways encased in 2″ of concrete or brick are considered outside the building. See other conditions No. (1) through No. (5).

SERVICES OVER 600 V — OVERCURRENT PROTECTION

230.205(B) The service disconnect shall open all ungrounded conductors simultaneously and shall have a fault-closing rating not less than the maximum short-circuit current available at its supply terminals. Fuses are permitted to contribute to the fault-closing rate of disconnect.

230.208 Short-circuit protective devices shall be installed on the load side or as an integral part of the service disconnect. The device shall protect all ungrounded conductors and be rated for interrupting values at its location. Equipment used for protection of service-entrance conductors shall be per Article 490. See also 230.208(A) and (B).

490.21(A)(1 – 4) If the CB is closed on an overload, it shall be free to open the circuit. This may be done by using trip-free CBs. They shall have ratings not less than the maximum available fault current at their terminals and clearly indicate if open or closed.

SERVICES OVER 600 V — ISOLATING SWITCHES

A. 230.204(A) Where the service disconnect is an oil switch or an air or oil CB, an air-break isolating switch is required on the supply side of the disconnect and all service equipment. See Exception.

230.204(C – D) Air-break isolators shall be accessible to qualified persons only and arranged for grounding on the load side.

SERVICES OVER 600 V — FUSE AS ISOLATING SWITCH

A. 230.204(B) Fuses shall be permitted to serve as an isolating switch if they can be operated as a disconnecting switch.

SERVICES OVER 600 V — DISCONNECTING MEANS

110.31(C)(1) Outdoor electrical installations shall comply with Article 225 if open to unqualified persons.

110.31(C)(2) Outdoor installations with exposed live parts shall be accessible to qualified persons only and shall also comply with 110.34, 110.36, and 490.24.

230.208(A – B) Disconnecting means for over 600 V services shall open all ungrounded conductors of the circuit simultaneously and shall have a sufficient fault-closing rating for the maximum available short-circuit current. See 230.205. Some acceptable forms of disconnecting means and fault protection are: (1) oil fuse cutout; (2) nonautomatic oil switch; (3) air load interrupter switch; and (4) automatic trip CB.

NUMBER OF OVERCURRENT DEVICES TO PROTECT CIRCUIT CONDUCTORS — 600 V OR LESS

LOAD
OVERCURRENT DEVICES
NEUTRAL CONDUCTOR
SUPPLY – 1φ, 3-WIRE

LOADS
NEUTRAL CONDUCTOR
SUPPLY – 3φ, 3-WIRE (UNGROUNDED)

SUPPLY – 3φ, 4-WIRE

240.15(A) For circuits of 600 V or less, overcurrent protection is required in series with each ungrounded conductor.

240.15(B)(1 – 4) For circuits of 600 V or less, CBs are required to open all conductors of the circuit. Single-pole breakers with identified handle ties are permitted for 3-wire DC or 1φ circuits, and lighting and appliance branch circuits connected to 4-wire, 3φ systems or 5-wire, 2φ systems if the circuits are supplied from a system having a grounded neutral, the voltage does not exceed that permitted by 210.6 and for 3-wire DC circuits.

240.100(A) For circuits over 600 V, overcurrent protection is required for each ungrounded conductor.

240.100(A)(1) (General Rule) On circuits over 600 V, where CBs are used for AC 3φ circuits, they shall have at least 3 overcurrent relays operated from 3 current transformers.

240.100(A)(2) For circuits over 600 V, where fuses are used for overcurrent protection, a fuse is required in each ungrounded conductor.

CAPACITORS — MOTOR CIRCUITS

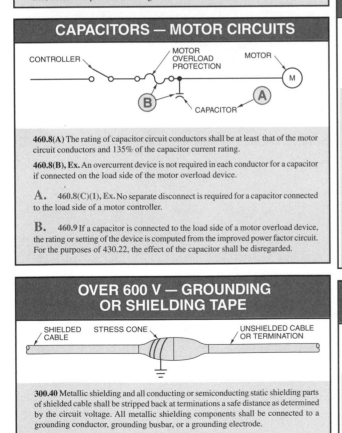

CONTROLLER
MOTOR OVERLOAD PROTECTION
MOTOR
M
B
A
CAPACITOR

460.8(A) The rating of capacitor circuit conductors shall be at least that of the motor circuit conductors and 135% of the capacitor current rating.

460.8(B), Ex. An overcurrent device is not required in each conductor for a capacitor if connected on the load side of the motor overload device.

A. **460.8(C)(1), Ex.** No separate disconnect is required for a capacitor connected to the load side of a motor controller.

B. **460.9** If a capacitor is connected to the load side of a motor overload device, the rating or setting of the device is computed from the improved power factor circuit. For the purposes of 430.22, the effect of the capacitor shall be disregarded.

OVER 600 V — GROUNDING OR SHIELDING TAPE

SHIELDED CABLE
STRESS CONE
UNSHIELDED CABLE OR TERMINATION

300.40 Metallic shielding and all conducting or semiconducting static shielding parts of shielded cable shall be stripped back at terminations a safe distance as determined by the circuit voltage. All metallic shielding components shall be connected to a grounding conductor, grounding busbar, or a grounding electrode.

CAPACITORS — OVER 600 V

3φ LINE, OVER 600 V
3φ OVERCURRENT DEVICE
3φ CAPACITOR BANK
3φ DELTA-CONNECTED CAPACITOR BANK OVER 600 V

460.2(A) Vaults or outdoor enclosures per Article 110, Part III are required for capacitors containing over 3 gal. of flammable liquid. This limit shall apply to any single unit in an installation of capacitors.

460.24(A)(1 – 4) A group-operated switch is rated to carry continuously at least 135% of the rated capacitor bank current, rated to interrupt load of bank, rated for minimum inrush currents, and rated to carry capacitor fault currents.

460.25(A – B) Overcurrent protection is required to detect and interrupt dangerous fault current. A 3φ device is permitted to protect a 3φ bank.

460.26 The capacitor's nameplate shall list the maker's name, rated voltage, kvar or A, frequency, phases, and type and amount of liquid.

460.27 If grounded, the capacitor enclosure and neutrals shall comply with Article 250, and be grounded to an EGC, except if on a structure operating at other than ground potential.

460.28(A – B) A discharge means is required to reduce the residual voltage to 50 V or less in 5 min or less. The discharge circuit shall be permanently connected or have an automatic connection after the capacitor bank is disconnected from the source.

CAPACITORS — 600 V OR LESS

CAPACITOR
A
B
C

460.2(B) Live parts shall be enclosed or guarded except when enclosed and accessible only to authorized and qualified personnel.

A. **460.8(A)** The ampacity of conductors shall be at least 135% of the capacitor current rating.

B. **460.8(B)** Overcurrent protection shall be set or rated as low as practicable.

C. **460.8(C)(1 – 3)** The disconnecting means shall open all ungrounded conductors simultaneously. Its continuous ampacity shall not be less than 135% of the rated capacity current.

460.10 Capacitor cases shall be connected to an EGC per Article 250, except if on a structure operating at other than ground potential.

460.28(A) For over 600 V, a discharge means is required to reduce the residual voltage to 50 V or less in 5 min or less.

OVER 600 V — BRAID-COVERED INSULATED CONDUCTORS, OPEN RUNS

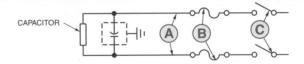

FLAME-RETARDANT BRAID
A
1"/kV
B

A. **300.39** Where braid-covered insulated conductors are used for open runs for circuits over 600 V, the braid shall be flame retardant.

B. **300.39** Braid shall be stripped back a safe distance and where practicable, shall not be less than 1" for each kilovolt to ground in circuit.

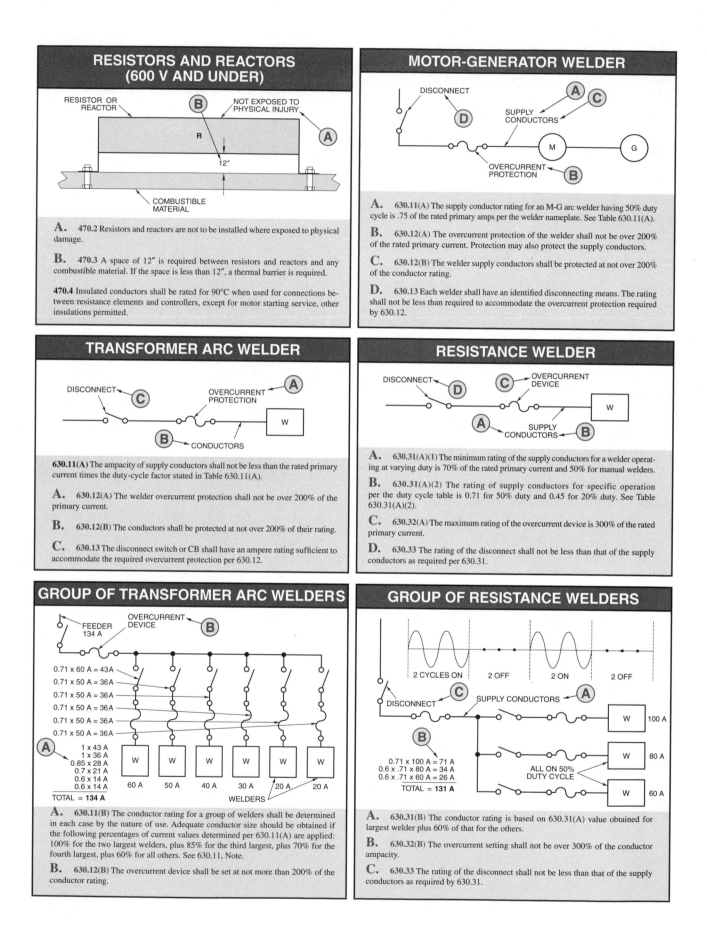

RESISTORS AND REACTORS (600 V AND UNDER)

A. 470.2 Resistors and reactors are not to be installed where exposed to physical damage.

B. 470.3 A space of 12″ is required between resistors and reactors and any combustible material. If the space is less than 12″, a thermal barrier is required.

470.4 Insulated conductors shall be rated for 90°C when used for connections between resistance elements and controllers, except for motor starting service, other insulations permitted.

MOTOR-GENERATOR WELDER

A. 630.11(A) The supply conductor rating for an M-G arc welder having 50% duty cycle is .75 of the rated primary amps per the welder nameplate. See Table 630.11(A).

B. 630.12(A) The overcurrent protection of the welder shall not be over 200% of the rated primary current. Protection may also protect the supply conductors.

C. 630.12(B) The welder supply conductors shall be protected at not over 200% of the conductor rating.

D. 630.13 Each welder shall have an identified disconnecting means. The rating shall be not less than required to accommodate the overcurrent protection required by 630.12.

TRANSFORMER ARC WELDER

630.11(A) The ampacity of supply conductors shall not be less than the rated primary current times the duty-cycle factor stated in Table 630.11(A).

A. 630.12(A) The welder overcurrent protection shall not be over 200% of the primary current.

B. 630.12(B) The conductors shall be protected at not over 200% of their rating.

C. 630.13 The disconnect switch or CB shall have an ampere rating sufficient to accommodate the required overcurrent protection per 630.12.

RESISTANCE WELDER

A. 630.31(A)(1) The minimum rating of the supply conductors for a welder operating at varying duty is 70% of the rated primary current and 50% for manual welders.

B. 630.31(A)(2) The rating of supply conductors for specific operation per the duty cycle table is 0.71 for 50% duty and 0.45 for 20% duty. See Table 630.31(A)(2).

C. 630.32(A) The maximum rating of the overcurrent device is 300% of the rated primary current.

D. 630.33 The rating of the disconnect shall not be less than that of the supply conductors as required per 630.31.

GROUP OF TRANSFORMER ARC WELDERS

0.71 x 60 A = 43A
0.71 x 50 A = 36A
0.71 x 50 A = 36A
0.71 x 50 A = 36A
0.71 x 50 A = 36A
0.71 x 50 A = 36A

1 x 43 A
1 x 36 A
0.85 x 28 A
0.7 x 21 A
0.6 x 14 A
0.6 x 14 A
TOTAL = 134 A

A. 630.11(B) The conductor rating for a group of welders shall be determined in each case by the nature of use. Adequate conductor size should be obtained if the following percentages of current values determined per 630.11(A) are applied: 100% for the two largest welders, plus 85% for the third largest, plus 70% for the fourth largest, plus 60% for all others. See 630.11, Note.

B. 630.12(B) The overcurrent device shall be set at not more than 200% of the conductor rating.

GROUP OF RESISTANCE WELDERS

0.71 x 100 A = 71 A
0.6 x .71 x 80 A = 34 A
0.6 x .71 x 60 A = 26 A
TOTAL = 131 A

A. 630.31(B) The conductor rating is based on 630.31(A) value obtained for largest welder plus 60% of that for the others.

B. 630.32(B) The overcurrent setting shall not be over 300% of the conductor ampacity.

C. 630.33 The rating of the disconnect shall not be less than that of the supply conductors as required by 630.31.

MOBILE HOMES — 550

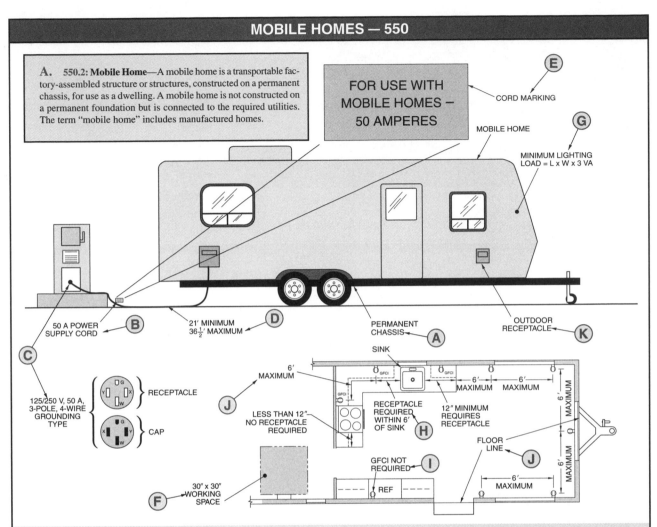

A. **550.2: Mobile Home**—A mobile home is a transportable factory-assembled structure or structures, constructed on a permanent chassis, for use as a dwelling. A mobile home is not constructed on a permanent foundation but is connected to the required utilities. The term "mobile home" includes manufactured homes.

FOR USE WITH MOBILE HOMES – 50 AMPERES

CORD MARKING — E

MOBILE HOME

MINIMUM LIGHTING LOAD = L x W x 3 VA — G

50 A POWER SUPPLY CORD — B

21' MINIMUM 36-½' MAXIMUM — D

PERMANENT CHASSIS — A

OUTDOOR RECEPTACLE — K

SINK

125/250 V, 50 A, 3-POLE, 4-WIRE GROUNDING TYPE — C

G Y X W RECEPTACLE

G X W Y CAP

6' MAXIMUM — J

LESS THAN 12" NO RECEPTACLE REQUIRED

RECEPTACLE REQUIRED WITHIN 6' OF SINK

6' MAXIMUM   6' MAXIMUM

12" MINIMUM REQUIRES RECEPTACLE — H

6' MAXIMUM

FLOOR LINE — J

30" x 30" WORKING SPACE — F

GFCI NOT REQUIRED — I

REF

6' MAXIMUM

550.4(A) Mobile homes not intended for use as dwelling units (e.g.: on-site construction office) are not required to have the number or capacity of circuits required by this article. All applicable provisions of Article 550 apply, however, when the service is 120/240 V, 3-wire.

B. 550.10(A) The power supply shall be a feeder assembly of not more than one listed 50 A power-supply cord or a permanently installed feeder.

550.10(A), Ex. 1 A 40 A listed power-supply cord is permitted for mobile homes factory-equipped with gas or oil-fired central heating equipment.

550.10(B) The power-supply cord shall be permanently connected to the distribution panelboard with adequate strain relief. The free end shall have an attachment plug cap. The listed cord shall have four conductors with the grounding conductor identified by a continuous green color or continuous green color with one or more yellow stripes.

C. 550.10(C) The attachment plug shall be a 125/250 V, 50 A, 3-pole, 4-wire, grounding type.

D. 550.10(D) The overall length of the power-supply cord shall be not less than 21' and not more than 36½'. The cord shall be not less than 20' in length from the face of the attachment plug cap to the point where the cord enters the mobile home.

E. 550.10(E) The power-supply cord shall be marked "For use with mobile homes – 40 amperes" or "For use with mobile homes – 50 amperes".

550.10(F) The point of entrance for the feeder assembly shall be in the exterior wall, floor, or roof.

550.10(G) The cord shall be protected at the point of entry by conduit, bushings, etc.

550.10(H) The attachment plug cap shall be protected against corrosion and mechanical damage.

550.10(I) For calculated loads over 50 A, the supply shall be by a mast weatherhead or either a metal or nonmetallic rigid raceway.

F. 550.11(A) A single disconnecting means is required inside a mobile home. The main CBs or fuses shall be clearly marked "Main". The neutral bar shall be insulated per 550.16(A). The distribution panelboard shall be located in an accessible location, but not in a bathroom or closet. A 30" wide – 30" deep clear working space shall be maintained in front of the panelboard from the floor to the top of the panelboard.

G. 550.12(A) The minimum lighting load shall be calculated by multiplying the area of the mobile home (using outside dimensions) by 3 VA per sq ft.

550.13(A) Except where supplying specific appliances, plug receptacles shall be the grounding type and installed per 406.4.

H. 550.13(B) Ex. All 120 V, 1φ, 15 A and 20 A receptacle outlets installed outdoors in compartments accessible from outside the unit and in bathrooms shall have GFCI protection. All receptacle outlets within 6' of a lavatory or sink or for serving countertops in kitchens shall have GFCI protection.

I. 550.13(B), Ex. The exceptions in 210.8(A) shall be permitted to be applied to mobile home GFCI requirements.

J. 550.13(D) Receptacles shall be installed in all rooms other than the bath, closet, and hallway areas. No point along the floor line shall be more than 6' from a receptacle outlet. Receptacles serving countertops shall be no more than 6' apart.

550.13(D)(1 – 9) Interrupted spaces at least 2' along the floor line and 12" for countertops shall be provided with receptacle outlets.

K. 550.13(D)(8) At least one receptacle outlet shall be located outdoors and be accessible at grade level, and be not more than 6' above grade.

RECREATIONAL VEHICLES

A. **551.2: Recreational Vehicle**—A vehicular type unit which is self-propelled or is mounted on or pulled by another vehicle and is primarily designed as temporary living quarters for camping, travel, or recreational use. Normally identified as Travel Trailer, Camping Trailer, Truck Trailer, Truck Camper, or Motor Home.

551.2: Travel Trailer—A vehicle that is mounted on wheels, has a trailer area less than 320 sq ft (excluding wardrobes, closets, cabinets, kitchen units, fixtures, etc.), is of such size and weight that a special highway use permit is not required, and is designed as temporary living quarters while camping or traveling.

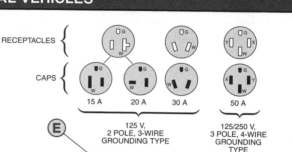

RECEPTACLES

CAPS

15 A   20 A   30 A   50 A

125 V, 2 POLE, 3-WIRE GROUNDING TYPE

125/250 V, 3 POLE, 4-WIRE GROUNDING TYPE

BRANCH CIRCUITS

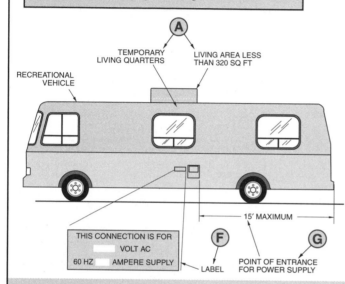

RECREATIONAL VEHICLE

TEMPORARY LIVING QUARTERS

LIVING AREA LESS THAN 320 SQ FT

15′ MAXIMUM

THIS CONNECTION IS FOR ___ VOLT AC
60 HZ ___ AMPERE SUPPLY

LABEL

POINT OF ENTRANCE FOR POWER SUPPLY

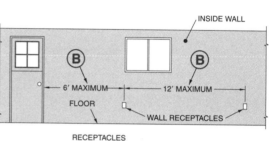

INSIDE WALL

6′ MAXIMUM   12′ MAXIMUM

FLOOR

WALL RECEPTACLES

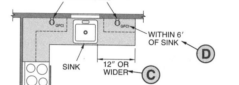

RECEPTACLES

SINK

WITHIN 6′ OF SINK

12″ OR WIDER

551.1 Article 551 covers the electrical conductors and equipment other than low-voltage and automotive vehicle circuits, installed in or on recreational vehicles, the conductors that provide the supply to RVs, and the equipment within an RV park.

551.40(A) The electrical equipment and material indicated for connection to a wiring system rated 120 V, 2-wire with ground, or 120/240 V, 3-wire with ground, shall be listed per Article 551, Parts I, III, IV, V and VI.

551.40(B) All electrical equipment and materials for the recreational vehicle shall be listed.

B. **551.41(A)** Receptacles shall be installed at wall spaces 2′ long or longer. No point along the floor line shall be more than 6′ from a receptacle. See Exceptions 1 and 2 for bath and hallway areas, occupied wall spaces, and wall spaces behind doors.

551.41(B) Receptacle outlets shall be installed in kitchens of recreational vehicles per 551.41(B)(1) through (3):

C. **551.41(B)(1)** Adjacent to countertops in kitchens with at least one on each side of the sink if there are 12″ or wider countertops on each side of the sink.

551.41(B)(2) Adjacent to the refrigerator and gas range space except if the appliances are gas-fired and require no electrical connection.

551.41(B)(3) Adjacent to countertop spaces 12″ or wider that cannot be reached by a 6′ cord without crossing a traffic area, sink, etc.

551.41(C) Each 125 V, 1ϕ, 15 A or 20 A receptacle outlet shall have GFCI protection per 551.41(C)(1) through (4):

551.41(C)(1) Adjacent to a bathroom lavatory.

D. **551.41(C)(2)** Where installed to serve countertops and within 6′ of any lavatory or sink. See Exceptions.

551.41(C)(3) In the area occupied by a toilet, shower, tub, or any combination of these.

551.41(C)(4) On the exterior of the vehicle except if located in the access panel. See Exception.

E. **551.42** Each recreational vehicle with a 120 V electrical system shall have circuit arrangements per 551.42(A) through (D):

551.42(A) A 15 A supply if one 15 A circuit.

551.42(B) A 20 A supply if one 20 A circuit.

551.42(C) A 30 A supply if two or more 15 A or 20 A circuits. See Exceptions.

551.42(D) A 50 A, 120/240 V supply if 6 or more circuits.

551.46(B) The power-supply cord shall be a minimum of 25′ when the point of entrance is on the side of the vehicle. It shall be a minimum of 30′ when the point of entrance is on the rear of the vehicle.

F. **551.46(D)** A label at least 1¾″ – 3″ shall be permanently affixed near the point of entrance. The label shall read, as appropriate, "THIS CONNECTION IS FOR 110 – 125 VOLT AC, 60 HZ _____ AMPERE SUPPLY." or "THIS CONNECTION IS FOR 208 Y/120 VOLT OR 120/240 VOLT AC, 3-POLE, 4 WIRE, 60 Hz _____ AMPERE SUPPLY."

G. **551.46(E)** The point of entrance for the power-supply assembly shall be located within 15′ of the rear, on the left (road) side, or at rear, and within 18″ of the outside wall. See Ex. 1, 2, and 3.

551.54(A) The grounding conductor in the power-supply cord shall be effectovely connected to the grounding bus or approved grounding means in the distribution panelboard.

551.55(A) All exposed metal parts, enclosures, frames, light fixture canopies, etc. shall be effectively bonded to grounding terminals or enclosure of the distribution panelboard.

X-RAY EQUIPMENT REQUIREMENTS

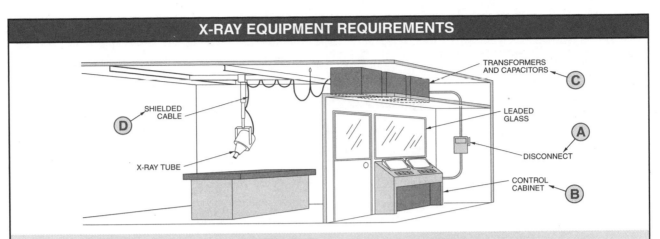

660.4(A) Hard-service cord and plug may be used to supply fixed and stationary unit not over 30 A. Over 30 A requires wiring which meets general NEC® requirements.

660.4(B) Individual branch circuits are not required for portable units with not over 60 A rating.

A. 660.5 A switch or CB at least 50% of the capacity of the momentary rating or 100% of the long time rating, whichever is greater, shall be readily accessible from the X-ray control station. Where 120 V, 30 A or less, a plug and receptacle may serve as the disconnect.

660.6(A) The ampacity of conductors and overcurrent devices shall be 50% of the momentary rating or 100% of the long time rating, whichever is greater.

B. 660.10 All new, reconditioned, or used equipment moved to a new location shall be of the approved type.

660.21 A manually-controlled device shall be located in or on equipment required for portable X-rays and shall comply with 660.20.

660.24 Where two or more pieces of equipment use the same high-voltage devices, each shall have a high-voltage switch or equivalent disconnecting means.

C. 660.35 Transformers and capacitors (part of X-ray) need not conform to Article 450 and Article 460.

660.36 Enclosures of grounded metal or insulating material shall be provided for capacitors.

D. 660.47(A) The connection from high-voltage equipment to X-ray tubes shall be made with high-voltage shielded cables.

660.48 Non-current-carrying metal parts shall be grounded. Portable equipment shall have an approved grounding-type plug unless battery-operated equipment is used.

MOTOR-GENERATOR EQUIPMENT

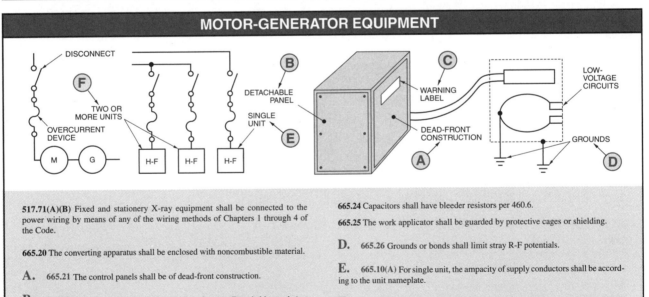

517.71(A)(B) Fixed and stationery X-ray equipment shall be connected to the power wiring by means of any of the wiring methods of Chapters 1 through 4 of the Code.

665.20 The converting apparatus shall be enclosed with noncombustible material.

A. 665.21 The control panels shall be of dead-front construction.

B. 665.22 Doors or panels are required for internal access. Detachable panels (not normally used for access) shall be difficult to remove. Doors and detachable panels giving access to more than 1000 V shall be interlocked with a circuit-interrupting device.

C. 665.23 Warning labels where necessary shall read: "DANGER—HIGH VOLTAGE—KEEP OUT".

665.24 Capacitors shall have bleeder resistors per 460.6.

665.25 The work applicator shall be guarded by protective cages or shielding.

D. 665.26 Grounds or bonds shall limit stray R-F potentials.

E. 665.10(A) For single unit, the ampacity of supply conductors shall be according to the unit nameplate.

F. 665.10(A) For two or more pieces of equipment, the feeder ampacity shall not be less than the sum of the nameplate currents, except where simultaneous operation is not possible.

665.7(A) "Local-Remote" switches shall be interlocked where multiple control points are used.

665.7(B) Foot switches shall be shielded from accidental contact.

SWIMMING POOLS — 680 . . .

680.2: Dry-Niche Lighting Fixture (Luminaire)—A lighting fixture installed in the floor or wall of a spa, pool, or fountain in a niche sealed against water entry.

Forming Shell—A structure mounted in a pool and designed to support a wet-niche lighting fixture assembly.

Hydromassage Bathtub—A permanently installed bathtub with a recirculating pump designed to accept, circulate, and discharge water for each use.

No-Niche Lighting Fixture (Luminare)—A lighting fixture installed above or below the water without a niche.

A. Permanently Installed Swimming Pool—A pool constructed in ground or partially above ground and designed to hold over 42″ of water, and all indoor pools regardless of depth.

Spa (Hot Tub)—An indoor or outdoor hydromassage pool or tub that is not designed to have the water discharged after each use.

Storable Swimming Pool—A pool constructed on or above ground and designed to hold less than 42″ of water, or a pool with nonmetallic, molded polymeric walls or inflatable fabric walls regardless of size.

Wet-Niche Lighting Fixture (Luminare)—A lighting fixture installed in a forming shell and completely surrounded by water.

680.1 All electric equipment installed in, on, or associated with pools, fountains, spas, hot tubs, and similar installations shall conform to Article 680.

680.23(A)(2) Transformers and power supplies that supply underwater fixtures shall be listed and shall be the two-winding type, with a grounded metal barrier between the primary and secondary windings or one that uses an approved system of double insulation between the windings.

B.   680.22(A)(3) At least one general use receptacle shall be located at least 6′ from the inside pool walls in dwelling units and not more than 20′.

C.   680.22(A)(3) Where a pool is installed for a dwelling, at least one receptacle is required between 6′ and 20′ from the pool.

D.   680.22(A)(4) All 15 A, 125 V and 20 A, 125 V receptacles located within 20′ of the pool shall have GFCI protection.

E.   680.22(B)(1) Lighting outlets, fixtures, and ceiling fans shall not be installed over the pool or over an area 5′ horizontally from the inside wall of the pool, unless the fixture is at least 12′ above the maximum water level.

F.   680.22(B)(3) Existing lighting fixtures and lighting outlets located less than 5′ horizontally from the inside wall of a pool and at least 5′ above the maximum water level shall be rigidly mounted and GFCI-protected.

680.22(B)(4) Lighting fixtures and lighting outlets installed between 5′ and 10′ horizontally from the inside wall of a pool shall be GFCI-protected unless at least 5′ above the maximum water level and rigidly mounted.

G.   680.22(B)(5) For cord-connected lighting fixtures installed within 16′ of the nearest water surface, the flexible cord shall be no more than 3′ in length, have a Cu EGC, and be no smaller than No. 12 with a grounding-type attachment plug.

H.   680.8 Overhead conductors shall meet minimum clearance requirements when routed over any of the following: 1. swimming or wading pools, including the area extending 10′ horizontally from the walls of the pool; 2. diving structures; or 3. towers, platforms, or observation stands.

I.   680.8(A) Table 680.8 lists overhead conductor clearances from the maximum water level.

J.   680.23(A)(3) Underwater fixtures, during normal use (not relamping), shall be free from shock hazard by design. If operating at more than the low-voltage contact limit, GFCI protection is required to eliminate shock hazard during relamping.

K.   680.23(A)(4) No underwater lighting fixture shall operate at over 150 V between conductors.

L.   680.23(A)(5) Lighting fixtures shall be installed in pool walls so that the top of the lens is at least 18″ below the normal water level. Lighting fixtures identified for the purpose are permitted 4″ below the normal water level.

M.   680.23(B)(1 – 2) Wet-niche fixtures shall have approved forming shells with threaded conduit entries. Metal conduit for wet-niche fixtures shall be of brass or other corrosion-resistant metal. Conduit shall be RMC, IMC, LFNC, or RNC. Where PVC is used, a No. 8 insulated Cu wire is required in the PVC with provisions to terminate in the forming shell, a junction box or transformer enclosure, or a GFCI enclosure. Termination of the No. 8 conductor in the forming shell be protected by covering or encapsulating with listed potting compound.

680.23(B)(4) Where flexible cord is used, the end of the cord jacket and wire terminals in the forming shell shall be similarly protected with a suitable potting compound.

680.23(B)(5) Noncurrent parts of wet-niche fixtures shall be bonded to the forming shell by a positive locking device. A tool shall be required to remove the fixture from the forming shell.

680.24(A)(1)(1 – 3) Junction boxes connected to conduit that extends to a forming shell shall be listed as a swimming pool JB.

680.24(A)(1, 3) Junction boxes connected to conduit that extends to a forming shell shall provide continuity by means of copper, brass, or other approved corrosion-resistant metal that is integral with the box.

680.24(A)(2)(a – b) Junction boxes connected to conduit that extends to a forming shell shall be not less than 4′ from the inside wall of the pool, unless separated by barrier and not less than 4″ from the inside bottom of the box above grade level or 8″ above the maximum water level, whichever is greater.

680.24(D) Junction boxes, transformers, power-supply enclosures, and GFCI enclosures shall have one more grounding terminal than conduit entries.

N.   680.26(B)(1 – 7) The following parts shall be bonded: All metal parts of electric pool water handling equipment and other metallic elements within or attached to pool structure, including reinforcing rods, coping stones, deck, piping systems, ladders, rails, diving board platforms, drains, skimmers, refill pipes, forming shells and all other metal piping or fixed metal parts within 5′ of the inside walls of the pool or within 12′ above the maximum water level of the pool or any observation stands, towers, or platforms (unless separated from pool by permanent barrier).

680.26(B)(1)(a) Steel tie wires are suitable for bonding the rebar.

680.26(B)(5) Isolated metal parts, no more than 4″ in any dimension and which do not penetrate into the pool structure more than 1″ do not require bonding.

680.26(B)(1) Poured concrete and concrete block are considered conductive materials due to water permeability and porosity.

680.26(B)(6) Ex. Metal parts of listed double-insulated equipment do not require bonding.

680.26(B) The parts required to be bonded by 680.26(B) (1) through (7) shall be connected to a common bonding grid with not less than No. 8 solid Cu wire, insulated, covered or bare, and connected by pressure connectors or clamps of brass, copper, or copper alloy. The common bonding grid is permitted to be: (1) reinforcing steel with usual tie wires; (2) the wall of a bolted or welded metal pool; or (3) No. 8 or larger solid Cu wire, insulated, covered or bare.

680.26(B)(6)(b) Pool heaters over 50 A shall be grounded and bonded per specific instructions.

680.6 The following electrical equipment shall be grounded: (1) through-wall and underwater lighting fixtures (2) electrical equipment within 5′ of the pool, (3) all associated recirculating equipment, (4) junction boxes, (5) transformer and power-supply enclosures, (6) GFCIs, and (7) panelboards that are not part of the service equipment.

680.23(F)(2) Wet-, dry-, or no-niche fixtures shall be connected to an insulated, Cu EGC sized per Table 250.122, but no smaller than No. 12. The insulated EGC shall be run in RMC, IMC, or RNC.

680.23(B)(3) Flexible cords to wet-niche lighting fixtures shall contain an insulated, Cu EGC not smaller than No. 16.

680.6(7) Panelboards that are not part of the service equipment and that supply pool equipment shall be grounded.

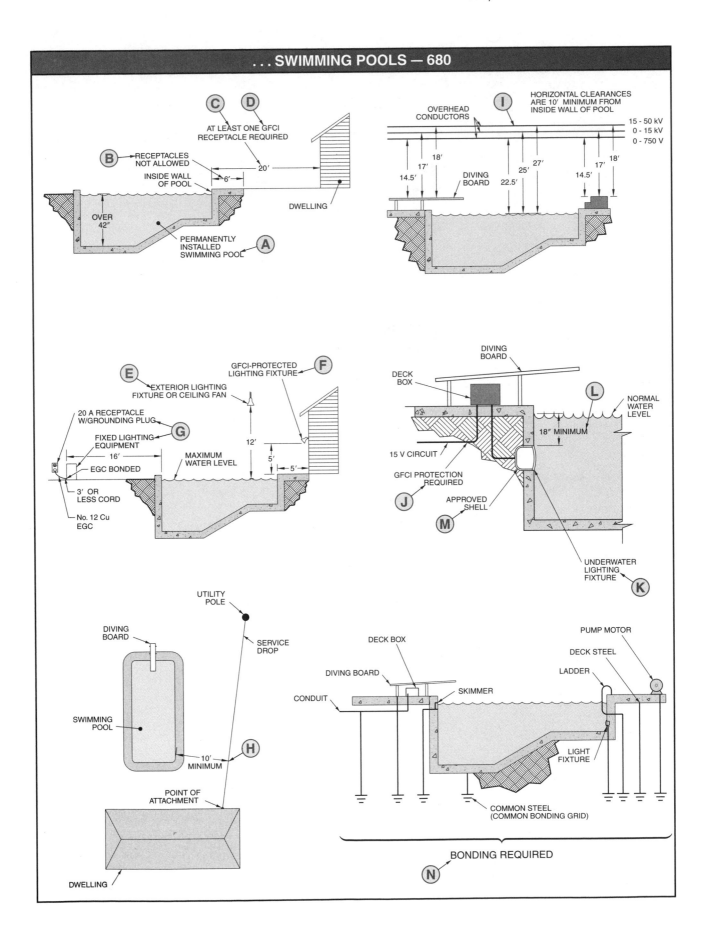

OFFICE FURNISHINGS — 605

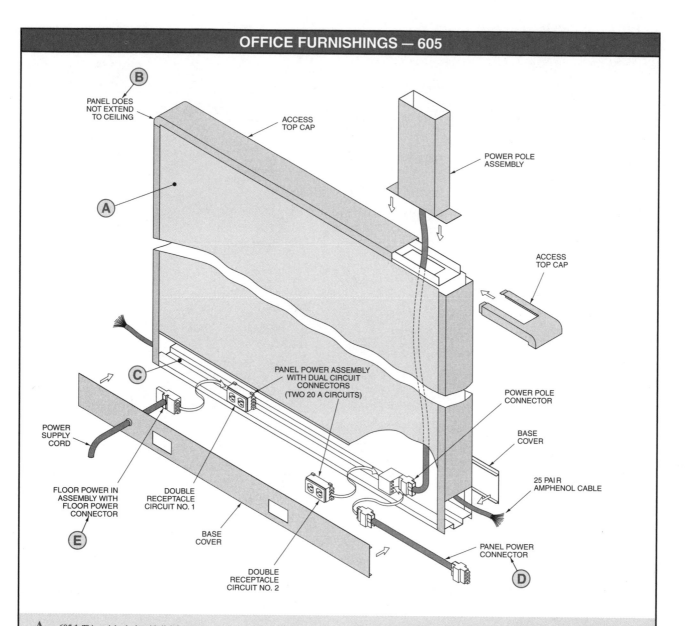

A. **605.1** This article deals with lighting accessories, electrical equipment, and wiring systems that are contained within or installed on relocatable wired partitions.

B. **605.2** Partitions are not permitted to extend from floor to ceiling. Wiring systems providing power for appliances and lighting accessories in wired partitions shall be identified for such use.

605.2, Ex. The AHJ can permit relocatable wired partitions to extend to the ceiling but not penetrate the ceiling.

C. **605.3** Wiring channels in partitions containing conductors and connections shall be free of projections that could damage conductor insulation.

D. **605.4** A flexible assembly, identified for use with wired partitions, shall be used for the electrical connection between partitions.

605.4(1 – 4) Flexible cord is permitted to interconnect partitions when partitions are mechanically contiguous. Cord is extra-hard usage type with 12 AWG or larger conductors and with an insulated EGC. Cord is as short as possible, but no longer than 2′, and cord terminates in an attachment plug and cord-connector with strain relief.

605.5 Lighting equipment shall be listed and identified for use with wired partitions and shall: (A) have adequate support; (B) where cord-and-plug connection shall not exceed 9′, be not smaller than No. 18, contain an EGC, and shall be of the hard usage type; and (C) lighting accessories shall not contain receptacle outlets.

605.6 Partitions secured to the building shall be permanently wired by wiring methods in Chapter 3. Multiwire branch circuits supplying power to a partition shall be provided with a means to disconnect simultaneously all ungrounded conductors.

605.7 Freestanding partitions shall be permitted to be permanently connected to the building electrical system by wiring methods in Chapter 3.

E. **605.8** Free standing individual partitions that are not over 30′ long and are mechanically contiguous are permitted to be cord-and-plug connected providing: (A) cord is at least No. 12 AWG with insulated grounding conductor of extra-hard usage type and not over 2′ long; (B) permanent building wiring connection is permitted by cord if the receptacle is on a separate circuit and located not more than 12″ from the partition; (C) not more than thirteen 15 A, 125 V receptacles are permitted in individual or group of interconnected partitions; and (D) multiwire circuits not permitted.

605.8(A) The flexible power supply cord shall be a No. 12 or larger extra-hard usage cord with insulated grounding conductor not over 2′ long.

Name_____ Date _____

NEC®	Answer	

_____   T   F   **1.** The secondary circuits of 460 V wound-rotor induction motors are permitted to be protected against overload by the motor-overload device.

_____   T   F   **2.** The primary windings of voltage transformers installed indoors shall be protected with fuses.

_____   _____   **3.** The conductors in a pull box having any dimension over 6′ shall be cabled or ___ in an approved manner.

_____   T   F   **4.** A transformer is a polyphase or single transformer, identified by a single nameplate.

_____   _____   **5.** Less-flammable, liquid-insulated transformers installed indoors shall be installed in vaults if rated over ___ V.
     A. 5000   C. 25,000
     B. 15,000   D. 35,000

_____   _____   **6.** In Article 430, the term "in sight" refers to a distance of not more than ___′.

_____   _____   **7.** In general, the motor disconnecting means shall disconnect both the motor and the ___.

_____   _____   **8.** Enclosures for CBs, switches, and fuses that are located in wet locations in agricultural buildings shall be listed and ___.

_____   _____   **9.** Except for polyphase wound-rotor units, AC motors ___ HP and larger shall have code letters or locked-rotor amps marked on their nameplates.

_____   T   F   **10.** The effect of capacitors used with motors shall be disregarded in determining the motor circuit conductor rating in 430.22.

_____   _____   **11.** The ampacity of supply conductors for a transformer arc welder with a 100% duty cycle and 30 A primary current is ___ A.
     A. 30   C. 60
     B. 37.5   D. 90

_____   T   F   **12.** A ¾ HP, 115 V motor on an automatic air compressor does not need overload protection.

_____   _____   **13.** Separate running overcurrent protection for continuous-duty motors more than 1 HP shall be based on the FLC, which is listed on the ___ of the motor.

_____   _____   **14.** Wired office partitions may extend from floor to ___ with permission of the AHJ.

_____   T   F   **15.** The rated load current of a sealed (hermetic) motor-compressor shall be marked on either the equipment or the nameplate.

_____ _____ **16.** The controller may be an attachment plug and receptacle or cord connector for portable motors rated at ___ HP or less.

_____ _____ **17.** On an ungrounded circuit, ___ thermal overcurrent units are needed to protect a 3φ induction motor from overload.

_____ _____ **18.** The running overcurrent protection for a DC motor with a 1.5 service factor and a 40°C temperature rise is ___%.

_____ _____ **19.** A suitable plug and hard-usage cord can supply a fixed or stationary nonmedical X-ray unit if the branch circuit is rated at no more than ___ A.

_____ _____ **20.** The maximum size of the overload relay used to protect a 50 A, 230 V, 40°C motor is ___ A when starting current is a problem.

      A. 50   C. 70

      B. 75   D. 100

_____ _____ **21.** Fuses used as motor overload protection devices shall be inserted in each ___ conductor.

_____ _____ **22.** The conductors of mineral-insulated, metal-sheathed cable (Type MI) are made of solid ___ nickel or nickel-clad copper.

_____ _____ **23.** FMC used in a manufactured wiring system shall contain at least a No. ___ AWG Cu grounding conductor.

_____ _____ **24.** The minimum clearance from the 0 V to 750 V insulated service conductors with grounded messenger wire, to the high-water line of a swimming pool is ___ ′.

_____ _____ **25.** Exposed live parts of motors and controllers operating at ___ V or more between terminals shall be guarded against accidental contact.

_____ _____ **26.** A general-use snap switch suitable only for use on AC may be used to disconnect a 2 HP, 240 V motor if the FLC rating of the motor does not exceed ___% of the ampere rating of the switch.

_____ _____ **27.** A 3 HP, 240 V, 3φ squirrel-cage motor with no code letter is started at full voltage. The size of a TW branch-circuit conductor shall be not less than No. ___.

      A. 8   C. 12

      B. 10   D. 14

_____ _____ **28.** Unless otherwise designed and marked, busways shall be supported at least every ___ ′.

_____ _____ **29.** The minimum length of a pull box for a straight-through pull of three No. 4/0 conductors in 2½″ conduit is ___ ″.

_____ _____ **30.** For stationary motors of 2 HP or less and 300 V or less, a general-use switch may be used as a controller if the rating is at least ___ times the full-load motor current.

_____ T   F **31.** Transformers rated over 600 V may have overcurrent protection devices in both primary and secondary circuits.

_____ T   F **32.** Voltage transformers shall be fused on the primary side when installed indoors.

Name _____ Date _____

NEC®	Answer	

_____ T F **1.** An overload in electrical apparatus includes short circuits or ground faults.

_____ _____ **2.** Control circuit transformers rated less than ___ VA require no additional protection.

_____ _____ **3.** A 30 HP, 230 V, 3φ squirrel-cage motor with code letter B is started with an autotransformer compensator. The initial rating of a dual element fuse for branch-circuit protection shall not exceed ___ A.
  A. 100   C. 150
  B. 125   D. 175

_____ _____ **4.** No underwater lighting fixture shall operate on supply circuits at a voltage greater than ___ V between conductors.

_____ _____ **5.** A(n) ___ building is of closed construction and is made or assembled in manufacturing facilities on or off the building site for installation or assembly and installation on the building site, other than manufactured homes, mobile homes, park trailers, or recreational vehicles.

_____ _____ **6.** A remote No. 14 AWG control circuit tapped from the line side of a magnetic starter requires no additional protection if the branch circuit is protected by a(n) ___ A device or less.

_____ _____ **7.** Sealed refrigeration units have ___ refrigerant motor-compressors.

_____ _____ **8.** In general, a controller used to stop and start a motor shall have a rating equivalent to the ___ of the motor.

_____ _____ **9.** The maximum overcurrent protection of a No. 6 THHN feeder that supplies two 30 A transformer arc welders rated for continuous duty shall be ___ A.

_____ T F **10.** A nonautomatically started, not permanently installed, 1 HP motor within sight of the controller location shall be permitted to be protected against overload by the branch-circuit, short-circuit, and ground-fault protective device.

_____ _____ **11.** Overhead 120/240 V insulated cables, supported by a grounded messenger that crosses a swimming pool shall have a clearance of ___' above water level.

_____ T F **12.** The overcurrent protection device for a 30 A transformer arc welder shall not exceed 30 A.

_____ _____ **13.** Conductors supplying a capacitor with a nameplate current of 50 A shall be rated not less than ___ A.
  A. 57.5   C. 67.5
  B. 62.5   D. 72.5

_____ T F **14.** A 277 V-to-ground circuit is permitted to supply lighting fixtures incorporating mogul-base screw-shell lampholders.

_____ _____ **15.** The basic rule requires an individual disconnecting means for ___ motor.

_____ _____ **16.** A manually-operated resistance welder draws 50 A according to the nameplate. Circuit conductors are permitted to be chosen on the basis of ___ A.
  A. 25   C. 75
  B. 50   D. 100

_____ _____ **17.** A busway may be reduced in size provided the smaller busway has a current rating of at least one-third of the rating of the main busway overcurrent protection device and does not extend more than ___'.

_____ _____ **18.** Pool heaters rated over ___ A are to be grounded and bonded per manufacture instructions.

_____ T   F **19.** An adjustable overload protection device protecting a 50°C motor with an FLC rating of 20 A shall be selected to trip at not more than 23 A.

_____ _____ **20.** Junction boxes housing conductors for swimming pool equipment shall have ___ more grounding terminal(s) than the number of conduit entries.

_____ T   F **21.** The disconnecting means for 460 V capacitors need not open all ungrounded conductors simultaneously.

_____ _____ **22.** The conductors to an adjustable speed drive shall be based on not less than 125% of the rated ___ current to the power conversion equipment.

_____ _____ **23.** ___ fiber cable is a factory assembly of one or more optical fibers with an overall covering.

_____ _____ **24.** A No. 14 AWG motor-control circuit conductor requires separate overcurrent protection when the control circuit extends beyond the control equipment enclosure and the motor branch-circuit protective device exceeds ___ A.
  A. 30   C. 60
  B. 45   D. 75

_____ _____ **25.** In general, the disconnecting means for a motor shall have an ampacity of at least ___% of the FLC rating of the motor.

_____ _____ **26.** The minimum height of the door sill or curb in an approved transformer vault is ___".

_____ _____ **27.** The controller can be an attachment plug and receptacle or cord connector for a(n) ___ motor rated at ⅓ HP or less.

_____ _____ **28.** The FLC table rating for a 10 HP, 230 V, 2φ AC motor is ___ A.

_____ _____ **29.** The minimum size of conductors permitted in a cablebus is No. ___.
  A. 1   C. 1/0
  B. 2   D. 2/0

_____ _____ **30.** It is mandatory that one of the disconnecting means to a motor be ___ accessible.

_____ _____ **31.** Transformers used for the supply of underwater fixtures shall be provided with a grounded metallic ___ between primary and secondary windings unless an approved system of double insulation is employed.

_____ _____ **32.** Overcurrent protection for a manually operated resistance welder that has a rated primary current of 50 A shall not exceed a maximum value of ___ A.
  A. 150   C. 200
  B. 175   D. 300

_____ _____ **33.** The allowable ampacities of wires used in crane circuits is ___ than for general electrical construction work.

_____ _____ **34.** The metal shell of wet niche fixtures shall be ___ and secured to the forming shell.

Name_____ Date _____

NEC®	Answer	
_____	_____	**1.** Flexible cords used to wire wet-niche lighting fixtures, other than listed low-voltage lighting systems, shall be equipped with an EGC not smaller than No. ___.
_____	_____	**2.** A stop-and-start station that can be locked open is not permitted to be used as the sole disconnecting means.
_____	T F	**3.** A ¼₀ HP, split-phase, automatically started motor does not require running overcurrent protection if protected by the branch-circuit, short-circuit, and ground-fault protection device.
_____	_____	**4.** Conductors rated over 35,000 V shall enter metal-enclosed switchgear or be installed in a transformer ___.
_____	T F	**5.** Motor overload protection shall never be shunted during the starting period.
_____	_____	**6.** Motors with no starting current problems and with a temperature rise of 50°C shall have overload protection provided at ___% of nameplate current.
_____	T F	**7.** Replacement parts may be stored in transformer vaults if at least 3′ from operating equipment.
_____	_____	**8.** Surge arresters rated at over 1 kV, shall be located so that they are ___ to unqualified personnel, unless listed for the installation.
_____	T F	**9.** The second largest X-ray unit on a feeder, in a health care facility, has a demand factor of 150% of the momentary rating.
_____	_____	**10.** The walls and roofs of transformer vaults shall have a minimum fire resistance of ___ hour(s). A. 1 C. 3 B. 2 D. none of the above
_____	_____	**11.** For other than industrial establishments, and under specific conditions, overcurrent protection shall be required where ___ are reduced in ampacity.
_____	T F	**12.** In general, an air-break isolating switch is required on the supply side of the disconnect and all service equipment for services over 600 V.
_____	T F	**13.** A motor branch-circuit, short-circuit protection device consisting of an NTDF shall never have a rating in excess of 400% of the full-load motor current.
_____	_____	**14.** A travel trailer has a maximum gross trailer area less than ___ sq ft.
_____	T F	**15.** Foreign pipes or ducts to the electrical installation shall not pass through a transformer vault.
_____	T F	**16.** Infrared heating units may be supplied from a 50 A branch circuit in an industrial occupancy.

_____ _____ **17.** Raceways routed in the building and encased in ___" of concrete or brick are considered to be outside the building.

_____ T F **18.** Single-phase conductors supplying a phase converter shall have an ampacity of not less than 125% of the 1ϕ input FLC rating for variable loads.

_____ _____ **19.** A feeder circuit supplying power to a mobile home shall be listed with a ___ conductor cord, with one conductor identified as the grounding conductor.

_____ T F **20.** An inherently short-time intermittent duty motor usually does not require separate running overload protection.

_____ _____ **21.** A ceiling fan without GFCI protection shall be hung no less than ___' above the maximum water level of a swimming pool.

_____ _____ **22.** The feeder circuit to a number of resistance welders shall not exceed ___% of the conductors' rating.

_____ T F **23.** In some cases, overload protection for a 50 A continuous duty motor may be set at 70 A.

_____ _____ **24.** The ampacity of branch-circuit conductors supplying power to X-ray equipment shall be ___% of the momentary rating or 100% of the long time rating, whichever is greater.

_____ _____ **25.** Enclosed overcurrent protection devices for services rated over 600 V are not required to be derated to ___% for continuous operation.

_____ _____ **26.** For electric-discharge lighting systems of more than 1000 V, a secondary open circuit voltage of 7500 V or less shall have an output not exceeding ___ mA.
    A. 240   C. 280
    B. 260   D. 300

_____ _____ **27.** All metal wiring methods and equipment within the walls of a swimming pool shall be bonded to the grid with a common copper conductor No. ___ or larger in size.

_____ _____ **28.** A resistance welder with a 50% duty cycle has a multiplier of ___%.

_____ _____ **29.** Each welder is required to have an identified disconnecting means either integral to the welder or in the ___ circuit.

_____ _____ **30.** Dry-type transformers rated over ___ V, primary or secondary, are required to be installed in a vault.

_____ _____ **31.** For capacitors rated over 600 V, the discharge means is required to be reduced to 50 V or less in ___ minutes or less.

_____ _____ **32.** The minimum rating of supply conductors for an automatically fed resistance welder operating at varying duty is ___% of the rated primary current.

_____ _____ **33.** The maximum rating of an OCPD for a No. 16 motor control circuit conductor is ___ amperes. Separate overcurrent protection is provided.
    A. 10   C. 20
    B. 15   D. 25

_____ _____ **34.** Conductors from surge arresters rated at over 1 kV to line and ground should not be any longer than ___ and shall avoid unnecessary bends.

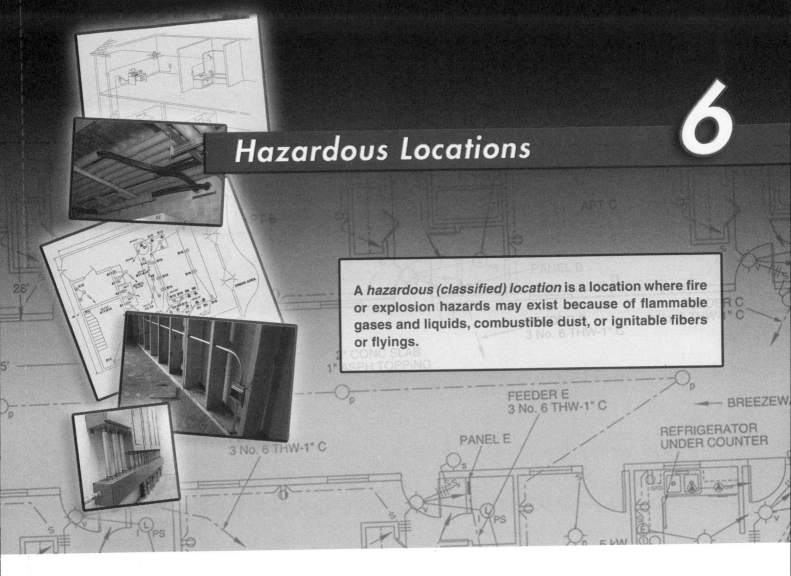

Hazardous Locations

6

> A *hazardous (classified) location* is a location where fire or explosion hazards may exist because of flammable gases and liquids, combustible dust, or ignitable fibers or flyings.

CLASSES

Some locations, because of their nature, call for higher standards with respect to materials or for methods of installation than are thought necessary in the general run of electrical wiring. The NEC® classifies hazardous locations under three headings: Class I, Class II, and Class III. Class I concerns activities that may cause fire or explosion from gases or vapors. Class II concerns the fire or explosion risk from dusts. Class III concerns the dangers of combustible fibers.

Each class has two divisions. In Division 1 locations, the particular danger is imminent at any or all times during the normal course of operation. In Division 2 locations, the danger is not present under normal conditions, but is likely to arise from a reasonable, foreseeable, accidental occurrence. For example, a building that houses a machine for compressing acetylene gas falls under Class I, Division 1. If high-pressure mains from this machine pass into another building where shut-off valves and pressure-reducing instruments are inserted in the lines, the adjacent building falls under Class I, Division 2.

Plants where combustible or electrically conductive dusts are present are in Class II, Division 1. Locations where readily-ignitable fibers are merely stored or handled, other than in the actual process of manufacture, are in Class III, Division 2. This classification applies mostly to warehouses where manufactured goods, in bales, are held for shipment to consumers.

Methods for Reducing Hazards

The NEC® requirements for proper seals in Class I, Division 1 and 2 locations are of utmost importance. Seals, when required, are installed within 18″ of an enclosure for switches, relays, CBs, fuses, or other equipment that could produce arcs, sparks, or high temperatures. The purpose of these seals is to restrict an explosion to one enclosure. Seals are also required in Class I, Division 1 and 2 areas when a conduit run of 2″ or larger enters an enclosure or fitting that contains terminals, splices, or taps. This seal is provided to limit the total volume that could be exposed to an explosion.

Seals are also required in conduits passing from a hazardous to a nonhazardous location. The purpose of this seal is to prevent an explosive mixture from being communicated through the conduit from a hazardous to a nonhazardous area.

In Class II areas, an alternate method of sealing is provided that permits a horizontal raceway not less than 10′ long or a vertical raceway extending not less than 5′ downward from a dust-ignition-proof enclosure. All sealing fittings and compound shall be approved for the purpose. At the present time, Underwriters Laboratories Inc. lists a sealing fitting, but does not list a compound except when a compound is used with the explosion-proof seal fittings of a specific manufacturer.

Intrinsically safe equipment and associated wiring that is approved for a specific location is permitted in hazardous areas. This equipment is such that under normal or abnormal conditions, it is not capable of releasing sufficient electrical or thermal energy to cause ignition of a specific explosive mixture.

The NEC® recommends that wherever possible, electrical equipment for hazardous locations be located in less hazardous areas. It also suggests that by adequate, positive-pressure ventilation from a clean source of outdoor air, the hazards may be reduced, or hazardous locations may be limited or eliminated. In many cases, the installation of dust collection systems can greatly reduce the hazards in a Class II area.

RMC, IMC, or Type MI cable are prescribed as the only acceptable wiring methods in the majority of hazardous locations. Grounding continuity shall be assured by means of bonding and not by double lockouts or locknut and bushing. Fittings shall be explosionproof, dust-ignition-proof, or dusttight, depending on the nature of the hazard.

Motors in the scope of Class I, Division 1 shall be explosionproof. Under Class I, Division 2, standard motors can be used if they have no sliding contacts. Motors in Classes II and III shall be totally enclosed or pipe-ventilated. Flexible cords of the three class areas shall be approved for extra hard use.

Specially Listed Occupancies

The NEC® considers all areas up to a level of 18″ in a commercial garage as Class I, Division 2 locations where ventilation that complies with 511.3(C)(1)(a) is not provided. A commercial garage where only exchange of parts and routine maintenance work are performed is considered a nonhazardous location. Aircraft hangars are given stricter, but similar protection. Space in the immediate vicinity of a gasoline dispensing island is Class I, Division 2 to the top of the dispenser enclosure extending 18″ horizontally to finished grade. Surrounding territory within a radius of 20′ of the island is Class I, Division 2 to a height of 18″ above grade. Bulk storage plants for gasoline are subject to comparable restrictions.

Finishing processes where paint, lacquer, or other flammable coatings are applied by spraying, dipping, brushing, or similar means, shall be carried out under prescribed safeguards for electrical wiring and equipment. Of particular interest here are electrostatic units that incorporate high-voltage transformers.

Anesthetizing locations of hospitals are Class I, Division 1 to a height of 5′ above floor level. Special rules apply to lighting and surgical equipment in the operating room, isolating transformers being generally required. Gas storage rooms are designated as Class I, Division 1 throughout.

Theaters and movie studios are given adequate consideration in the NEC®. Unless theaters are provided with sufficient emergency lighting, there is a likelihood of panic in case of fire. Fires may occur suddenly from temporary wiring used in connection with stage scenery or from projection room equipment. Movie studios also have a threat of fire from temporary wiring, arc lamps, and portable devices. Radio or television studios and receiving stations shall be guarded from the danger of high-voltage electronic wiring systems or equipment.

MATERIAL	CLASS I		CLASS II			CLASS III	
	Division 1	Division 2	Division 1	Division 2	Metal dust	Division 1	Division 2
RACEWAY†							
Rigid	✓	✓	✓	✓	✓	✓	✓
IMC	✓	✓	✓	✓	✓	✓	✓
EMT				✓			
Type PVC conduit						✓	✓
BOXES — Fittings							
Threaded	✓	✓	✓	✓	✓	✓	✓
Explosionproof	✓						
Dust-ignition-proof			✓ ‖		✓		
Dusttight			✓ #				
Dust-minimal				✓		✓	✓
FLEXIBLE CONDUIT							
Explosionproof	✓						
Standard							
Extra-hard-usage cord	✓	✓	✓	✓	✓	✓	✓
SWITCHES							
Explosionproof	✓	✓					
General purpose — hermetic‡		✓					
General purpose — oil§		✓					
Dust-ignition-proof			✓		✓		
Dust-minimal				✓		✓	✓

Title: **HAZARDOUS LOCATIONS GENERAL WIRING METHODS***

* See Exceptions for special applications for various metal cables, such as Types MI, MC-HL, ITC-HL, ITC, ITC-ER, PLTC, and PLTC-ER cable
† See Exceptions for special applications for Type PRC conduit and Type RTRC conduit
‡ General purpose enclosure – contacts hermetically sealed
§ General purpose enclosure – contacts immersed in oil
‖ Where joints or terminals
Where no joints or terminals

HAZARDOUS LOCATIONS — 500

Hazardous Location – A location where there is an increased risk of fire or explosion due to the presence of flammable gases, vapors, liquids, combustible dusts, or easily ignitable fibers or flyings.

Location – A position or site.

Flammable – Capable of being easily ignited and of burning quickly.

Gas – A fluid (such as air) that has no independent shape or volume but tends to expand indefinitely.

Vapor – A substance in the gaseous state as distinguished from the solid or liquid state.

Liquid – A fluid (such as water) that has no independent shape but has a definite volume. A liquid does not expand indefinitely and is only slightly compressible.

Combustible – Capable of burning.

Ignitable – Capable of being set on fire.

Fiber – A thread or piece of material.

Flyings – Small particles of a material.

Dust – Fine particles of matter.

Classes	Likelihood that a flammable or combustible concentration is present
I	Sufficient quantities of flammable gases and vapors present in air to cause an explosion or ignite hazardous materials.
II	Sufficient quantities of combustible dust are present in air to cause an explosion or ignite hazardous materials
III	Easily-ignitable fibers or flyings are present in air, but not in a sufficient quantity to cause an explosion or ignite hazardous materials

Divisions	Location containing hazardous substances
1	Hazardous location in which hazardous substance is normally present in air in sufficient quantities to cause an explosion or ignite hazardous materials
2	Hazardous location in which hazardous substance is not normally present in air in sufficient quantities to cause an explosion or ignite hazardous materials

Groups	Atmosphere containing flammable gases or vapors or combustible dust	
Class I	Class II	Class III
A B C D	E F G	none

DIVISION I EXAMPLES

Class I:
- Spray booth interiors
- Areas adjacent to spraying or painting operations using volatile flammable solvents
- Open tanks or vats of volatile flammable liquids
- Drying or evaporation rooms for flammable solvents
- Areas where fats and oil extraction equipment using flammable solvents are operated
- Cleaning and dyeing plant rooms that use flammable liquids
- Gas generator rooms
- Pump rooms for flammable gases or volatile flammable liquids that do not contain adequate ventilation
- Refrigeration or freezer interiors that store flammable materials
- All other locations where sufficient ignitable quantities of flammable gases or vapors are likely to occur during routine operations

Class II:
- Grain and grain products
- Pulverized sugar and cocoa
- Dried egg and milk powders
- Pulverized spices
- Starch and pastes
- Potato and woodflour
- Oil meal from beans and seeds
- Dried hay
- Any other organic materials that may produce combustible dusts during their use or handling

Class III:
- Portions of rayon, cotton, or other textile mills
- Manufacturing and processing plants for combustible fibers, cotton gins, and cotton seed mills
- Flax processing plants
- Clothing manufacturing plants
- Woodworking plants
- Other establishments involving similar hazardous processes or conditions

HAZARDOUS (CLASSIFIED) LOCATIONS

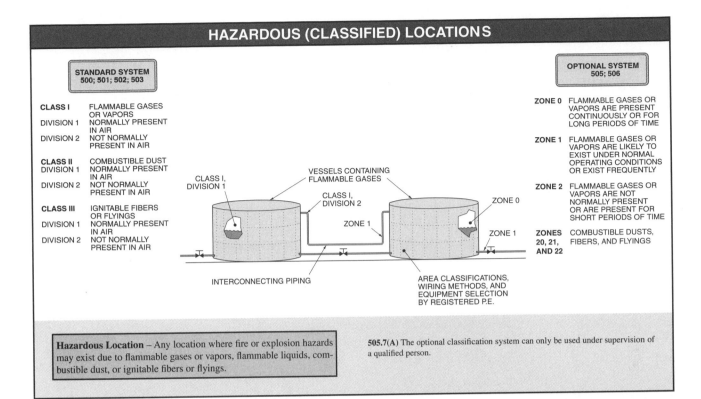

STANDARD SYSTEM
500; 501; 502; 503

CLASS I FLAMMABLE GASES OR VAPORS
DIVISION 1 NORMALLY PRESENT IN AIR
DIVISION 2 NOT NORMALLY PRESENT IN AIR

CLASS II COMBUSTIBLE DUST
DIVISION 1 NORMALLY PRESENT IN AIR
DIVISION 2 NOT NORMALLY PRESENT IN AIR

CLASS III IGNITABLE FIBERS OR FLYINGS
DIVISION 1 NORMALLY PRESENT IN AIR
DIVISION 2 NOT NORMALLY PRESENT IN AIR

CLASS I, DIVISION 1
VESSELS CONTAINING FLAMMABLE GASES
CLASS I, DIVISION 2
ZONE 0
ZONE 1
ZONE 1
INTERCONNECTING PIPING
AREA CLASSIFICATIONS, WIRING METHODS, AND EQUIPMENT SELECTION BY REGISTERED P.E.

OPTIONAL SYSTEM
505; 506

ZONE 0 FLAMMABLE GASES OR VAPORS ARE PRESENT CONTINUOUSLY OR FOR LONG PERIODS OF TIME

ZONE 1 FLAMMABLE GASES OR VAPORS ARE LIKELY TO EXIST UNDER NORMAL OPERATING CONDITIONS OR EXIST FREQUENTLY

ZONE 2 FLAMMABLE GASES OR VAPORS ARE NOT NORMALLY PRESENT OR ARE PRESENT FOR SHORT PERIODS OF TIME

ZONES 20, 21, AND 22 COMBUSTIBLE DUSTS, FIBERS, AND FLYINGS

Hazardous Location – Any location where fire or explosion hazards may exist due to flammable gases or vapors, flammable liquids, combustible dust, or ignitable fibers or flyings.

505.7(A) The optional classification system can only be used under supervision of a qualified person.

INTRINSICALLY SAFE SYSTEMS — 504

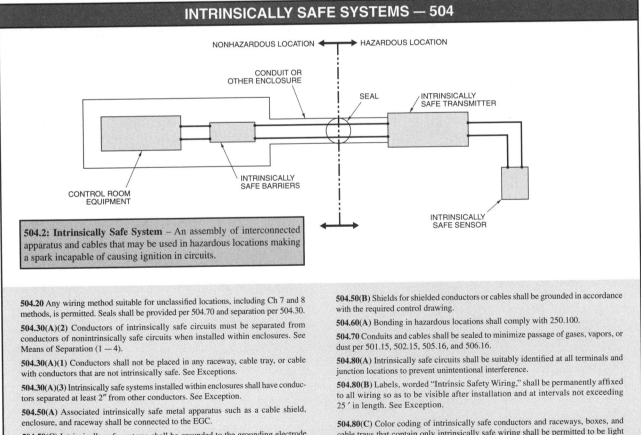

NONHAZARDOUS LOCATION ◄———► HAZARDOUS LOCATION

CONDUIT OR OTHER ENCLOSURE
SEAL
INTRINSICALLY SAFE TRANSMITTER
CONTROL ROOM EQUIPMENT
INTRINSICALLY SAFE BARRIERS
INTRINSICALLY SAFE SENSOR

504.2: Intrinsically Safe System – An assembly of interconnected apparatus and cables that may be used in hazardous locations making a spark incapable of causing ignition in circuits.

504.20 Any wiring method suitable for unclassified locations, including Ch 7 and 8 methods, is permitted. Seals shall be provided per 504.70 and separation per 504.30.

504.30(A)(2) Conductors of intrinsically safe circuits must be separated from conductors of nonintrinsically safe circuits when installed within enclosures. See Means of Separation (1 — 4).

504.30(A)(1) Conductors shall not be placed in any raceway, cable tray, or cable with conductors that are not intrinsically safe. See Exceptions.

504.30(A)(3) Intrinsically safe systems installed within enclosures shall have conductors separated at least 2″ from other conductors. See Exception.

504.50(A) Associated intrinsically safe metal apparatus such as a cable shield, enclosure, and raceway shall be connected to the EGC.

504.50(C) Intrinsically safe systems shall be grounded to the grounding electrode per 250.52(A)(1 — 4) and shall comply with 250.30(A)(4).

504.50(B) Shields for shielded conductors or cables shall be grounded in accordance with the required control drawing.

504.60(A) Bonding in hazardous locations shall comply with 250.100.

504.70 Conduits and cables shall be sealed to minimize passage of gases, vapors, or dust per 501.15, 502.15, 505.16, and 506.16.

504.80(A) Intrinsically safe circuits shall be suitably identified at all terminals and junction locations to prevent unintentional interference.

504.80(B) Labels, worded "Intrinsic Safety Wiring," shall be permanently affixed to all wiring so as to be visible after installation and at intervals not exceeding 25 ′ in length. See Exception.

504.80(C) Color coding of intrinsically safe conductors and raceways, boxes, and cable trays that contain only intrinsically safe wiring shall be permitted to be light blue provided no other light blue conductors are used.

CLASS I, DIVISION 1 LOCATIONS — 501

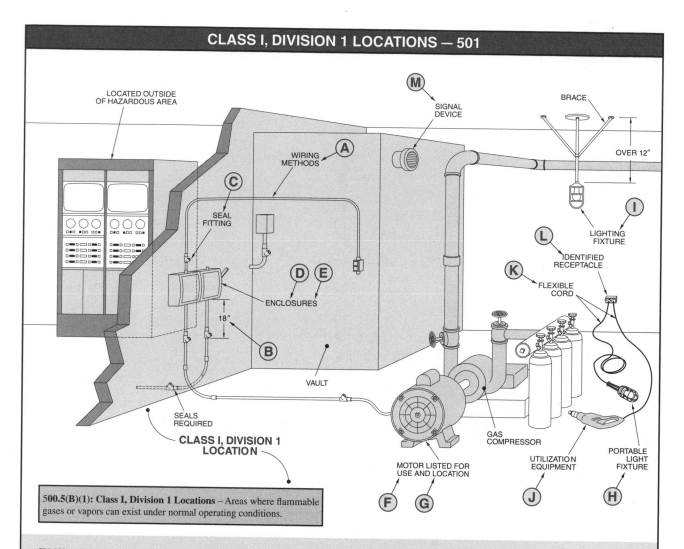

500.5(B)(1): Class I, Division 1 Locations – Areas where flammable gases or vapors can exist under normal operating conditions.

501.100(A)(1) Transformers or capacitors containing a liquid that will burn shall be installed in vaults per 450.41 through 450.48.

501.105(A) Meters, instruments, and relays shall be in enclosures identified for Class I, Division 1 locations.

A. **501.10(A)(1 – 2)** Wiring methods permitted for Class I, Division 1 locations are RMC, steel IMC, or Type MI cable with listed termination fittings. See Special Provisions for PVC (c) MC-HL cable and (d) ITC-HL cable. Explosionproof flexible connection with fittings shall be listed for Class I, Division 1 locations.

B. **501.15(A – F)** A seal shall be installed within 18″ of enclosures for spark- or heat-producing apparatus also in each run of 2″ or larger conduit where splices are made and within 18″ thereof.

502.10(A)(4) Seals are required where a conduit leaves a Class I, Division 1 location. Seals may be installed on either side and within 10′ of the boundary.

C. **501.15(C)(1)** Seal fittings shall be accessible and listed for one or more specific compounds.

501.15(C)(2, 3) Specifications for sealing compound are detailed in the NEC®.

501.17 Canned pumps, process connections for flow, pressure, or analysis measurement, etc. that depend upon a single seal, diaphragm, or tube to keep fluid from entering the electrical raceway or conduit system require an additional approved seal, barrier, or other means to keep fluids from entering the conduit system.

D. **501.115(A)** Switches, CBs, motor controllers, and fuses shall be in enclosures identified as a complete assembly for Class I, Division 1 locations.

E. **501.120(B)(1)** Switching mechanisms associated with transformers, impedance coils, and resistors shall be in enclosures identified for Class I, Division 1 locations. See 501.120(B)(1 – 3).

F. **501.125(A)(1)** Motors, generators, and other rotating machinery shall be identified for use in Class I, Division 1 locations.

G. **501.125(A)(1 – 2)** Motors and generators in Class I, Division 1 locations shall be identified for location. One alternative is to use totally enclosed types of equipment with positive-pressure ventilation that ensures enclosures are purged with ten volumes of air before equipment is energized.

H. **501.130(A)(1)** Portable light fixtures used as a complete assembly shall be listed for use in Class I, Division 1 locations.

I. **501.130(A)(3)** Pendant lighting fixtures shall be identified as complete units for use in Class I, Division 1 locations. Stems over 12″ shall be braced.

J. **501.135(A)** Utilization equipment shall be identified for use for Class I, Division 1 locations.

K. **501.140(B)** Flexible cord, listed for extra-hard usage type, with grounding conductor shall be terminated in approved manner, suitably supported, installed with suitable seals where necessary and be of a continuous length.

L. **501.145** Receptacles and attachment plugs shall be identified for use in Class I, Division 1 locations.

M. **501.150(A)** Signal devices shall be identified for use in Class I, Division 1 locations.

501.25 No uninsulated exposed parts are permitted in Class I, Division 1 locations that operate at more than 30 V (15 V in wet locations).

501.35(A) Surge arrestors shall be in an enclosure identified for use in Class I, Division 1 locations.

CLASS I, DIVISION 2 LOCATIONS — 501

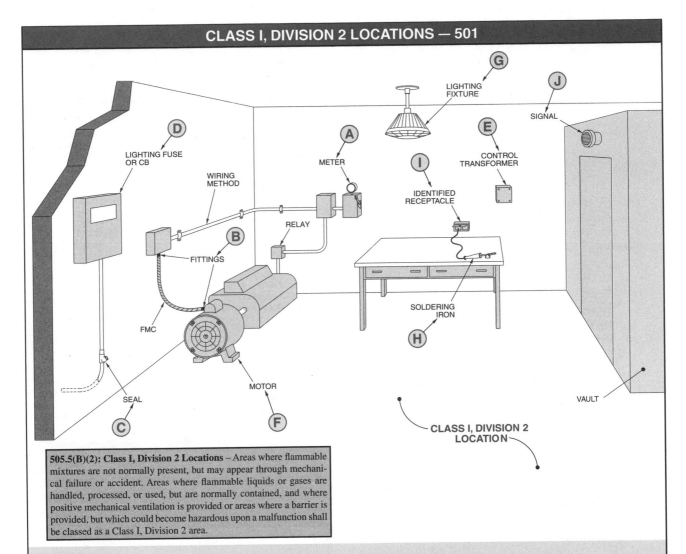

505.5(B)(2): Class I, Division 2 Locations – Areas where flammable mixtures are not normally present, but may appear through mechanical failure or accident. Areas where flammable liquids or gases are handled, processed, or used, but are normally contained, and where positive mechanical ventilation is provided or areas where a barrier is provided, but which could become hazardous upon a malfunction shall be classed as a Class I, Division 2 area.

501.100(B) Transformers are permitted to be installed as in nonhazardous locations and shall comply with 450.21–450.27. Capacitors shall comply with 460.2 through 460.28.

A. 501.105(B) (1 – 6) Meters, relays, resistors, etc., without Class I enclosures shall have make-and-break contacts immersed in oil or in hermetically sealed chambers unless on circuits of negligible energy (nonincendive circuits) and listed for Division 2.

501.100(B)(5) Instrument fuses are permitted in general-purpose enclosures per 501.100(B)(1 – 4) if not subject to overloading and preceded by a switch conforming to 501.105(B)(1).

B. 501.10(B)(1 – 4) FMC, LFMC, LFNC, or extra-hard usage cord with listed fittings are permitted. Wiring methods permitted in Class I, Division 2 locations include: threaded RMC, IMC, enclosed gasketed wireways or busways, Type PLTC cable per 725, Type ITC, MI, MC, MV, or TC cable. Boxes and fittings generally need not be explosionproof, except as required by 501.105(B)(1), 501.115(B)(1), and 501.150(B)(1).

C. 501.15(B)(2) A sealing fitting is required in the conduit run passing to the nonhazardous area. The sealing fitting is permitted to be located on either side of the boundry within 10′ of the boundary.

501.115(B)(1) Motor controllers, switches, and CBs shall be listed as Class I, Division 1 equipment per 501.105(A) unless make-and-break contacts are immersed in oil or in a chamber hermetically sealed against the entrance of gas or vapors, in a factory-sealed explosionproof chamber, or the device is solid state.

D. 501.115(B)(4) Listed cartridge fuses are permitted for supplementary protection with lighting fixtures.

E. 501.120(B)(2) General-purpose enclosures are permitted for transformers, solenoids, or impedance coils in Class I, Division 2 locations.

F. 501.125(B) Standard squirrel cage induction motors may be used in Class I, Division 2 locations. If the motor has sliding contacts, it must be identified for Class I, Division 2 locations, unless the contact arrangements are especially identified for the location.

G. 501.130(B)(1 – 6) Fixed fixtures shall have protection from physical damage. Note surface temperature limitation. Portable lamps shall be specifically approved for the purpose. Pendants to have reinforcing supports if over 12″ long. Fixture switches shall comply with 501.115(B)(1). Electric discharge starting and control equipment shall be per 501.120(B), except the potted thermal protector in the ballast of fluorescent lighting shall be approved for hazardous locations.

H. 501.135(B)(1) Electrically-heated appliances shall be identified for Class I, Division 2 locations unless they comply with 501.135(B)(1)(1).

I. 501.145 Plug receptacles shall be identified for the location and provide for connection to the EGC.

J. 501.150(B)(1) Signal unit shall be explosionproof, or contacts immersed in oil, or hermetically sealed unless on circuits of negligible energy. See Exception.

501.25 No uninsulated exposed live parts are permitted in Class I, Division 2 locations that operate at more than 30 V (15 V in wet locations).

501.30(A – B) Locknut-bushing type contacts are not approved for general bonding purposes in hazardous locations. Bonding jumpers are required for FMC. External bonding jumpers shall be per 250.102. See Exception.

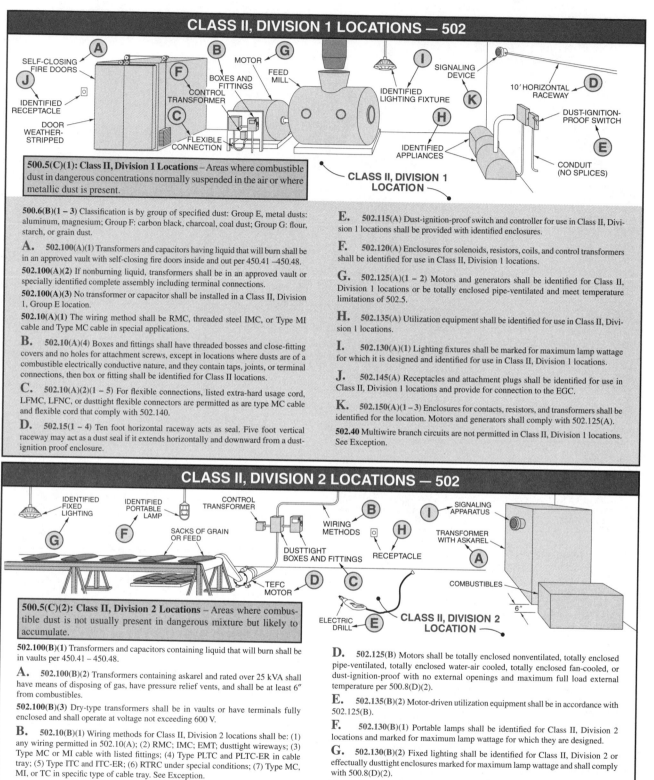

CLASS II, DIVISION 1 LOCATIONS — 502

500.5(C)(1): Class II, Division 1 Locations – Areas where combustible dust in dangerous concentrations normally suspended in the air or where metallic dust is present.

500.6(B)(1 – 3) Classification is by group of specified dust: Group E, metal dusts: aluminum, magnesium; Group F: carbon black, charcoal, coal dust; Group G: flour, starch, or grain dust.

A. **502.100(A)(1)** Transformers and capacitors having liquid that will burn shall be in an approved vault with self-closing fire doors inside and out per 450.41 –450.48.

502.100(A)(2) If nonburning liquid, transformers shall be in an approved vault or specially identified complete assembly including terminal connections.

502.100(A)(3) No transformer or capacitor shall be installed in a Class II, Division 1, Group E location.

502.10(A)(1) The wiring method shall be RMC, threaded steel IMC, or Type MI cable and Type MC cable in special applications.

B. **502.10(A)(4)** Boxes and fittings shall have threaded bosses and close-fitting covers and no holes for attachment screws, except in locations where dusts are of a combustible electrically conductive nature, and they contain taps, joints, or terminal connections, then box or fitting shall be identified for Class II locations.

C. **502.10(A)(2)(1 – 5)** For flexible connections, listed extra-hard usage cord, LFMC, LFNC, or dusttight flexible connectors are permitted as are type MC cable and flexible cord that comply with 502.140.

D. **502.15(1 – 4)** Ten foot horizontal raceway acts as seal. Five foot vertical raceway may act as a dust seal if it extends horizontally and downward from a dust-ignition proof enclosure.

E. **502.115(A)** Dust-ignition-proof switch and controller for use in Class II, Division 1 locations shall be provided with identified enclosures.

F. **502.120(A)** Enclosures for solenoids, resistors, coils, and control transformers shall be identified for use in Class II, Division 1 locations.

G. **502.125(A)(1 – 2)** Motors and generators shall be identified for Class II, Division 1 locations or be totally enclosed pipe-ventilated and meet temperature limitations of 502.5.

H. **502.135(A)** Utilization equipment shall be identified for use in Class II, Division 1 locations.

I. **502.130(A)(1)** Lighting fixtures shall be marked for maximum lamp wattage for which it is designed and identified for use in Class II, Division 1 locations.

J. **502.145(A)** Receptacles and attachment plugs shall be identified for use in Class II, Division 1 locations and provide for connection to the EGC.

K. **502.150(A)(1 – 3)** Enclosures for contacts, resistors, and transformers shall be identified for the location. Motors and generators shall comply with 502.125(A).

502.40 Multiwire branch circuits are not permitted in Class II, Division 1 locations. See Exception.

CLASS II, DIVISION 2 LOCATIONS — 502

500.5(C)(2): Class II, Division 2 Locations – Areas where combustible dust is not usually present in dangerous mixture but likely to accumulate.

502.100(B)(1) Transformers and capacitors containing liquid that will burn shall be in vaults per 450.41 – 450.48.

A. **502.100(B)(2)** Transformers containing askarel and rated over 25 kVA shall have means of disposing of gas, have pressure relief vents, and shall be at least 6″ from combustibles.

502.100(B)(3) Dry-type transformers shall be in vaults or have terminals fully enclosed and shall operate at voltage not exceeding 600 V.

B. **502.10(B)(1)** Wiring methods for Class II, Division 2 locations shall be: (1) any wiring permitted in 502.10(A); (2) RMC; IMC; EMT; dusttight wireways; (3) Type MC or MI cable with listed fittings; (4) Type PLTC and PLTC-ER in cable tray; (5) Type ITC and ITC-ER; (6) RTRC under special conditions; (7) Type MC, MI, or TC in specific type of cable tray. See Exception.

502.10(B)(4) Boxes and fittings to resist entrance of dust shall have telescoping or close-fitting covers to contain sparks. No holes in the box shall be left exposed after installation. All boxes and fittings shall be dusttight.

502.15 Sealing is the same as for Class II, Division 1 areas.

C. **502.10(B)(4)** Dusttight boxes and fittings are required in Class II, Division 2 locations.

D. **502.125(B)** Motors shall be totally enclosed nonventilated, totally enclosed pipe-ventilated, totally enclosed water-air cooled, totally enclosed fan-cooled, or dust-ignition-proof with no external openings and maximum full load external temperature per 500.8(D)(2).

E. **502.135(B)(2)** Motor-driven utilization equipment shall be in accordance with 502.125(B).

F. **502.130(B)(1)** Portable lamps shall be identified for Class II, Division 2 locations and marked for maximum lamp wattage for which they are designed.

G. **502.130(B)(2)** Fixed lighting shall be identified for Class II, Division 2 or effectually dusttight enclosures marked for maximum lamp wattage and shall comply with 500.8(D)(2).

H. **502.145(B)** Receptacle and attachment plug shall provide connection to EGC and be of type that supply circuit cannot be broken while live parts are exposed.

I. **502.150(B)** Enclosures for signaling apparatus for Class II, Division 2 locations shall be dusttight or identified for the location.

502.30 Equipment in Class II, Division 1 and 2 locations shall be grounded per Article 250. For bonding and types of EGCs, see 502.30(A – B).

CLASS III, DIVISION 1 LOCATIONS — 503

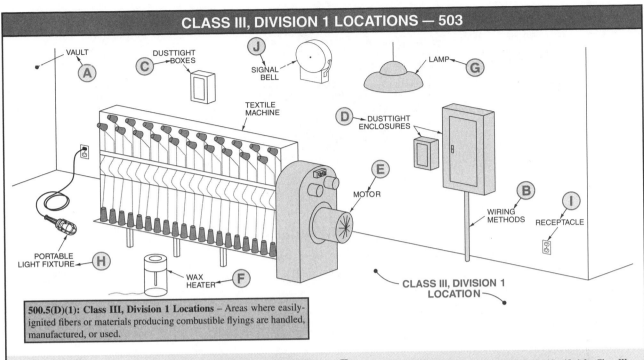

CLASS III, DIVISION 1 LOCATION

500.5(D)(1): Class III, Division 1 Locations – Areas where easily-ignited fibers or materials producing combustible flyings are handled, manufactured, or used.

A. 503.100 Transformers and capacitors containing liquid that will burn shall be in a vault per 450.41 – 450.48. They shall comply with 502.100(B).

B. 503.10(A) Wiring methods shall be RMC, IMC, approved Types MI or MC cables, or PVC, EMT, dusttight wireways, Type PLTC, PLTC-ER, ITC, and ITC-ER cables.

C. 503.10(A)(2) All boxes and fittings shall be dusttight.

D. 503.115 Switches and controllers shall be in dusttight enclosures.

E. 503.125, Ex. (1 – 3) Motors shall be totally enclosed or pipe-ventilated. Self-cleaning squirrel cage textile motors may be locally acceptable. Partially enclosed or splashproof motors shall not be installed in Class III, Division 1 locations.

F. 503.135(A) Electrically heated equipment shall be identified for Class III locations.

G. 503.130(A) Lamps shall be protected against entrance of lint and fibers. They shall also be marked for their maximum wattage. Their surface temperature shall not exceed 165°C.

H. 503.130(D) Portable light fixtures shall be unswitched. Exposed metal parts shall be grounded.

I. 503.145 Receptacles and plugs shall be the grounding type and designed to minimize entry or accumulation of fibers or filings and prevent escape of sparks. See Exception.

J. 503.150 Signals, remote-control, and intercommunications equipment shall comply with the requirements of Article 503.

CLASS III, DIVISION 2 LOCATIONS — 503

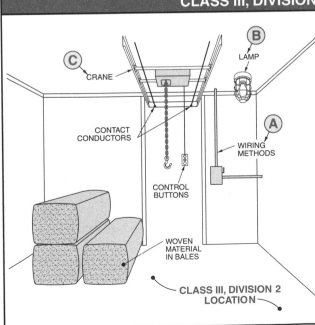

CLASS III, DIVISION 2 LOCATION

500.5(D)(2) Class III, Division 2 Locations – Areas where easily-ignited fibers are stored or handled other than in the process of manufacture.

A. 503.10(B) The same wiring methods as required for Class II, Division 1 are acceptable, except open wiring on insulators per Article 398 is permissible in storage areas that have no machinery. See Exception.

B. 503.130(A) Lamps shall be protected against entrance of lint and fibers. They shall also be marked for their maximum wattage. Their surface temperature shall not exceed 165°C.

C. 503.155 Cranes operating over combustible fibers, etc. shall be specially wired: (A) power supply to contact conductors isolated and ungrounded; (B) contact conductors to be guarded against accidental contact; (C) current collectors to be guarded against escape of hot particles or sparks; and (D) control equipment per 503.115 and 503.120.

503.25 No uninsulated exposed live parts are permitted in Class III, Division 2 locations that operate at over 30 V (15 V in wet locations). See 503.155 for an exception to the general rule.

503.30(A – B) Equipment and wiring shall be grounded per Article 250. Bonding and EGCs shall conform to 503.30(A)(B).

Note: In general, provisions for wiring in Class III, Division 1 locations apply in Class III, Division 2 locations.

COMMERCIAL GARAGES — 511

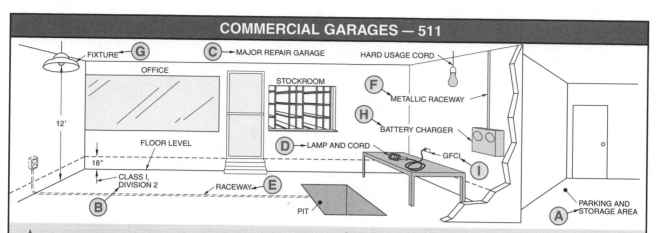

A. 511.3(A – E) Where the garage is used for parking and storage and only routine maintenance requiring no electrical tools, or where exchange of parts is performed in adequate ventilation, it is not classified as a hazardous area.

B. 511.3(B) Areas in which flammable fuel is dispensed into vehicle tanks shall be classified per Table 514.3(B)(1).

C. 511.2 A major repair garage is a building or part of a building where major repairs, such as engine overhaul, are performed. Such repairs include the draining of the fuel tank.

D. 511.4(B)(2) If a lamp and cord can be used in a hazardous area, it shall be identified for a Class I, Division 1 area. The outlet shell shall be molded composition or other suitable material.

E. 511.4(A) Wiring located within Class I locations, as classified in 511.3, shall conform to Article 501.

F. 511.7(A)(1) Fixed wiring above hazardous areas shall be in metallic raceways or RNC, MI, TC, FMC, LFMC, LFNC, or MC cable. PLTC, TC, and ITC cables are

also permitted where installed in accordance with their respective articles. Conductors are not permitted in any cell, header, or duct containing foreign pipes.

511.7(B)(1)(a) Arc or spark producing equipment, such as switches or motors (excluding receptacles, lamps, and lampholders), mounted less than 12' from the floor, shall be the totally enclosed type.

G. 511.7(B)(1)(b) Fixtures shall be at least 12' above the driveway unless they are the totally enclosed type.

H. 511.10 Battery chargers and their control equipment shall be placed outside hazardous area.

511.10(B)(1) Electric vehicle charging equipment shall be installed per Article 625.

I. 511.12 GFCIs are required in commercial repair garages for all 125 V, 1φ, 15 A and 20 A receptacles where electrical automotive diagnostic equipment, electrical hand tools, and portable lighting are used.

AIRCRAFT HANGARS — 513

A. 513.3(A) Pit shall be designated as a Class I, Division 1 location.

B. 513.3(B) The general hazardous area throughout the hangar is 18" above the floor and classified as a Class I, Division 2 location.

C. 513.3(C)(1) The area within 5' horizontally from aircraft fuel tanks and from the floor to 5' above aircraft wings is a Class I, Division 2 area.

513.3(D) Adjacent areas are unclassified areas where adequately ventilated and physically isolated by walls or partitions.

D. 513.8(A) Wiring under a hangar floor is a Class I, Division 1 location. Where wiring is located in volts, pits, or ducts, drainage means shall be provided.

513.7(A) Fixed wiring not within the hazardous area shall be in metallic raceway, Type MI, TC, or Type MC cable. See Exception.

513.7(B) For pendants, hard or extra-hard usage cord with a separate EGC shall be used for portable equipment.

E. 513.7(C) Equipment less than 10' above wings and engine enclosures shall be of the totally enclosed type. See Exception.

F. 513.7(D) No metal-shell, fiber-lined lampholders are permitted in Class I locations.

G. 513.16(A) Metal raceways, metal-jacketed cables, and non-current-carrying metal parts of fixed or portable electrical equipment shall be grounded per Article 250.

513.12 GFCI protection is required for all 125 V, 1φ, 15 A and 20 A receptacles used for testing, diagnostics, or lighting.

MOTOR FUEL DISPENSING FACILITIES — 514

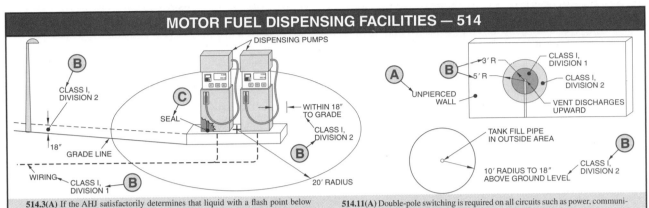

514.3(A) If the AHJ satisfactorily determines that liquid with a flash point below 100°F will not be handled, the location shall not be required to be classified.

A. **514.3(B)(1)** The hazardous area shall not extend beyond an unpierced wall, roof, or other solid partition.

B. **Table 514.3(B)(1)** The area to a height of 18″ above grade for a distance of 20′ horizontally in every direction from the dispensing pump is a Class I, Division 2 location. Wiring below the surface of Class I, Division 1 and 2 locations shall be Class I, Division 1 locations.

Spherical volume with a 3′ radius at the point of discharge is considered a Class I, Division 1 location. Volume between a 3′ and 5′ radius is considered a Class I, Division 2 location.

Within 18″ horizontally in all directions extending to grade from the dispenser enclosure or the portion of the enclosure with liquid handling components is a Class I, Division 2 location.

Area within 10′ radius of tank fill pipe is considered a Class I, Division 2 location to a height of 18″ above ground level.

514.7 Wiring above Class I areas shall be in metallic raceways, RNC, or Types MI, TC, MC cables and Types PLTC, and ITC cables per 511.7(A)(1).

514.11(A) Double-pole switching is required on all circuits such as power, communications, and date and video circuits leading to or through dispensing pump including the grounded conductor, if any. The use of handle ties is not permitted.

514.11(B – C) Attended or unattended self-service station shall have emergency control disconnects with the required distances. Location for attended station is within 100′ from the dispenser.

C. **514.9(A)** Seal fittings are required on all conduits entering or leaving the dispensing area. The seal shall be the first fitting after the conduit emerges from the earth or concrete.

514.8 Conduit runs below the surface of hazardous area are considered Class I, Division 1 to point of emergence above grade. Underground wiring is permitted in RMC or threaded steel IMC. Where PVC or Type RTRC is used, it shall be at least 2′ deep and, if emerging from the underground, the last 2′ of underground to emergence above ground shall be RMC or IMC with threaded fittings. Where PVC or Type RTRC is used, an EGC is required. All wiring systems shall comply with Article 300 including Table 300.5, which requires 24″ depth under areas subject to heavy vehicular traffic.

514.16 Metal raceways, metal-jacketed cables, and non-current-carrying metal parts of equipment are grounded and bonded per Article 250. Grounding in Class I locations shall comply with 501.130.

BULK STORAGE PLANTS — 515

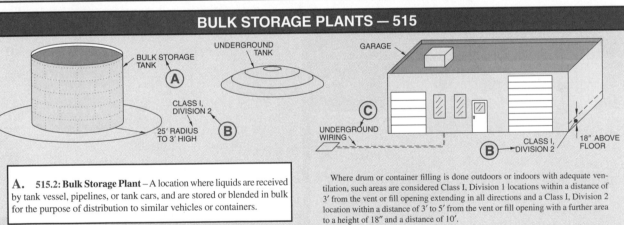

A. **515.2: Bulk Storage Plant** – A location where liquids are received by tank vessel, pipelines, or tank cars, and are stored or blended in bulk for the purpose of distribution to similar vehicles or containers.

B. **Table 515.3** When loading tank trucks or cars, the area extending 3′ in all directions from open domes or atmospherically-ventilated closed domes is a Class I, Division 1 location and the distance from 3′ to 15′ is a Class I, Division 2 location. The area within 3′ of a fixed connection used in bottom or in dome loading is a Class I, Division 2 location and where such loading is from a bottom connection, the area shall extend to a height of 18″ within a radius of 10′.

Pumps, bleeders, withdrawal fittings, meters, and similar devices located indoors where flammable liquids are handled under pressure shall be considered Class I, Division 2 locations within a distance of 5′ in all directions from any surface of handling equipment and to a height of 3′ for a distance of 25′.

Where Class I liquids are stored, handled, or dispensed, Table 515.2 shall be used to delineate and classify bulk storage plants. A Class I location shall not extend beyond an unpierced wall, roof, or other substantial partition.

For above-ground tanks, the area above the roof and within the shell of a floating roof tank as a Class I, Division 1 location, while the area above the roof of a closed-roof tank is a Class I, Division 2 location. Areas within 10′ of a tank shell and inside a dike are Class I, Division 2 locations. Areas within 5′ of a vent opening are Class I, Division 1 locations and within 5′ to 10′, Class I, Division 2 locations.

Where drum or container filling is done outdoors or indoors with adequate ventilation, such areas are considered Class I, Division 1 locations within a distance of 3′ from the vent or fill opening extending in all directions and a Class I, Division 2 location within a distance of 3′ to 5′ from the vent or fill opening with a further area to a height of 18″ and a distance of 10′.

Garages for tank vehicles are Class I, Division 2 locations to a height of 18″ above the floor unless special hazard exists.

Pits within Division 1 or Division 2 locations are considered Class I, Division 1 locations unless adequately ventilated, then they become Class I, Division 2 locations. Those outside such locations but containing piping or valves are Class I, Division 2 locations.

515.7(A – B) Fixed wiring above hazardous area shall be in metallic raceways, PVC Sch. 80, Type RTRC-XW, MI, TC, or Type MC cable. Sparking equipment shall be totally enclosed or guarded. Type PLTC and PLTC-ER are permitted as are Type ITC and ITC-ER, where installed in accordance with the provisions of their respective articles.

C. **515.8(A – C)** Underground wiring is permitted in RMC, threaded steel IMC, PVC, or approved MI cable. Where PVC or Type RTRC is used, it shall be at least 2′ underground, and, if emerging from underground in a Class I area, the last 2′ of underground to emergence aboveground shall be in RMC or steel IMC with threaded fittings. Where PVC or Type RTRC is used, an EGC is required. Where cable is used, it shall be in RMC or steel IMC from the lowest buried cable to the connection point aboveground. All conductor insulation shall comply with 501.20.

SPRAYING, DIPPING, AND COATING PROCESSES — 516

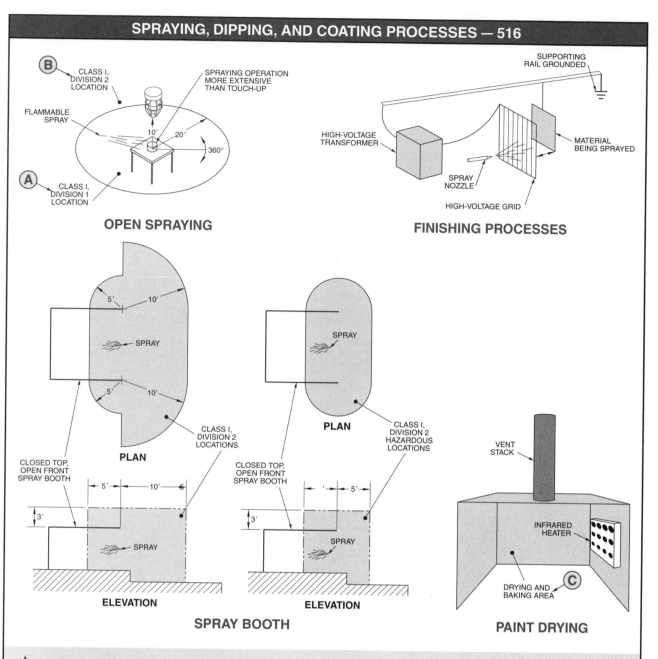

OPEN SPRAYING

FINISHING PROCESSES

SPRAY BOOTH

PAINT DRYING

A. **516.3(A)(1 – 2)** The area containing paint spray is considered a Class I, Division 1 location. This includes (1) the interior of any open or closed container of flammable liquid and (2) the interior of any dip tank or coating tank.

B. **516.3(C)(1 – 7)** The area outside of the Class I, Division 1 location per Figure 516.3(C)(1) is classified as a Class I, Division 2 location.

516.3(C)(3) If open top spray booth, the 3' above the booth and the 3' from the other openings are considered as Class I, Division 2 locations.

516.3(C)(4) If the booth is enclosed, 3' in all directions from all openings are Class I, Division 2 locations.

C. **516.3(F)** Drying and baking areas are generally considered unclassified if adequately ventilated, protective devices of the ventilating system are interlocked with the heating circuit, and acceptable to the AHJ.

516.4(B) Electrical equipment shall not be installed where subject to vapor or deposit of residue unless listed for the location.

516.4(C) Illumination through glass panels or equivalent permitted if installed per 516.4(C)(1 – 5).

516.10 Fixed electrostatic spray equipment shall comply with 516.10(A).

516.10(A)(2) Electrodes and electrostatic atomizing heads shall be adequately supported and insulated from ground.

516.10(A)(10)(a) The separation between grid and work shall be at least twice the sparking distance.

516.10(B) Electrostatic hand-spraying equipment shall be listed. Transformers, etc. shall be located outside the hazardous area. The handles of spray guns and other conductive objects shall be grounded. Target objects shall maintain contact with a grounded support. The area shall be properly ventilated and interlocked with spraying equipment. Hand guns and connections to power supply are permitted in the hazardous area.

516.7(A) Fixed wiring above hazardous areas shall be in metal raceways; ENT; PVC; Type RTRC conduit; or Type MI, TC, or MC cable.

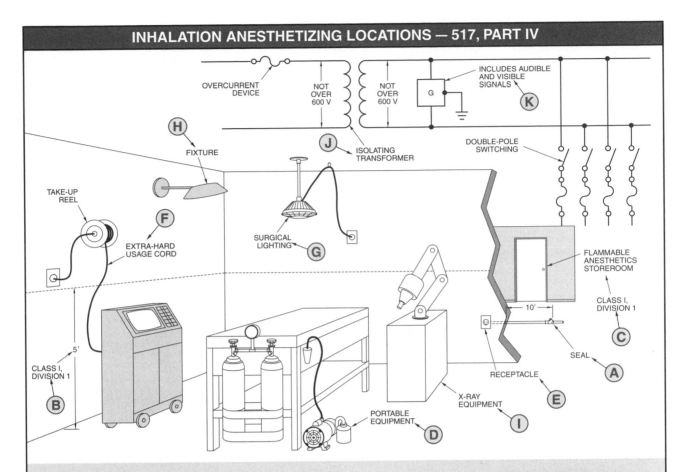

INHALATION ANESTHETIZING LOCATIONS — 517, PART IV

A. 501.15(A)(4) Seals may be located within 10′ of the point where conduit emerges from the boundary wall provided the junction box contains a seal-off device and the conduit between the box and the fitting is one continuous piece.

B. 517.60(A)(1) Anesthetizing locations shall be Class I, Division 1 locations to a height of 5′ above the floor.

C. 517.60(A)(2) A storeroom for combustible gas or other volatile agents shall be a Class I, Division 1 location throughout.

517.60(B) An inhalation anesthetizing location designated only for use of nonflammable anesthetizing agents shall be considered to be a nonhazardous area.

D. 517.61(A)(3) Portable equipment shall be approved for use in Class I, Division 1 locations if more than 10 V. A suction unit shall be approved for Class I, Division 1 locations and also specifically approved for hazardous atmospheres involved. Fixed wiring and equipment, and portable equipment over 10 V shall conform to Class I requirements per 501.1 through 501.25, 501.100 through 501.150, and 501.30(A) and (B), and shall be specifically approved for the location.

517.61(A)(4) Boxes, fittings, or enclosures partially within the hazardous area shall be identified for a Class I, Division 1 location.

E. 517.61(A)(5) Receptacles and attachment plugs shall provide connection for a grounding conductor and be listed for Class I, Group C locations.

F. 517.61(A)(6) Cords supplying equipment of more than 8 V shall be of a type approved for extra-hard usage and shall include an additional conductor for grounding. See Table 400.4.

517.61(B)(1) Wiring above the hazardous area shall be per 517.60 in RMC, EMT, IMC, or Type MI or MC cable with a continuous gastight/vaportight metal sheath.

517.61(B)(2) Spark-producing devices, except wall-mounted receptacles above hazardous areas, shall be the totally enclosed type or constructed to prevent escape of sparks. See Exception.

G. 517.61(B)(3) Surgical lighting units shall be identified for Class I, Division 2 locations, except there are no surface temperature limitations. See 501.130(B).

H. 517.61(B)(3), Ex. 1 Fixtures above the hazardous area shall conform to the same requirements as fixtures in Class I, Division 2 locations per 501.130(B)(1), except for temperature limitations and integral or pendant switches.

517.61(B)(3), Ex. 2 An integral or pendant surgical switch is not required to be explosionproof if it is not lowerable into the hazardous area.

517.61(B)(4) Seals are required when any part of the system extends into a hazardous area per 501.15 and 501.15(A)(4).

517.61(B)(5) Single-phase, 120 V attachment plugs and receptacles above the hazardous area shall be listed for hospital use and voltage, frequency, rating, number of conductors, and provide for connection of the grounding conductor.

517.64(A)(1 – 3) Equipment in contact with persons shall not be over 10 V or shall be approved as intrinsically safe or double-insulated and be moisture-resistant.

517.64(C)(2) Isolating transformer for low-voltage circuits shall have the core and case connected to an EGC.

517.64(F) Receptacles and attachment plugs on low voltage shall be noninterchangeable with higher voltages.

I. 517.75 All new, used, or reconditioned X-ray equipment moved to a new location shall be an approved type.

517.160(A)(1) Circuits to equipment over 10 V shall be isolated and have a switch in each conductor. Isolated power system may be transformer, M-G set, or batteries.

J. 517.160(A)(2) In general, circuits in anesthetizing locations shall be supplied by isolating transformers rated at not more than 600 V primary or secondary. The primary winding shall have overcurrent protection. Secondary circuits shall have overcurrent protection, double-pole switching, and be ungrounded.

K. 517.160(B)(1) The secondary circuit shall have a ground detector, a green light when under 3.7 mA, a red light and audible when 5.0 mA or over. Lights shall be conspicuously visible to persons in the room. See Exception.

THEATERS AND SIMILAR LOCATIONS – 520

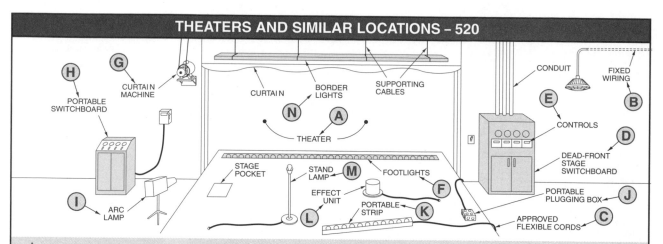

A. **520.1** This article includes the building or part of the building used for presentation, musical, dramatic, motion picture projection, and assembly areas of motion picture and TV studios.

B. **520.5(A)** Fixed wiring methods shall be metal raceways, nonmetallic raceways encased in 2″ of concrete, or Type MI, MC, or AC cable with insulated EGC. See Exception.

C. **520.5(B)** Portable equipment is permitted to be wired with approved flexible cords and cables.

520.5(C) Nonrated areas are permitted to be wired with Type NM, AC cable, or RNC or ENT.

520.7 No exposed live parts are permitted on stage. Live parts shall be enclosed or guarded.

D. **520.21** Stage switchboard shall be dead-front and per 408, Part IV unless suitable as a stage switchboard and approved by AHJ.

520.22 Exposed live parts on back shall be enclosed by walls, grills, or other approved means and the entrances of enclosures shall have self-closing doors.

E. **520.23** Controls and overcurrent devices for portable branch circuits are required at stage switchboards.

520.24 A metal hood is required unless the stage switchboard is dead-front and dead-rear (or recessed).

520.41(A – B) Circuits for footlights, border lights, and side lights shall not be over 20 A unless they have heavy-duty lamps. See 210 for heavy-duty lampholders.

520.42 Conductors shall have at least 125°C insulation. Ampacity shall be based on that of 60°C conductors.

F. **520.43(B)** Footlights shall be a metal trough or individual outlets with RMC, FMC, IMC, or Type MI or MC cable. Circuit conductors shall be soldered to lampholder terminals.

520.43(C) An automatic disconnect switch is required for disappearing footlights.

520.44(C)(1) Electric cables for border lights, drop boxes, and connector strips are permitted only where flexibility is necessary. They shall be listed for extra-hard usage and suitably supported.

520.45 Stage receptacles for equipment or fixtures shall be rated in amperes. Conductors shall be rated per Article 310 and Article 400.

520.46 Receptacles for portable stage lighting are required to be in suitable pockets or enclosures and shall comply with 520.45.

520.47 Lamps in scene locks shall be guarded and 2″ away from combustible material. See Exception.

G. **520.48** Curtain machines shall be listed.

520.49 For smoke ventilators, there shall be two required externally operable switches in the closed circuit flue. The damper control shall be set at the electrician's station.

H. **520.51** A fused switch or CB shall control a portable switchboard.

520.53(G) A pilot light shall be provided in the enclosure.

520.53(H)(1) Supply cords or cables shall be listed for extra-hard usage.

I. **520.61** Arc lamps including lamps and ballasts are required to be listed.

J. **520.62 (A – B)** Portable plugging box receptacles shall have overcurrent protection per 520.45 and be totally enclosed.

K. **520.64** Portable strips shall be constructed per the specifications for border lights per 520.44(A).

L. **520.66** "Effect units" shall be guarded against contact with combustible materials.

M. **520.68(A)(2)** A listed, hard-usage cord is acceptable for a stand lamp if the circuit protection is not over 20 A.

N. **520.81** All metal raceways, metal frames, and enclosures of equipment (including border lights and portable fixtures) shall be connected to an EGC per Article 250.

CABLE TRAYS – 392

A. **Table 392.10(A)** Cable trays shall be permitted to support: MI cable, ENT, AC, MC, NM, SE, UF, power and control tray cable, ITC, power-limited tray cable, and other factory-assembled tray cables, IMC, RMC, RNC, EMT, FMT, FMC, optical fiber cables, LFMC, LFNC, fire alarm cable, and communication cables. Cable tray installations shall not be limited to industrial establishments. See Table 392.10 (A).

392.10(B) In industrial establishments where supervision and maintenance is present, other single conductors and multiconductor cables are permitted in cable trays.

392.10(B)(1)(c) The EGC is permitted to be No. 4 or larger.

392.18(B) Each run of cable tray shall be a complete system before cables are installed.

B. **392.20(B)** Cables rated 600 V or less are not permitted with cables rated over 600 V except: (1) where separated by a fixed barrier of material; (2) Type MC cable is used.

392.18(D) Cable trays may pass through partitions, walls, floors, or platforms where the possible spread of fire is prevented per 300.21.

392.18(F) Adequate working space is required about cable trays to permit installation and maintenance of cables.

C. **392.60(A)** Metal cable trays shall be grounded in the same manner as conductor enclosures per 250.96 if they support electrical conductors.

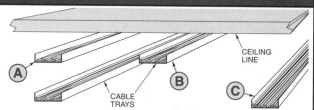

392.60(A – B) Metal trays shall be grounded per 250, and, if used as EGCs, shall comply with 392.60(B). See Table 392.60(A).

392.56 Approved cable splices are permitted in the cable tray if they are accessible and insulated by approved methods. Splices are not permitted to extend above the side rails if they are subject to physical damage.

392.80(A)(1 – 3) Multiconductor cables shall have ampacity per Tables 310.15(B)(16) and 310.15(B)(18) except they may be derated to 95% if more than 6′ of trays have an unventilated solid cover. Derating shall be limited to the number of current-carrying conductors in the cable and not to the number of conductors in the cable tray.

BRACKET FIXTURE WIRING — 520.63

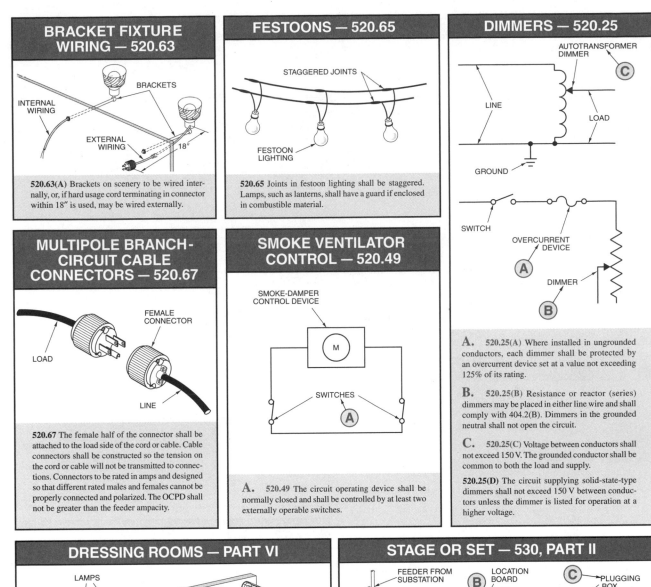

520.63(A) Brackets on scenery to be wired internally, or, if hard usage cord terminating in connector within 18″ is used, may be wired externally.

FESTOONS — 520.65

520.65 Joints in festoon lighting shall be staggered. Lamps, such as lanterns, shall have a guard if enclosed in combustible material.

DIMMERS — 520.25

A. **520.25(A)** Where installed in ungrounded conductors, each dimmer shall be protected by an overcurrent device set at a value not exceeding 125% of its rating.

B. **520.25(B)** Resistance or reactor (series) dimmers may be placed in either line wire and shall comply with 404.2(B). Dimmers in the grounded neutral shall not open the circuit.

C. **520.25(C)** Voltage between conductors shall not exceed 150 V. The grounded conductor shall be common to both the load and supply.

520.25(D) The circuit supplying solid-state-type dimmers shall not exceed 150 V between conductors unless the dimmer is listed for operation at a higher voltage.

MULTIPOLE BRANCH-CIRCUIT CABLE CONNECTORS — 520.67

520.67 The female half of the connector shall be attached to the load side of the cord or cable. Cable connectors shall be constructed so the tension on the cord or cable will not be transmitted to connections. Connectors to be rated in amps and designed so that different rated males and females cannot be properly connected and polarized. The OCPD shall not be greater than the feeder ampacity.

SMOKE VENTILATOR CONTROL — 520.49

A. **520.49** The circuit operating device shall be normally closed and shall be controlled by at least two externally operable switches.

DRESSING ROOMS — PART VI

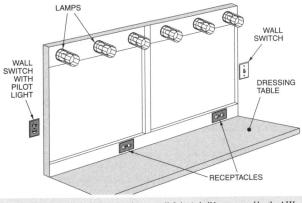

110.2 Equipment (including dressing room lighting) shall be approved by the AHJ. Also see 90.4 and 100 –Definitions, approved for purpose.

520.71 Pendant lampholders are not permitted in dressing rooms.

520.72 Lamps within 8′ of the floor shall have riveted, open-end guards or otherwise be sealed or locked in place.

520.73 Lamps and receptacles located adjacent to mirrors and above the dressing table shall be controlled by wall switches. Switch-controlling receptacles shall have pilot lights located outside the dressing room.

STAGE OR SET — 530, PART II

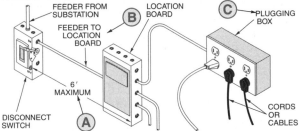

530.11, 530.12, 530.13 Studio wiring requirements generally follow those for theater stage.

A. **530.13** A single externally operable switch may be used to disconnect all contactors on a location board that is not more than 6 ′ away.

B. **530.18(B)** Overcurrent devices for feeders may be set at not over 400% of the conductor rating.

C. **530.18(E)** Where plugging box does not have overcurrent devices, every cord or cable smaller than No. 8 supplied through it shall have its own overcurrent device.

530.19(A) See Table 530.19(A) for feeder demand factors for stage set lighting.

530.20 Pendant and portable lamps and special portable equipment need not be grounded if the operating supply voltage is not over 150 V DC to ground.

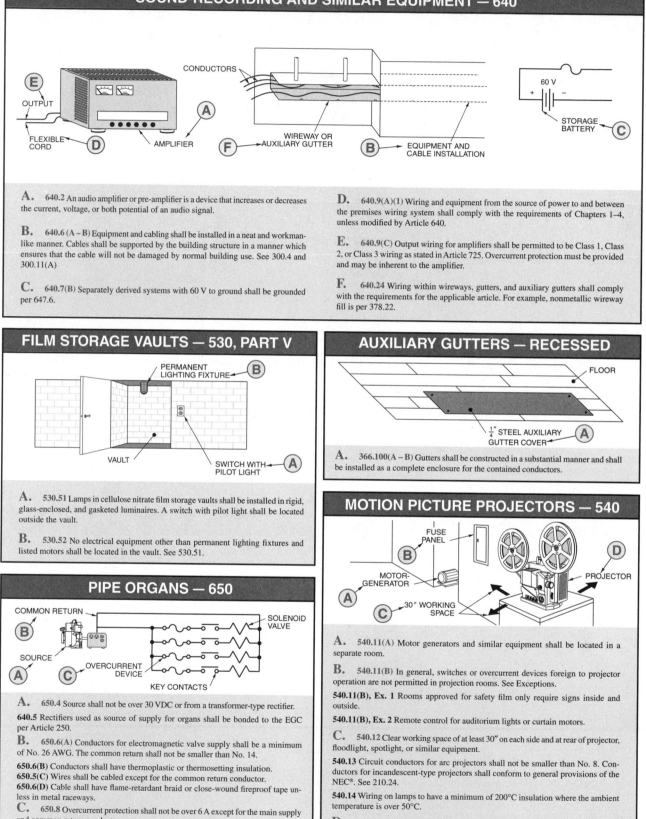

SOUND-RECORDING AND SIMILAR EQUIPMENT — 640

A. 640.2 An audio amplifier or pre-amplifier is a device that increases or decreases the current, voltage, or both potential of an audio signal.

B. 640.6 (A – B) Equipment and cabling shall be installed in a neat and workmanlike manner. Cables shall be supported by the building structure in a manner which ensures that the cable will not be damaged by normal building use. See 300.4 and 300.11(A)

C. 640.7(B) Separately derived systems with 60 V to ground shall be grounded per 647.6.

D. 640.9(A)(1) Wiring and equipment from the source of power to and between the premises wiring system shall comply with the requirements of Chapters 1–4, unless modified by Article 640.

E. 640.9(C) Output wiring for amplifiers shall be permitted to be Class 1, Class 2, or Class 3 wiring as stated in Article 725. Overcurrent protection must be provided and may be inherent to the amplifier.

F. 640.24 Wiring within wireways, gutters, and auxiliary gutters shall comply with the requirements for the applicable article. For example, nonmetallic wireway fill is per 378.22.

FILM STORAGE VAULTS — 530, PART V

A. 530.51 Lamps in cellulose nitrate film storage vaults shall be installed in rigid, glass-enclosed, and gasketed luminaires. A switch with pilot light shall be located outside the vault.

B. 530.52 No electrical equipment other than permanent lighting fixtures and listed motors shall be located in the vault. See 530.51.

AUXILIARY GUTTERS — RECESSED

A. 366.100(A – B) Gutters shall be constructed in a substantial manner and shall be installed as a complete enclosure for the contained conductors.

MOTION PICTURE PROJECTORS — 540

A. 540.11(A) Motor generators and similar equipment shall be located in a separate room.

B. 540.11(B) In general, switches or overcurrent devices foreign to projector operation are not permitted in projection rooms. See Exceptions.

540.11(B), Ex. 1 Rooms approved for safety film only require signs inside and outside.

540.11(B), Ex. 2 Remote control for auditorium lights or curtain motors.

C. 540.12 Clear working space of at least 30″ on each side and at rear of projector, floodlight, spotlight, or similar equipment.

540.13 Circuit conductors for arc projectors shall not be smaller than No. 8. Conductors for incandescent-type projectors shall conform to general provisions of the NEC®. See 210.24.

540.14 Wiring on lamps to have a minimum of 200°C insulation where the ambient temperature is over 50°C.

D. 540.20 Projectors and enclosures for associated equipment shall be listed.

PIPE ORGANS — 650

A. 650.4 Source shall not be over 30 VDC or from a transformer-type rectifier.

640.5 Rectifiers used as source of supply for organs shall be bonded to the EGC per Article 250.

B. 650.6(A) Conductors for electromagnetic valve supply shall be a minimum of No. 26 AWG. The common return shall not be smaller than No. 14.

650.6(B) Conductors shall have thermoplastic or thermosetting insulation.
650.5(C) Wires shall be cabled except for the common return conductor.
650.6(D) Cable shall have flame-retardant braid or close-wound fireproof tape unless in metal raceways.

C. 650.8 Overcurrent protection shall not be over 6 A except for the main supply and common return conductors.

EMERGENCY SYSTEMS — 700

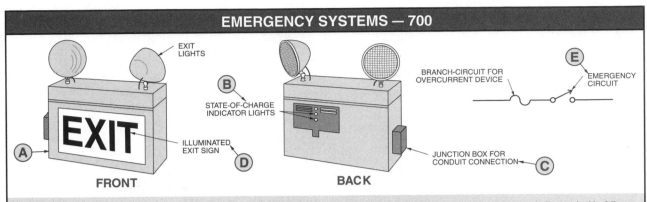

FRONT

BACK

EXIT LIGHTS

B — STATE-OF-CHARGE INDICATOR LIGHTS

A

ILLUMINATED EXIT SIGN — D

E — EMERGENCY CIRCUIT

BRANCH-CIRCUIT FOR OVERCURRENT DEVICE

JUNCTION BOX FOR CONDUIT CONNECTION — C

A. **700.1** Emergency systems are normally installed in buildings where large numbers of persons assemble or occupy such buildings, such as theaters, assembly halls, hotels, sports arenas, hospitals, and similar institutions. Emergency systems also provide power for hospital equipment that is essential to maintain life and for illumination and power for operating rooms, fire alarm systems, fire pumps, critical public address systems, industrial processes where interruption would cause hazards, and similar functions. See Life Safety Code, NFPA 101, 2006 (ANSI) for critical emergency illumination and exit lights and lighting. 700.1 applies when emergency systems or circuits are legally required by municipal, state, federal, or other codes or by any governmental agency having jurisdiction.

B. **700.6** Audible and visual derangement signals shall be provided to ensure constant readiness of emergency system. See 700.6(A – D).

C. **700.10(B)** Wiring to be entirely separate from all other wiring and boxes and raceways to be identified. See 700.10(A – D).

700.12(F) Unit requirement for emergency lighting to consist of: (1) rechargeable battery; (2) means to charge battery; (3) one or more lamps on unit (terminals are permitted for remote lamp location); (4) a relay automatically energized by failure of general lighting system.

Battery shall be able to maintain at least 87% of nominal voltage at full lamp load for at least 1 hours or maintain at least 60% of rated illumination for at least 1 hours. Unit shall be approved for emergency service. Unit not to be portable and to be wired per Ch 3. Flexible cord and plug permitted if not over 3′ long. Unit wired to same branch circuit as other normal lighting in area and connected ahead of any local switches. Remote location unit equipment lamps shall be wired per Ch 3. Branch circuit feeding unit equipment shall be clearly identified at distribution panel.

700.15 Only emergency light outlet may be served by emergency lighting system.

D. **700.16** Emergency lighting includes exit lights and illuminated exit signs.

E. **700.20** Only authorized persons have access to emergency switch. Series, 3-, or 4-way switches not permitted. See Ex. 1, 2.

700.22 Exterior emergency light may be switched by light-actuated device where there is sufficient daylight available.

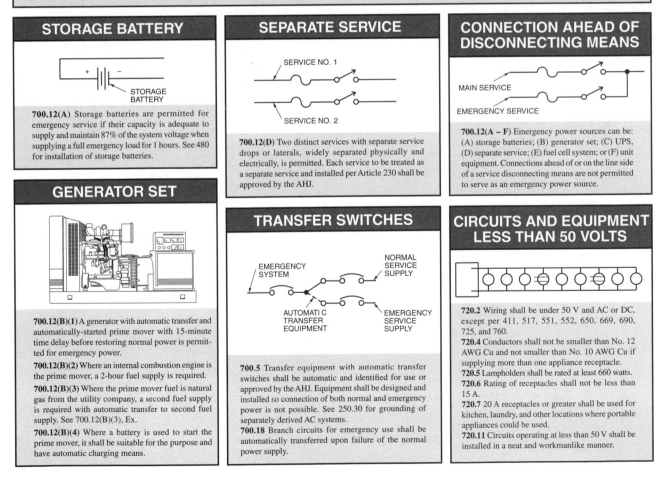

STORAGE BATTERY

STORAGE BATTERY

700.12(A) Storage batteries are permitted for emergency service if their capacity is adequate to supply and maintain 87% of the system voltage when supplying a full emergency load for 1 hours. See 480 for installation of storage batteries.

GENERATOR SET

700.12(B)(1) A generator with automatic transfer and automatically-started prime mover with 15-minute time delay before restoring normal power is permitted for emergency power.

700.12(B)(2) Where an internal combustion engine is the prime mover, a 2-hour fuel supply is required.

700.12(B)(3) Where the prime mover fuel is natural gas from the utility company, a second fuel supply is required with automatic transfer to second fuel supply. See 700.12(B)(3), Ex.

700.12(B)(4) Where a battery is used to start the prime mover, it shall be suitable for the purpose and have automatic charging means.

SEPARATE SERVICE

SERVICE NO. 1

SERVICE NO. 2

700.12(D) Two distinct services with separate service drops or laterals, widely separated physically and electrically, is permitted. Each service to be treated as a separate service and installed per Article 230 shall be approved by the AHJ.

TRANSFER SWITCHES

EMERGENCY SYSTEM

NORMAL SERVICE SUPPLY

AUTOMATIC TRANSFER EQUIPMENT

EMERGENCY SERVICE SUPPLY

700.5 Transfer equipment with automatic transfer switches shall be automatic and identified for use or approved by the AHJ. Equipment shall be designed and installed so connection of both normal and emergency power is not possible. See 250.30 for grounding of separately derived AC systems.

700.18 Branch circuits for emergency use shall be automatically transferred upon failure of the normal power supply.

CONNECTION AHEAD OF DISCONNECTING MEANS

MAIN SERVICE

EMERGENCY SERVICE

700.12(A – F) Emergency power sources can be: (A) storage batteries; (B) generator set; (C) UPS, (D) separate service; (E) fuel cell system; or (F) unit equipment. Connections ahead of or on the line side of a service disconnecting means are not permitted to serve as an emergency power source.

CIRCUITS AND EQUIPMENT LESS THAN 50 VOLTS

720.2 Wiring shall be under 50 V and AC or DC, except per 411, 517, 551, 552, 650, 669, 690, 725, and 760.

720.4 Conductors shall not be smaller than No. 12 AWG Cu and not smaller than No. 10 AWG Cu if supplying more than one appliance receptacle.

720.5 Lampholders shall be rated at least 660 watts.

720.6 Rating of receptacles shall not be less than 15 A.

720.7 20 A receptacles or greater shall be used for kitchen, laundry, and other locations where portable appliances could be used.

720.11 Circuits operating at less than 50 V shall be installed in a neat and workmanlike manner.

CLASS 1 CIRCUITS

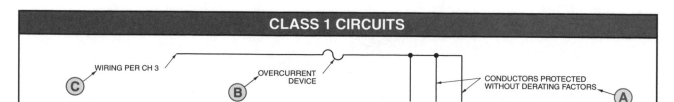

WIRING PER CH 3

Ⓒ

OVERCURRENT
DEVICE

Ⓑ

CONDUCTORS PROTECTED
WITHOUT DERATING FACTORS

Ⓐ

725.41(A – B) Class 1 Circuits–Power Limitations: Source rated output not over 30 V and 1000 VA and having overcurrent protection not over 167% of VA rating of source divided by source-rated volts. Overcurrent devices to be noninterchangeable and approved for purpose. May be integral part of power supply. (1) Article 450 applies where transformers are used to supply Class 1 circuits. (2) If source is other than transformer, it shall have maximum output of 2500 VA and maximum amps times maximum volts shall not exceed 10,000 VA with overcurrent protection by-passed. (B) Class 1 remote-control and signaling circuits not over 600 V but power output need not be limited.

A. **725.43** Conductors (No. 14 or larger) to be protected per Tables 310.15(B)(16) through 310.15(B)(19) without derating factors. Maximum overcurrent protection for No. 18 is 7 A and 10 A for No. 16. See Ex. and Note.

B. **725.45(A)** Overcurrent device at point where conductor receives supply.

725.45(B) Class 1 circuit conductors are permitted to be tapped where the overcurrent device protecting the larger conductor also protects the smaller conductor.

725.45(C) Branch-circuit taps require only short-circuit and ground-fault protection.

C. **725.46** Wiring installed per Ch 3 except where permitted by 725.48 through 725.51 and other articles of the NEC®. See Ex. 1 and 2.

725.48 Two or more Class 1 circuits may be in the same box cable or raceway if all wires are insulated for the maximum voltage. Also, Class 1 and power supply conductors shall be intermingled only when supplying equipment to which Class 1 conductors are functionally associated.

725.49(A) No. 18 and No. 16 conductors permitted if in raceways or listed cables and if leads do not exceed ampacity per 402.5.

725.49(B) Conductors rated 600 V. Conductors No. 16 and larger per Article 310. No. 16 and No. 18 shall be TF or similar. See Types listed in 725.49(B).

725.51(A – B) Ampacity of Class 1 wires shall be reduced for over 3 conductors if continuous load. Ampacity of all reduced when power supply and Class 1 in raceway and more than three Class 1 carry continuous loads or ampacity of power supply conductors only when more than three.

CLASS 2 AND CLASS 3 CIRCUITS

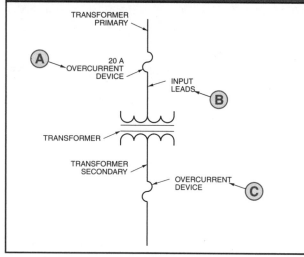

TRANSFORMER
PRIMARY

Ⓐ 20 A
OVERCURRENT
DEVICE

INPUT
LEADS
Ⓑ

TRANSFORMER

TRANSFORMER
SECONDARY

OVERCURRENT
DEVICE
Ⓒ

725.48(B)(1 – 4) Class 1 circuits and power supply circuits can only occupy the same cable, enclosure, or raceway where the equipment powered is functionally associated. See (B)(2) for factory or field-assembled control centers and (B)(3) for conductors in access holes and (B)(14) for cable tray.

725.121(B) Power supplies shall not be interconnected or paralleled unless listed for such interconnection.

A. **725.127** Transformer is protected (maximum) by a 20 A overcurrent device.

B. **725.127, Ex.** Input leads shall not be smaller than No. 14 except may be No. 18 if not over 12″ long and insulation per 725.49(B).

725.139(E) Class 2 and Class 3 circuits may be run in same cable as communication circuits, but shall be classed as communication circuits and conform to Article 800.

C. **Ch 9, Table 11(A), Notes** Overcurrent devices shall not be interchangeable with devices of higher ratings and they may be an integral part of power supply. Overcurrent devices shall be located at a point where conductor to be protected receives its supply.

COMBINING CLASS 2 AND CLASS 3 CIRCUITS

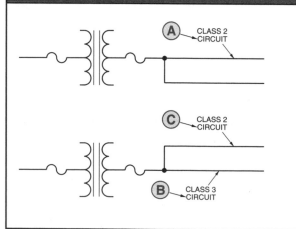

Ⓐ CLASS 2
CIRCUIT

Ⓒ CLASS 2
CIRCUIT

Ⓑ CLASS 3
CIRCUIT

A. **725.139(A)** Two or more Class 2 circuits are permitted in the same raceway, cable, or enclosure.

B. **725.139(B)** Two or more Class 3 circuits are permitted in the same raceway, cable, or enclosure.

C. **725.139(C)** One or more Class 2 circuits are permitted in same raceway, cable, or enclosure with Class 3 circuits if Class 2 conductor insulation is at least equal to that required for Class 3 circuits.

725.141 Where Class 2 or Class 3 circuits extend beyond building and are subject to possible contact with over 300 V to ground, light, or power circuits, they shall conform to 800.44, 800.50, 800.53, 800.93, 800.100, 800.170(A), and 800.170(B) for other than coaxial conductors and 820.44, 820.93, and 820.100 for coaxial conductors.

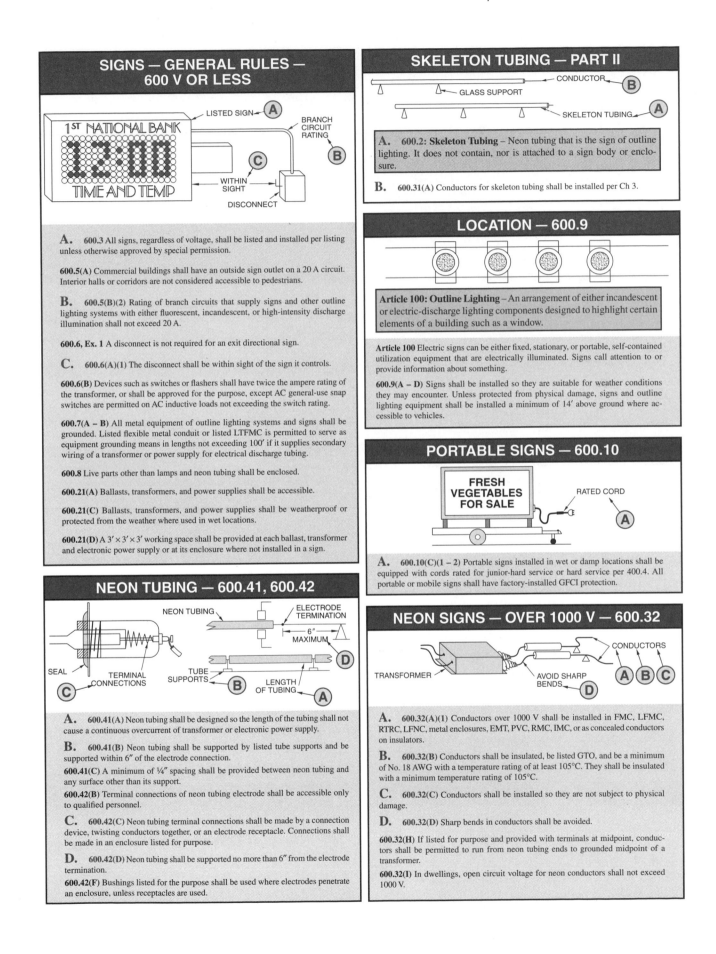

SIGNS — GENERAL RULES — 600 V OR LESS

A. **600.3** All signs, regardless of voltage, shall be listed and installed per listing unless otherwise approved by special permission.

600.5(A) Commercial buildings shall have an outside sign outlet on a 20 A circuit. Interior halls or corridors are not considered accessible to pedestrians.

B. **600.5(B)(2)** Rating of branch circuits that supply signs and other outline lighting systems with either fluorescent, incandescent, or high-intensity discharge illumination shall not exceed 20 A.

600.6, Ex. 1 A disconnect is not required for an exit directional sign.

C. **600.6(A)(1)** The disconnect shall be within sight of the sign it controls.

600.6(B) Devices such as switches or flashers shall have twice the ampere rating of the transformer, or shall be approved for the purpose, except AC general-use snap switches are permitted on AC inductive loads not exceeding the switch rating.

600.7(A – B) All metal equipment of outline lighting systems and signs shall be grounded. Listed flexible metal conduit or listed LTFMC is permitted to serve as equipment grounding means in lengths not exceeding 100′ if it supplies secondary wiring of a transformer or power supply for electrical discharge tubing.

600.8 Live parts other than lamps and neon tubing shall be enclosed.

600.21(A) Ballasts, transformers, and power supplies shall be accessible.

600.21(C) Ballasts, transformers, and power supplies shall be weatherproof or protected from the weather where used in wet locations.

600.21(D) A 3′ × 3′ × 3′ working space shall be provided at each ballast, transformer and electronic power supply or at its enclosure where not installed in a sign.

NEON TUBING — 600.41, 600.42

A. **600.41(A)** Neon tubing shall be designed so the length of the tubing shall not cause a continuous overcurrent of transformer or electronic power supply.

B. **600.41(B)** Neon tubing shall be supported by listed tube supports and be supported within 6″ of the electrode connection.
600.41(C) A minimum of ¼″ spacing shall be provided between neon tubing and any surface other than its support.
600.42(B) Terminal connections of neon tubing electrode shall be accessible only to qualified personnel.

C. **600.42(C)** Neon tubing terminal connections shall be made by a connection device, twisting conductors together, or an electrode receptacle. Connections shall be made in an enclosure listed for purpose.

D. **600.42(D)** Neon tubing shall be supported no more than 6″ from the electrode termination.
600.42(F) Bushings listed for the purpose shall be used where electrodes penetrate an enclosure, unless receptacles are used.

SKELETON TUBING — PART II

A. **600.2: Skeleton Tubing** – Neon tubing that is the sign of outline lighting. It does not contain, nor is attached to a sign body or enclosure.

B. **600.31(A)** Conductors for skeleton tubing shall be installed per Ch 3.

LOCATION — 600.9

Article 100: Outline Lighting – An arrangement of either incandescent or electric-discharge lighting components designed to highlight certain elements of a building such as a window.

Article 100 Electric signs can be either fixed, stationary, or portable, self-contained utilization equipment that are electrically illuminated. Signs call attention to or provide information about something.

600.9(A – D) Signs shall be installed so they are suitable for weather conditions they may encounter. Unless protected from physical damage, signs and outline lighting equipment shall be installed a minimum of 14′ above ground where accessible to vehicles.

PORTABLE SIGNS — 600.10

A. **600.10(C)(1 – 2)** Portable signs installed in wet or damp locations shall be equipped with cords rated for junior-hard service or hard service per 400.4. All portable or mobile signs shall have factory-installed GFCI protection.

NEON SIGNS — OVER 1000 V — 600.32

A. **600.32(A)(1)** Conductors over 1000 V shall be installed in FMC, LFMC, RTRC, LFNC, metal enclosures, EMT, PVC, RMC, IMC, or as concealed conductors on insulators.

B. **600.32(B)** Conductors shall be insulated, be listed GTO, and be a minimum of No. 18 AWG with a temperature rating of at least 105°C. They shall be insulated with a minimum temperature rating of 105°C.

C. **600.32(C)** Conductors shall be installed so they are not subject to physical damage.

D. **600.32(D)** Sharp bends in conductors shall be avoided.

600.32(H) If listed for purpose and provided with terminals at midpoint, conductors shall be permitted to run from neon tubing ends to grounded midpoint of a transformer.

600.32(I) In dwellings, open circuit voltage for neon conductors shall not exceed 1000 V.

OUTSIDE BRANCH CIRCUITS AND FEEDERS — 225

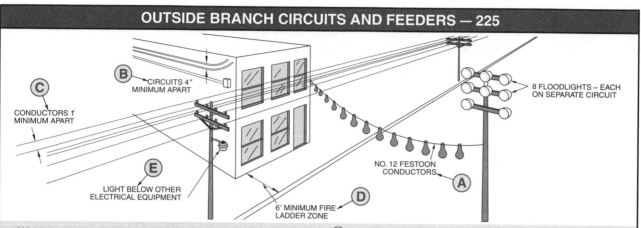

225.4 Festoons or open conductors within 10' of building shall be insulated or covered.

225.4, Ex. EGC and grounded conductor where elsewhere permitted.

225.6(A)(1) Overhead spans for 600 V or less and not over 50', No. 10 Cu or No. 8 Al; over 50', No. 8 Cu or No. 6 Al.

225.6(A)(2) Overhead spans for over 600 V, No. 6 Cu or No. 4 Al for separate conductors and No. 8 Cu or No. 6 Al for cable.

A. **225.6(B)** Festoon conductors shall be not less than No. 12 unless messenger supported. Messenger wire is required where the span exceeds 40'.

225.7(B) A common neutral may be used for circuits on a pole installation. The neutral size shall not be less than the total load connected to any one phase of the supply circuit.

B. **225.14(C)** Open conductors shall be separated from open conductors of other circuits by at least 4".

C. **225.14(D)** Conductors on poles shall be at least 1' apart where not on racks or brackets. For horizontal climbing space, see 225.14(D).

D. **225.19(E)** Fire ladder zone of 6' adjacent to building or beginning not more than 8' therefrom if building is over 3 stories or 50' high.

225.24 Stranded conductors are required where pin-type outdoor lampholders are used.

E. **225.25** Lights shall be below other electrical equipment unless safeguards or clearances are provided for relamping or equipment is controlled by lockable disconnect.

800.44(A)(1 – 4) Aerial communication conductors shall be below power lines and not on the same crossarm with light or power.

800.44(B) Communication conductors shall be 8' minimum over roofs.

COMMUNICATION CIRCUITS — 800

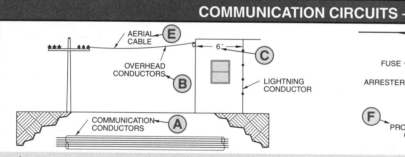

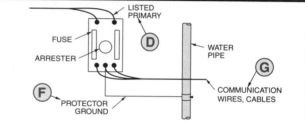

A. **800.47(A)** Underground communication conductors shall be in separate compartment from light, power conductors, Class I, or nonpower-limited fire alarm conductors.

B. **800.44** Overhead communications wires and cables entering buildings shall comply with 800.44 (A) and (B).

800.50(C) Where protector is installed in a building, wires shall enter (in general) through a noncombustible, nonabsorbent insulating bushing or metal raceway.

C. **800.53** Six foot separation from lightning conductors where practical.

D. **800.90(A)(1)** Fuseless type protectors may be used under special conditions outlined in 800.90(A)(1)(a – e).

800.90(A)(2) Protector has fuse in each line wire and arrester between each wire and ground.

800.90(B) Primary protectors shall be located in, on, or where the conductors enter the building.

800.90(C) Protectors shall not be located in any hazardous location.

E. **800.93(A)** Metal sheath of aerial or underground cables entering buildings shall be grounded or interrupted at the building by an insulating joint or equivalent device.

F. **800.100(A)(1 – 6)** The protector bonding conductor or GEC shall be at least No. 14 Cu, 30 mil rubber insulation, run in straight line to the nearest accessible grounding electrode location per 800.100(B)(2)(1-7).

G. **800.133(A)(1 – 2)** Electric light, power, Class I circuits shall not be placed in any raceway outlet box, junction box with communication conductors, or wiring.

SHOW WINDOW SIGNS — OVER 600 V

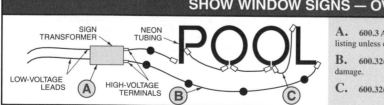

A. **600.3** All electric signs, regardless of voltage, shall be listed and installed per listing unless excepted by special permission or as in 600.3(A) or (B).

B. **600.32(C)** Conductors shall be installed as they are not subject to physical damage.

C. **600.32(D)** Sharp bends in installed conductors shall be avoided.

RECEIVING EQUIPMENT — ANTENNA SYSTEMS — 810, PART II

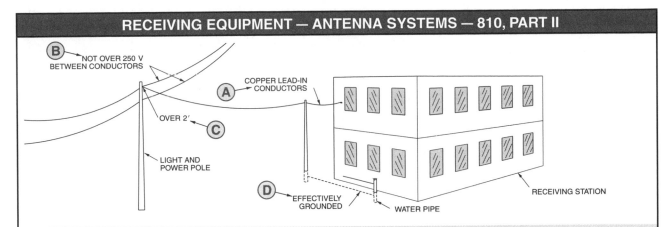

A. 810.11 Antenna and lead-in conductors shall be hand-drawn copper, bronze, aluminum alloy, copper-clad steel, or other high strength corrosion-resistant material.

810.11, Ex. Soft- or medium-drawn copper shall be used for lead-in if the span is not over 35′.

B. 810.12 Antenna and associated conductors shall not be attached to a pole carrying light and power wires over 250 V between conductors or to electric service masts.

C. 810.13 Antenna and lead-in conductors shall not cross over light and power conductors or come within 2′ of them if 250 V or less.

D. 810.15 Masts shall be effectively grounded per 810.21.

810.16(A) Antenna conductors are sized per Table 810.16(A).

810.18(A) Lead-in conductors shall be supported so they cannot swing closer than 2′ to 250 V wires or 10′ to conductors over 250 V. Lead-ins shall not be closer than 6′ to the lightning rod system unless bonded per 250.60. Underground conductors shall be separated at least 12″ from light or power or Class I circuits.

810.18(B) Indoor antennas and lead-ins shall not be closer than 2″ to other wiring systems. See Exceptions.

810.18(B), Ex. 1 Unless in metal raceway or cable armor.

810.18(B), Ex. 2 Permanently separated therefrom by continuous and firmly secured nonconductor such as porcelain tubes or flexible tubing.

810.20(A) Lead-in shall have a listed antenna discharge unit.

810.20(A), Ex. Listing is not required where lead-in wires are enclosed in a continuous metal shield that is effectively grounded or protected by the antenna discharge unit.

810.21(B) Bonding conductors or GECs may be uninsulated.

810.21(E) Grounding conductor shall be run in a straight line as practicable from the mast or discharge unit.

810.21(F) The grounding conductor shall be connected to the grounding means per 810.21(F)(1 – 3).

810.21(G) Bonding conductor or GEC for receiving station may be inside or outside of the building.

810.21(H) The bonding conductors or GECs shall not be smaller than No. 10 Cu, No. 8 Al, or No. 17 Cu-clad steel or bronze.

810.21(I) Bonding conductors or GECs may be used for both protective and operating purposes.

AMATEUR TRANSMITTING AND RECEIVING STATION — ANTENNA SYSTEMS — PART III

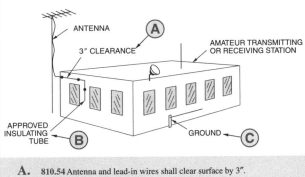

A. 810.54 Antenna and lead-in wires shall clear surface by 3″.

810.54, Ex. Lead-in may be attached to surface if encased in grounded shield. See 810.58.

B. 810.55 Except when shielded, lead-in conductors shall enter buildings through: (1) a rigid, noncombustible, nonabsorbent insulating tube; (2) a hole giving 2″ of clearance all around; or (3) a drilled window pane.

810.57 An antenna discharge unit is required in transmitting stations or other suitable means to drain static charges. See Exceptions.

810.57, Ex. 1 Where continuous metal shield is effectively grounded.

810.57, Ex. 2 Where antenna is effectively and permanently grounded per 810.58.

C. 810.58(B) The protective bonding conductors or GECs for transmitting stations shall be as large as the lead-in but not less than No. 10 Cu, bronze, or Cu-clad steel.

810.58(C) The operating bonding conductors or GECs shall be not less than No. 14 Cu or equal.

INTERIOR INSTALLATION TRANSMITTING STATION — PART IV

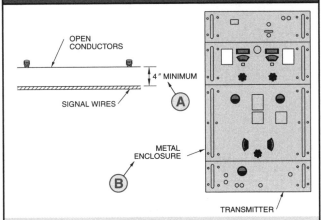

A. 810.70 Conductors inside the building shall be separated from other open conductors at least 4″ except: 1. as provided in Article 640; and 2. where conductors are separated by a fixed nonconductor or are in conduit. See exceptions.

B. 810.71(A) The transmitter shall be enclosed by metal or by installing a grounded barrier to separate the transmitter from operating space.

810.71(A – B) All noncurrent metal parts including handles shall be effectively connected to an EGC.

810.71(C) All doors giving access to voltages in excess of 350 V shall have interlocks.

Name _____ Date _____

NEC®	Answer	

_____ (T) F **1.** Seals shall be installed in conduit runs passing between two adjacent Class I, Division 2 rooms.

_____ T (F) **2.** Under some conditions, askarel-filled transformers may be used in Class II, Division 2 locations.

_____ _____ **3.** Locations where combustible dust is normally in the air in quantities sufficient to produce an explosive mixture are designated as ___.
    A. Class I, Division 2   C. Class II, Division 2
    B. Class II, Division 1   D. Class III, Division 1

_____ T F **4.** Locations where concentrations of combustible dust of an electrically conductive nature may be present in sufficient quantities to be hazardous are designated as Class II, Division 1.

_____ _____ **5.** Emergency lighting includes exit lights and illuminated exit ___.

_____ T F **6.** Open squirrel-cage induction motors that have no sliding contacts may be used in Class I, Division 1 locations if they are identified for the location.

_____ _____ **7.** Sealing compound is used with mineral-insulated cable in a Class I location for the purpose of ___.
    A. preventing passage of   C. limiting the extent of a
       gas or vapor   possible explosion
    B. excluding moisture   D. preventing the escape of insulating powder

_____ T F **8.** Hazardous concentrations of gas or vapor are not present under normal operating conditions in Class I, Division 1 locations.

_____ _____ **9.** Mobile aircraft energizers shall carry a sign reading "WARNING–KEEP ___ FEET CLEAR OF AIRCRAFT ENGINES AND FUEL TANK AREAS."

_____ _____ **10.** In Class I, Division 1 locations, conduit seals shall be placed no farther than ___″ from an explosionproof enclosure.
    A. 12   C. 24
    B. 18   D. 30

_____ _____ **11.** In a conduit run leaving a Class I, Division 1 location, the seal may be placed on either side of the ___ within 10″ from the boundary.

_____ _____ **12.** Where gases and dust are present at the same time, both shall be considered when determining the safe operating ___ of equipment.

_____ _____ **13.** Equipment that has been approved for a Division 1 location shall be permitted in a Division 2 location of the same class, ___, and temperature class.

_____ _____ **14.** For limited flexibility of motor connections in a Class I, Division 2 location, flexible metal conduit with ___ fittings is permitted.

_____ _____ **15.** In a completed seal in a Class I location, the minimum thickness of the sealing compound for other than listed cable sealing fittings shall not be less than the trade size of the conduit and, in no case, less than ___″.

_____ _____ **16.** Fused isolating switches for transformers not intended to interrupt current in a Class I, Division 2 location ___.

    A. shall be explosionproof   C. shall have interlocking devices

    B. may be installed in a general-purpose type enclosure   D. may not have doors or openings in area

_____ _____ **17.** In Class I, Division 2 locations, concentrations of explosive gases are normally ___ in closed containers or systems.

_____ _____ **18.** Anesthetizing locations shall be Class I, Division 1 locations to a height of ___′ above the floor.

_____ _____ **19.** A flexible cord may be used in a Class I location between a portable appliance and the ___ portion of its supply circuit.

_____ _____ **20.** In Class I, Division 1 locations, pendant fixtures with rigid conduit stems over 12″ long shall be permanently and effectively___.

_____ _____ **21.** Locations where combustible fibers are stored are designated as ___.

    A. Class II, Division 2   C. Class III, Division 2

    B. Class III, Division 1   D. nonhazardous

_____ _____ **22.** In a Class II location, a horizontal run of conduit not less than ___′ long may be used as the sealing means.

_____ _____ **23.** Single-pole ___ utilizing handle ties shall not be permitted to serve as a circuit disconnecting means for gasoline dispensing equipment.

_____ _____ **24.** A parking garage where only storage or parking occurs is considered to be a(n) ___ area.

_____ T   F **25.** A location where cloth is woven is designated as Class III, Division 1.

_____ _____ **26.** Flexible cords suitable for the type of service and listed for ___ usage may be used as pendants in a commercial garage.

_____ _____ **27.** Underground conduit runs within 20′ of a gasoline dispensing location are considered to be in a Class I, Division ___ location.

_____ _____ **28.** Class III locations are those where ___ fibers are present.

_____ _____ **29.** Where electrodes penetrate an enclosure, ___ listed for the purpose shall be provided unless receptacles are installed.

_____ _____ **30.** The secondary voltage of an isolating transformer used in an anesthetizing location may not exceed ___ V between conductors of each circuit.

_____ _____ **31.** All threaded NPT joints in conduit used in Class I, Division 1 locations shall be made with at least ___ threads fully engaged.

_____ _____ **32.** Disappearing stage footlights shall have provisions for ___ disconnecting the current supply when the footlights are replaced in the storage recess designed for them.

_____ _____ **33.** Incandescent lamps less than 8′ from the floor of a theatrical dressing room shall be equipped with ___ guards.

_____ _____ **34.** In Class I, Division 1 locations, conduit seals are required adjacent to boxes containing splices if the conduit size is equal to or larger than ___″.

    A. ¾   C. 1½

    B. 1   D. 2

_____ _____ **35.** A switch controlling a plug receptacle in a theatrical dressing room shall be equipped with a(n) ___ located outside the dressing room.

Name _____ Date _____

NEC®	Answer		

_____ T F **1.** Receptacles and attachment plugs in Class II, Division 1 locations shall be listed for Class II locations.

_____ T F **2.** In general, explosionproof portable electric lamps may be used inside a paint spray booth during operation.

_____ _____ **3.** Where magnesium, aluminum, or aluminum-bronze powders may be present, ___ and capacitors are not permitted to be installed in Class II, Division 1 locations.

_____ T F **4.** Portable appliances used inside an aircraft hangar shall be suitable for Class I, Division 2 locations or zone 2 locations.

_____ _____ **5.** Class 2 and 3 circuits can occupy the same raceway as communication circuits if they are ___ as such and installed per Article 800.

_____ _____ **6.** For general wiring in Class I, Division 1 locations, ___ may be used.
  A. RMC   C. FMC
  B. EMT   D. all of the above

_____ T F **7.** Floor areas in minor repair garages without pits or below grade work areas shall be unclassified.

_____ _____ **8.** Unit equipment to be used as emergency lighting can be cord-and-plug connected if the cord is not over ___′ long.

_____ _____ **9.** Each branch circuit supplying an electric sign shall be controlled by a(n) ___ operable switch or circuit breaker that opens all ungrounded conductors.

_____ _____ **10.** Areas within 18″ horizontally of gas-dispensing pumps are considered Class I, Division 2 locations to the top of the ___.
  A. island   C. concrete
  B. curb   D. dispenser

_____ _____ **11.** All 120 V, 1ϕ, 15 A and 20 A circuits within ___′ of a wet bar or sink in a mobile home shall have GFCI protection.

_____ _____ **12.** Markings for maximum allowable lamp wattage for signs and outline lighting systems with incandescent lamp holders shall be ___ during relamping.

_____ _____ **13.** Where the AHJ so decides, locations with dried, flammable paints that have ventilating equipment interlocked with the electrical equipment may be designated as ___.
  A. Class I, Division 2   C. Class II, Division 2
  B. Class II, Division 1   D. unclassified

_____ _____ **14.** Portable lamps used in commercial garages shall be equipped with a handle, lampholder, hook, and substantial ___.

_____ _____ **15.** Neon tubing that is ___ to pedestrians shall be protected from physical damage.

_____ T F **16.** Open spaces within 20′ horizontally of a gas-dispensing pump are designated as Class I, Division 1 locations.

_____ T F **17.** Audible and visual signal devices shall be used, where practical, to give warning of derangement of emergency power.

_____ _____ **18.** If acceptable to the AHJ, a separate ___ is suitable for use as an emergency power source.

_____ _____ **19.** Internal combustion engines used as prime movers shall have at least a(n) ___ hour supply of fuel.

_____ T F **20.** Open outside areas within 20′ horizontally from the fill pipe of an underground gasoline tank shall be considered as a Class I, Division 2 location to a height of 18″.

_____ _____ **21.** Two or more Class 1 circuits can be in the same box, cable, or raceway where all wires are insulated for the ___ voltage.

_____ T F **22.** Under some conditions, spray booths may be illuminated through suitable glass panels.

_____ _____ **23.** Power supply for Class 2 and 3 circuits shall not be connected in ___ unless they are listed for such interconnection.

_____ _____ **24.** Effectively isolated storerooms and similar areas adjacent to aircraft hangars shall be designated as ___ where adequately ventilated.
   A. Class I, Division 2   C. Class II, Division 2
   B. Class II, Division 1   D. unclassified

_____ _____ **25.** The output to Class I remote-control and signaling circuits is unlimited for systems rated at ___ V or less.

_____ _____ **26.** Wiring to lamps in a motion picture projector shall have insulation with high ambient temperature if exposed to over ___ °C.

_____ _____ **27.** Lampholders operating on systems rated less than 50 V are required to be at least ___ W.

_____ _____ **28.** When a separate service is used as the emergency power source, the service conductors shall be sufficiently electrically and physically ___ from the normal service.

_____ _____ **29.** In a Class II location where electrically conducting dust is present, flexible connections at motors may be made with ___.
   A. FMC   C. hard-usage cord
   B. Type AC armored cable   D. none of the above

_____ _____ **30.** Branch circuits used for emergency lighting shall be ___ transferred upon failure of normal service.

_____ _____ **31.** OCPDs for feeders can be set at a maximum of ___% their conductor rating in buildings used for motion picture production.

_____ _____ **32.** Where a pool is installed for a dwelling, at least one receptacle is required between ___′ and 20′ from the pool.

_____ _____ **33.** No electrical equipment, other than permanent lighting fixtures, shall be located in a film storage ___ with cellulose nitrate film.

_____ _____ **34.** Where lighter-than-air gaseous fuels will not be transferred, such locations in a commercial garage shall be ___.

_____ _____ **35.** For other than simple apparatus, all intrinsically safe apparatus and associated apparatus shall be ___.

Name _____ Date _____

NEC®   **Answer**

_____   T   F   **1.** Resistance and reactor dimmers for stage lighting may be placed in the grounded conductor of a circuit.

_____   _____   **2.** If the AHJ can determine that flammable liquids with a flash point below 100°F, such as gasoline, will not be handled, the location is not required to be ___.

_____   T   F   **3.** Branch circuits that supply only neon tubing installations shall be rated at 30 A or less.

_____   _____   **4.** A well-ventilated room where flammable anesthetics are stored in approved containers shall be considered a ___ area.
     A. Class I, Division 1   C. Class II, Division 2
     B. Class I, Division 2   D. nonhazardous

_____   T   F   **5.** Storage batteries, motor-generator sets, or isolating transformers may be used to provide ungrounded circuits in anesthetizing locations.

_____   _____   **6.** Electric cranes operating over combustible fibers shall be supplied by a(n) ___ power source.

_____   _____   **7.** For calculating the load for the required branch circuit for signs or outline lighting in commercial occupancies, ___ VA is used.

_____   _____   **8.** Conduits installed below the surface of a gasoline dispenser are considered to be in a(n) ___ location.

_____   T   F   **9.** A low-voltage isolating transformer supplying circuits within an anesthetizing location shall have the tie core and case connected to an EGC.

_____   _____   **10.** Pendant lampholders are not permitted in ___ rooms of theaters.

_____   _____   **11.** Each commercial building and occupancy with ground floor footage shall have at least one outside sign outlet rated at ___ A.
     A. 15   C. 25
     B. 20   D. 30

_____   _____   **12.** Boxes and fittings used in Class III, Division 1 locations must be ___.

_____   _____   **13.** A feeder supplying power to stage lighting units with a load of 70,000 VA can have a demand factor of ___ applied.

_____   T   F   **14.** Lamps in cellulose nitrate film vaults may be installed in general-use nongasketed fixtures provided they are equipped with suitable guards.

_____   _____   **15.** The circuit supplying an autotransformer dimmer used in theaters shall not exceed ___ V between conductors.

_____   _____   **16.** Dimmers used for theater lighting shall have protection set at ___% of their rating.

_____ _____ **17.** In motion picture studios, the ampacity of feeder conductors to the stage may be protected at a maximum value of ___%.
      A. 200   C. 400
      B. 250   D. 500

_____ _____ **18.** Cable tray runs shall be ___ before conductors are installed.

_____ T   F **19.** Surgical fixtures more than 5' above the floor of an operating room may be installed on a grounded circuit.

_____ _____ **20.** All circuits routed through a(n) ___ pump shall be provided with a means to disconnect simultaneously from the source of supply of all conductors.

_____ _____ **21.** Lamps used in Class III, Division 2 locations shall be protected against the ___ of lint and fibers.

_____ _____ **22.** In aircraft hangars, lighting fixtures less than ___' above wings and engine enclosures shall be totally enclosed.
      A. 10   C. 20
      B. 15   D. none of the above

_____ _____ **23.** All metal parts around a dispensing pump shall be ___ and bonded.

_____ _____ **24.** Switches and controllers used in Class III locations shall be installed in ___ enclosures.

_____ _____ **25.** In general, live parts of electric signs and outline lighting other than lamps and neon tubing shall be ___.

_____ _____ **26.** The ___ fitting is the first fitting after a conduit emerges from the earth or concrete in a dispenser pump.

_____ _____ **27.** A storage battery supplying emergency lighting and power shall maintain not less than 87½% of full voltage at total load for a period of at least ___ hour(s).
      A. ¼   C. 1½
      B. 1   D. 2

_____ _____ **28.** In a Class II location, a 50 kVA transformer containing askarel shall be located at least ___" from combustible material.

_____ _____ **29.** Storeroom office and control rooms in aircraft hangers are considered ___ if proper ventilation is provided.

_____ _____ **30.** Portable lamps used in Class III locations shall be of the ___ type with no provisions for raceway attachment plugs.

_____ _____ **31.** Motor-driven utilization equipment shall be dust-___ if installed in a Class II location.

_____ _____ **32.** Dry-type transformers with open terminals that are installed in Class II locations are required to be installed in ___.

_____ _____ **33.** The primary voltage of an isolating transformer used in an anesthetizing location shall not exceed ___ V.
      A. 150   C. 300
      B. 250   D. 600

_____ _____ **34.** A 10' or longer length of horizontal conduit can serve as a(n) ___ in Class II locations.

_____ _____ **35.** The entire floor area in a major repair garage adjacent to classified locations in which flammable vapors are not likely to be released shall not be classified where mechanical ventilation providing ___ air changes per hour is provided.

Name _____ Date _____

NEC®	Answer	

_____   T   F   **1.** A plug receptacle used exclusively for cleaning purposes may be tapped from the emergency circuit wires.

_____   _____   **2.** For open conductor spacing of outdoor feeders, conductors on poles, where not placed on racks or brackets, shall be separated by no less than ___″.
     A. 6   C. 18
     B. 12   D. 24

_____   T   F   **3.** The area containing a dip tank or coating tank is considered a Class I, Division 1 location.

_____   _____   **4.** The color ___ can be used to identify intrinsically safe conductors.

_____   T   F   **5.** In patient care areas, metal raceways shall not be required where AC cable is used as a wiring method when the metal jacket itself qualifies as an EGC.

_____   _____   **6.** In hospitals, if the normal power source is lost, the emergency system shall be automatically restored to operation within ___ seconds.

_____   _____   **7.** An anesthetizing location is considered a Class I, Division 1 location to a height of ___′ above the floor.

_____   T   F   **8.** Testing each level is required when equipment ground fault protection (GFPE) is first installed in a health care facility.

_____   _____   **9.** The rating of a lampholder on a circuit operating at less than 50 V shall be at least ___ VA.
     A. 220   C. 550
     B. 330   D. 660

_____   T   F   **10.** A connection on the supply side of the service disconnect may be permitted as an emergency service.

_____   _____   **11.** On circuits of 600 V or less, overhead feeder spans up to 50′ in length shall have copper conductors no smaller than No. ___.
     A. 6   C. 10
     B. 8   D. 12

_____   _____   **12.** Lighting fixtures installed over ___′ above finished grade in commercial garages can be of the general purpose type.

_____   T   F   **13.** Business offices in health care facilities are not considered as patient care areas.

_____   _____   **14.** Electrical equipment to be installed in hazardous locations shall be marked to show the ___, group, and operating temperature.

_____   _____   **15.** Areas in which flammable fuel is dispensed into vehicle fuel tanks shall be classified in accordance with Table ___.

_____ _____ **16.** In operating rooms using flammable anesthetics, the entire area is a Class I, Division 1 location up to a level of ___′ above the floor.

    A. 4   C. 6

    B. 5   D. 7

_____ _____ **17.** Surge arrestors installed in Class I, Division 1 locations are to be in ___ enclosures for Class I, Division I.

_____ _____ **18.** Each branch of the essential electrical system shall be served by ___ or more transfer switches.

    A. one   C. three

    B. two   D. four

_____ _____ **19.** Class I locations in bulk storage plants shall not extend beyond a(n) ___ partition, such as walls, floors, or roofs, that has no communicating opening.

_____ T   F **20.** In patient care areas, the grounding terminals of all receptacles shall be connected to an insulated copper equipment grounding conductor.

_____ _____ **21.** Receptacles and attachment plugs installed in Class I, Division 1 locations shall be ___ for the location.

_____ _____ **22.** Signaling apparatus used in Class II, Division 2 locations shall be in ___ metal enclosures or other enclosures identified for the location.

_____ _____ **23.** Open-type generators shall be permitted to be used in Class I, Division 2 locations if they have no ___ devices exposed.

_____ _____ **24.** Boxes and fittings used in Class II, Division 1 locations shall be ___.

_____ _____ **25.** In unattended self-service stations, emergency controls shall be located at least ___′ from dispensers.

_____ _____ **26.** Flexible cord used for flexible connections in a Class II, Division 1 location shall be listed and be of the ___ type and terminated with listed dusttight fittings.

_____ _____ **27.** The isolated circuit conductors for a 3ϕ system supplying an anesthetizing location shall have the third conductor identified as ___ with at least one distinctive colored stripe other than white, green, or gray.

    A. yellow   C. red

    B. black   D. blue

_____ _____ **28.** A minimum of ___″ spacing shall be provided between neon tubing and any surface other than its support.

_____ _____ **29.** In general, the space adjacent to enclosed coating and dipping operations shall be considered ___.

_____ _____ **30.** Portable lamps shall be identified for Class II locations and marked for ___ lamp wattage.

_____ _____ **31.** No transformer or capacitor is permitted in a Class II, Division 1, Group ___ location.

_____ _____ **32.** A Class I, Division 1 location is an area where flammable gases and vapors can exist under ___ operating conditions.

_____ T   F **33.** Communication wires and cables are permitted on the same crossarm with light or power wires if separated by at least 2′.

_____ _____ **34.** Explosionproof apparatus is capable of withstanding an explosion without rupturing and of ___ the ignition of a gas or vapor surrounding the enclosure.

_____ _____ **35.** Where a Class II, Group E motor operates without being subject to overload, its operating temperature is ___°C.

Name_____ Date _____

NEC®	Answer	

230.03 B)   T   (F)   **1.** The smallest permissible hard-drawn copper overhead service conductor is No.10.

_____   _____   **2.** No. ___ Al EGC is required for an automatic OCPD set at 200 A.
  A. 2   C. 6
  B. 4   D. 8

_____   T   F   **3.** The GEC for a small, one-family dwelling shall be permitted to be spliced by the exothermic welding process.

_____   _____   **4.** The grounded system conductor for a high-impedance grounded system shall be at least No. ___ Cu.

_____   _____   **5.** Gaps in plaster, drywall, or plasterboard and other noncombustible surfaces shall be repaired so that they do not exceed ___ at the edge of the box.

_____   T   F   **6.** Any available wall space in a dwelling shall be within 7' of a plug receptacle.

_____   T   F   **7.** Under certain conditions, switch enclosures may be used as junction boxes.

_____   T   F   **8.** GFCI protection is required for all receptacles serving countertop surfaces and those located within 6' of the wet bar sink in areas other than a kitchen.

_____   _____   **9.** The service disconnecting means rating shall not be less than ___ A 3-wire for a one-family dwelling.
  A. 50   C. 70
  B. 60   D. 100

_____   _____   **10.** Conductors of dissimilar metals shall not be intermixed in a terminal or splicing connector where physical contact occurs, unless the device is ___ for the purpose and conditions of use.
  A. marked   C. listed
  B. approved   D. identified

_____   T   F   **11.** The service disconnecting means may consist of not more than six switches or CBs per service grouped in any one location.

_____   _____   **12.** The service or feeder capacity necessary for four 10 kVA electric ranges is ___ kVA.
  A. 15   C. 19
  B. 17   D. 21

_____   _____   **13.** When initially installed and permitted by 400.7, flexible cords shall be used only in ___ lengths.

_____   _____   **14.** The clearance of 120/240 V service-drop conductors over residential driveways not subject to truck traffic shall be no less than ___ ' where the voltage does not exceed 300 V to ground.

_____   _____   **15.** The first 20,000 VA of computed load in a hotel shall be assessed at ___%.

_____ _____ **16.** Unless otherwise specified in the Code, conductors used to carry current shall be of ___.
    A. copper   C. copper-clad aluminum
    B. aluminum   D. none of the above

_____ _____ **17.** The Cu GEC for a 350 kcmil Cu, 120/240 V service installation, if run to a waterpipe electrode, shall be no smaller than No. ___.

_____ _____ **18.** The smallest permissible driven nonlisted ground rod electrode made of copper material is ___″ in diameter.
    A. ⅜   C. ⅝
    B. ½   D. ¾

_____ T   F **19.** It is permissible to install ten No.12 conductors in a 4″ by 1½″ square box that also contains one switch.

_____ _____ **20.** The minimum clearance from ground of service drop conductors operating at 600 V or less, which pass over public streets, alleys, roads, and driveways, on other-than-residential property, shall be ___′.
    A. 10   C. 15
    B. 12   D. 18

_____ _____ **21.** Under the optional calculation for a one-family dwelling, the electric space heating load is assessed at ___% if less than four separately controlled units.

_____ _____ **22.** In general, Type AC cable shall be supported at intervals not exceeding ___′.

_____ _____ **23.** Heating cables installed on drywall ceilings shall be secured at intervals not exceeding ___″ by approved means.
    A. 10   C. 14
    B. 12   D. 16

_____ _____ **24.** The first 3000 VA of computed load in a multifamily dwelling shall be assessed at ___%.

_____ _____ **25.** Receptacles for a 20 A circuit shall be rated no less than ___ A.

_____ _____ **26.** The leads of a 120 V heating cable are marked with the color identification ___ to indicate the circuit voltage on which it is to be used.

_____ _____ **27.** Except in dwellings, each ___′ or fraction of multioutlet assembly strip is usually considered as a load of 180 VA when it is unlikely that appliances will be used simultaneously.

_____ T   F **28.** The cross-sectional area of all conductors in any part of a metal auxiliary gutter shall not exceed 15% of the cross-sectional area of the gutter.

_____ _____ **29.** A 5 HP, 230 V, 3φ squirrel-cage motor with no code letter is marked 40°C and has a nameplate rating of 15 A. In no case shall the setting of an adjustable overload device ___ A.
    A. 19   C. 21
    B. 20   D. 25

_____ _____ **30.** The ampacity of No. 2 THWN-THHN, dual-rated copper conductors in rigid conduit and installed in a wet location is ___ A.

_____ T   F **31.** A header used with precast cellular concrete floor raceways shall not be run at right angles to the cells.

_____ T   F **32.** An AC general-use snap switch controlling an inductive load shall have an ampere rating equal to 150% of the load.

_____ _____ **33.** In general, LFNC other than Type LFNC-B is limited to lengths not longer than ___′.

_____ _____ **34.** Metal surface raceways used for signaling and lighting circuits shall have two separate compartments, each identified by a stamp, imprint, or color coding of the ___ finish.

_____ _____ **35.** Splices and taps in underfloor raceway installations shall be made only in ___.

_____ _____ **36.** Nonmetallic underground conduit with conductors, of Type NUCC, is permitted to be used ___.
  A. for direct burial C. encased or embedded in concrete
  B. in cinder fill D. all of the above

_____ _____ **37.** Sheet metal auxiliary gutters installed in wet locations shall be ___ for the location.

_____ _____ **38.** The motor ___ means shall be in sight of the motor and the driven machinery location or shall be individually capable of being locked in the open position.

_____ T F **39.** The maximum ampere rating of a 250 V cartridge fuse is 400 A.

_____ T F **40.** A No. 18 Cu conductor may be used to ground a portable device connected to a 30 A circuit.

_____ T F **41.** Conductors in underfloor raceway systems shall be permitted to be looped to connect individual outlets.

_____ _____ **42.** The minimum size Cu GEC required for a 120/240 V service consisting of one 250 kcmil conductor per phase is No. ___. The grounding electrode is a ground rod and it is a sole connection to the electrode.

_____ _____ **43.** The allowable ampacity of a conductor is 55 A. It is considered protected by a(n) ___ A fuse or CB.

_____ T F **44.** Secondary circuits of wound-rotor induction motors require separate overload protection.

_____ _____ **45.** Two or more services are permitted when the capacity requirements for the service are over ___ A.
  A. 1500 C. 2500
  B. 2000 D. 3000

_____ T F **46.** Transformers are never permitted to have overcurrent protection devices in both primary and secondary circuits.

_____ _____ **47.** A wood building, without water, is built on the earth without footings. If a 3000 A, 460 V service is installed and grounded solely by ground rods, the minimum size copper GEC is No. ___ AWG.
  A. 3/0 C. 6
  B. 4 D. 8

_____ _____ **48.** The conductor fill of an underfloor raceway shall not exceed ___ of the interior area of the raceway.
  A. 20 C. 75
  B. 40 D. 100

_____ T F **49.** A motor branch-circuit, short-circuit device may sometimes be rated in excess of 300% of the full load motor current.

_____ T F **50.** The ampacity of the ungrounded conductors from the generator to the first overcurrent device shall not be less than 115% of the nameplate current of the generator.

_____ _____ **51.** Branch-circuit conductors supplying a 50 HP, 460 V, 3φ induction motor shall be rated for at least ___ A.

    A. 65   C. 91

    B. 82   D. 130

_____ T   F **52.** A ½ HP permanently installed pump motor controlled by a float switch requires running overcurrent protection.

_____ T   F **53.** The primary windings of potential transformers installed indoors shall always be fused.

_____ _____ **54.** The largest size conductor permitted in an underfloor raceway is ___.

    A. 250 kcmil   C. 500 kcmil

    B. No. 0   D. determined by design of the underfloor raceway

_____ T   F **55.** An 8 A, 115 V, 1φ sealed (hermetic) motor compressor shall show the LRC on its nameplate.

_____ T   F **56.** The minimum size of Type THW Cu conductors supplying a 70 A, 480 V transformer arc welder used at 30% duty application is No. 8.

_____ _____ **57.** A lighting and power service has two No. 3/0 ungrounded copper conductors (one per phase) and a No. 1/0 neutral. A Cu GEC attached to a waterpipe electrode shall not be smaller than No. ___.

    A. 2   C. 6

    B. 4   D. 8

_____ T   F **58.** All receptacles on the property shall be located a minimum of 10′ from the inside wall of a permanently installed swimming pool.

_____ _____ **59.** The maximum initial rating of branch circuit NTDFs for a 5 HP, 230 V, 3φ squirrel-cage motor with no code letter is ___ A.

    A. 40   C. 50

    B. 45   D. 60

_____ T   F **60.** The color coding permitted for intrinsically safe conductors is light green.

_____ _____ **61.** A CB disconnecting means for a 50 HP, 460 V, 3φ motor shall have an ampere rating of at least ___ A.

    A. 63   C. 91

    B. 75   D. 26

_____ _____ **62.** The maximum size of NTDFs for branch circuit overcurrent protection for a 50 HP, 460 V, 3φ motor is ___ A.

    A. 90   C. 250

    B. 200   D. 300

_____ _____ **63.** Duct heaters must include a fan circuit interlock, which ensures that the fan circuit is ___ when any heater circuit is energized.

    A. locked out   C. energized

    B. locked in   D. none of the above

_____ _____ **64.** A 5 HP, 230 V, 3φ squirrel-cage motor with no code letter is started at full voltage. The minimum ampacity of the branch-circuit conductor is ___ A.

    A. 15.2   C. 19

    B. 18   D. 20

Name _____ Date _____

NEC®	Answer	

_____   T   F   **1.** When the power for a fire pump is provided by a tap ahead of the service disconnecting means, the tap is permitted to be made within the service disconnecting means compartment.

_____   _____   **2.** A demand factor of ___% may be applied to the neutral feeder load for electric ranges.
     A. 40   C. 70
     B. 60   D. 80

_____   T   F   **3.** FMC used to provide flexibility at equipment shall be equipped with an EGC.

_____   T   F   **4.** A tap for an individual lampholder or fixture that is less than 12″ long may not be smaller than the branch circuit conductors.

_____   T   F   **5.** Under certain conditions, the neutral feeder conductor is permitted to be smaller than the ungrounded conductors.

_____   _____   **6.** In general, wiring for emergency systems shall be kept entirely ___ of all other wiring and equipment and shall not occupy the same raceway, box, cable or cabinet with other wiring.

_____   T   F   **7.** SE conductors, are permitted to be spliced if specific conditions are met.

_____   _____   **8.** The ampacity of a No. 3/0 THW Cu conductor at a room temperature of 50°C is ___ A.
     A. 125   C. 175
     B. 150   D. 200

_____   _____   **9.** The rating of conductors supplying power to a capacitor shall be at least one-third of the motor circuit conductors and ___% of the FLC rating of the capacitor.

_____   _____   **10.** Conduit bodies shall not contain splices, taps, or devices unless they are marked by the manufacturer to indicate their ___ .

_____   _____   **11.** For 600 V or less, electrical equipment rated over ___ A or more than 6′ wide, that contains overcurrent devices, there shall be at least one entrance 24″ wide and 6½′ high at each end of the work space.

_____   T   F   **12.** Embedded electric space heating cables shall not be spliced.

_____   T   F   **13.** Dry-type transformers installed indoors and rated at 15,000 V shall be installed in vaults.

_____   _____   **14.** A 480 V transformer with a primary FLC rating of less than 2 A shall be protected at ___% of its primary current where "primary only" protection is provided.

_____   _____   **15.** The ampacity of conductors supplying a 20 A oscillator induction heating unit shall not be less than ___ A.

_____ _____ **16.** The primary OCPD, when properly selected, may protect the secondary conductors only if the secondary of the transformer is 1φ and feeds a(n) ___ wire circuit.

_____ _____ **17.** The feeder load for three electric ranges (13 kVA, 18 kVA, and 20 kVA) is ___ kVA.
A. 15   C. 19
B. 17.5   D. 21

_____ _____ **18.** Where thermal cutouts are used for motor overload protection for a 3φ motor, a total of ___ thermal cutouts are required.

_____ _____ **19.** Class I equipment shall not have any ___ surface that operates at a temperature in excess of the ignition temperature of the specific gas or vapor.

_____ _____ **20.** Suitable disconnecting switches or ___ connectors shall be installed to permit the disconnection of all ungrounded conductors of each temporary circuit in a temporary installation.
A. 5   C. 15
B. 10   D. 20

_____ _____ **21.** To supply a feeder for night lights in an apartment, the required size of THW Cu conductors for a continuous 100 A load shall not be smaller than No. ___ AWG.
A. 0   C. 2
B. 1   D. 3

_____ T   F **22.** The screw shell of a fuseholder shall not be connected to the line side of the circuit.

_____ _____ **23.** Inside locations of spray booths are classified as a Class I, ___ location.

_____ _____ **24.** The general lighting load for an office building is calculated at ___ VA per sq ft.
A. 2   C. 3.5
B. 2.5   D. 5

_____ _____ **25.** Transformers shall not be installed in a Group ___ location.

_____ _____ **26.** In general, raceways shall not be used as a means of support for other ___, cables, or nonelectric equipment unless they are identified or associated control conductors.

_____ _____ **27.** The maximum rating of a cord-and-plug-connected appliance used on a 20 A branch circuit shall be ___ A.
A. 10   C. 16
B. 12   D. 20

_____ _____ **28.** The area within ___' horizontally from aircraft fuel tanks is classified as a Class I, Division 2 location.

_____ _____ **29.** The electrical systems in aircraft shall be ___ when they are stored in hangars or undergoing maintenance.

_____ _____ **30.** PVC and RTRC conduit can be used to supply power to a dispenser pump if buried at least ___' in the earth.

_____ _____ **31.** Under certain conditions, an unbroken run of conduit can pass through a Class I, Division 1 location without ___ being provided if the termination is in an unclassified location against the passage of gas or vapor.

_____ _____ **32.** Sealing ___ shall be used to seal fittings in a conduit run.

_____ _____ **33.** Receptacles in marinas and boatyards that provide shore power to boats rated 30 A and 50 A shall be of the locking and ___ types.

_____ _____ **34.** Non-current-carrying metal parts of X-ray and associated equipment including ___ cables shall be grounded.

_____ _____ **35.** Receptacles that supply power to spas and hot tubs with associated electrical components shall be ___-protected.

_____ _____ **36.** The screw shell of a plug-type fuseholder shall be connected to the ___side of the circuit.

_____ _____ **37.** The full load current of a 5 HP, 208 V, 3ϕ induction motor is ___ A.

_____ _____ **38.** The radius of bends in mineral-insulated cable with a ¾″ external diameter shall be not less than ___ times the cable diameter.

_____ _____ **39.** All exposed noncurrent-carrying metal parts of information technology equipment shall be ___ to the EGC.

_____ _____ **40.** Horsepower rated switches are not required as a disconnecting means for 300 V stationary motors of ___ HP or less.

_____ _____ **41.** In general, the controller disconnect shall be in sight of the ___ location.

_____ T   F **42.** Open wiring on insulators is generally permitted to be used for branch circuit temporary wiring on construction sites.

_____ _____ **43.** Open transformers installed indoors and rated less than 112½ kVA shall have a clearance of at least ___″ from combustible material unless a fire-resistant, heat-insulated barrier is provided.

_____ _____ **44.** A psychiatric hospital is used exclusively for such care on a 24-hour basis for ___ or more inpatients.

_____ _____ **45.** The isolated circuit conductors for a 2-wire system supplying an anesthetizing location shall be identified by using ___ conductors with at least one stripe other than white, green, or gray.
    A. orange and brown   C. purple and pink
    B. black and red   D. none of the above

_____ _____ **46.** Flexible cords and cables shall not be used as a substitute for the ___wiring of a structure.

_____ _____ **47.** Intrinsic safety wiring methods shall be marked with labels and identified "Intrinsic Safety Wiring" every ___′.

_____ _____ **48.** Unshielded lead-in conductors of amateur transmitting stations shall clear the building surface that is wired over by a distance of at least ___″.
    A. 1   C. 3
    B. 2   D. 4

_____ _____ **49.** General purpose enclosures can be used in Class I, Division 2 locations if the contacts are immersed in ___.

_____ _____ **50.** A basic parking garage is not considered a(n) ___ location if it is used only for parking and storage.

_____ _____ **51.** Flexible cords feeding 120 V equipment within a hazardous anesthetizing location shall be of the ___ type and shall include one additional conductor for grounding.

_____ _____ **52.** A 10′ or longer length of horizontal conduit can serve as a(n) ___ in Class II locations.

_____ _____ **53.** The entire floor area in a commercial garage up to a level of 18″ can be classified as nonhazardous with ventilation providing ___ air changes per hour.

_____ _____ **54.** Ten 10 kVA ovens located in an apartment complex can have a demand factor of ___ kVA.

_____ _____ **55.** In general, junction boxes for swimming pools shall be located at least 4′ from the inside wall of the pool and they shall be located not less than 4″ from the pool deck or 8″ from the maximum level of the pool water, whichever provides the ___ elevation.

_____ _____ **56.** With no overcurrent protection in the secondary, the primary overcurrent device for a 600 V dry-type transformer with a rated primary current of 100 A shall not exceed ___ A.

    A. 125   C. 250
    B. 150   D. 600

_____ _____ **57.** The use of a continuous white or gray covering on a conductor or termination is reserved for the ___ circuit conductor.

_____ _____ **58.** A bathroom is an area with a(n) ___ and one or more of the following: toilet, tub, shower, bidet, urinal, or similar plumbing fixture.

_____ _____ **59.** Receptacles installed over kitchen countertops 12″ or wider shall be installed so that no point on the wall is more than ___″ from a receptacle outlet.

_____ _____ **60.** A service with phase conductors rated at 300 kcmil per phase shall have a minimum grounded conductor of No. ___ Cu.

_____ T   F **61.** Two to six disconnecting means can serve to disconnect the service equipment of a multifamily dwelling.

_____ _____ **62.** Fixtures over ___ lb shall be supported independently of the box unless listed for the weight to be supported.

_____ _____ **63.** The lowest point of attachment for a service drop to a building is 12′.

_____ _____ **64.** THHN Cu conductors in a circuit routed through an ambient temperature of 121°F shall be derated ___%.

_____ _____ **65.** A No. 6 Cu conductor rated at 60°C terminated to an OCPD has an allowable ampacity of ___ A.

_____ _____ **66.** The smallest size EGC permitted in cable trays is No. ___ or larger.

_____ T   F **67.** The ungrounded conductors of a 3-pole OCPD used as a service disconnecting means shall be disconnected simultaneously from the premises wiring system.

_____ _____ **68.** Each length of PVC conduit shall be clearly and durably marked within every ___′.

_____ _____ **69.** A 6′ fence with 1′ of barbed wire on top is equivalent to a(n) ___′ fence for guarding equipment of over 600 V.

_____ _____ **70.** A(n) ___ VA rating is required for each linear foot of show window.

_____ _____ **71.** A control transformer with a primary FLC of 1 A can be protected with a(n) ___ A fuse.

_____ _____ **72.** A disconnecting means located within ___′ and within sight can serve as the disconnecting switch for a motor.

_____ _____ **73.** A 4¹¹⁄₁₆″ by 1¼″ box has a(n) ___ cu in. volume rating.

THREE-PHASE VOLTAGE VALUES

For 208 V × 1.732, use 360
For 230 V × 1.732, use 398
For 240 V × 1.732, use 416
For 440 V × 1.732, use 762
For 460 V × 1.732, use 797
For 480 V × 1.732, use 831
For 2400 V × 1.732, use 4157
For 4160 V × 1.732, use 7205

POWER FORMULA ABBREVIATIONS AND SYMBOLS

P = Watts	V = Volts
I = Amps	VA = Volt Amps
A = Amps	φ = Phase
R = Ohms	√ = Square Root
E = Volts	

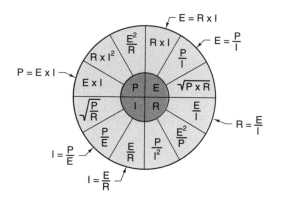

VALUES IN INNER CIRCLE
ARE EQUAL TO VALUES
IN CORRESPONDING
OUTER CIRCLE

OHM'S LAW AND POWER FORMULA

POWER FORMULAS—1φ, 3φ

Phase	To Find	Use Formula	Example		
			Given	Find	Solution
1φ	I	$I = \dfrac{VA}{V}$	32,000 VA, 240 V	I	$I = \dfrac{VA}{V}$ $I = \dfrac{32,000\ VA}{240\ V}$ $I = \textbf{133 A}$
1φ	VA	$VA = I \times V$	100 A, 240 V	VA	$VA = I \times A$ $VA = 100\ A \times 240\ V$ $VA = \textbf{24,000 VA}$
1φ	V	$V = \dfrac{VA}{I}$	42,000 VA, 350 A	V	$V = \dfrac{VA}{I}$ $V = \dfrac{42,000\ VA}{350\ A}$ $V = \textbf{120 V}$
3φ	I	$I = \dfrac{VA}{V \times \sqrt{3}}$	72,000 VA, 208 V	I	$I = \dfrac{VA}{V \times \sqrt{3}}$ $I = \dfrac{72,000\ VA}{360\ V}$ $I = \textbf{200 A}$
3φ	VA	$VA = I \times V \times \sqrt{3}$	2 A, 240 V	VA	$VA = I \times V \times \sqrt{3}$ $VA = 2 \times 416$ $VA = \textbf{832 VA}$

AC/DC FORMULAS

To Find	DC	AC		
		1φ, 115 or 220 V	1φ, 208, 230, or 240 V	3φ – All Voltages
I, HP known	$\dfrac{HP \times 746}{E \times E_{ff}}$	$\dfrac{HP \times 746}{E \times E_{ff} \times PF}$	$\dfrac{HP \times 746}{E \times E_{ff} \times PF}$	$\dfrac{HP \times 746}{1.73 \times E \times E_{ff} \times PF}$
I, kW known	$\dfrac{kW \times 1000}{E}$	$\dfrac{kW \times 1000}{E \times PF}$	$\dfrac{kW \times 1000}{E \times PF}$	$\dfrac{kW \times 1000}{1.73 \times E \times PF}$
I, kVA known		$\dfrac{kVA \times 1000}{E}$	$\dfrac{kVA \times 1000}{E}$	$\dfrac{kVA \times 1000}{1.763 \times E}$
kW	$\dfrac{I \times E}{1000}$	$\dfrac{I \times E \times PF}{1000}$	$\dfrac{I \times E \times PF}{1000}$	$\dfrac{I \times E \times 1.73 \times PF}{1000}$
kVA		$\dfrac{I \times E}{1000}$	$\dfrac{I \times E}{1000}$	$\dfrac{I \times E \times 1.73}{1000}$
HP (output)	$\dfrac{I \times E \times E_{ff}}{746}$	$\dfrac{I \times E \times E_{ff} \times PF}{746}$	$\dfrac{I \times E \times E_{ff} \times PF}{746}$	$\dfrac{I \times E \times 1.73 \times E_{ff} \times PF}{746}$

HORSEPOWER FORMULAS

To Find	Use Formula	Example		
		Given	Find	Solution
HP	$HP = \dfrac{I \times E \times E_{ff}}{746}$	240 V, 20 A, 85% E$_{ff}$	HP	$HP = \dfrac{I \times E \times E_{ff}}{746}$ $HP\ \dfrac{20\text{ A} \times 240\text{ V} \times 85\%}{746}$ $HP = \mathbf{5.5}$
I	$I = \dfrac{HP \times 746}{E \times E_{ff} \times PF}$	10 HP, 240 V, 90% E$_{ff}$, 88% PF	I	$I = \dfrac{HP \times 746}{E \times E_{ff} \times PF}$ $I\ \dfrac{10\text{ HP} \times 746}{240\text{ V} \times 90\% \times 88\%}$ $I = \mathbf{39\ A}$

VOLTAGE DROP FORMULAS—1φ, 3φ

Phase	To Find	Use Formula	Example		
			Given	Find	Solution
1φ	VD	$VD = \dfrac{2 \times R \times L \times I}{1000}$	240 V, 40 A, 60' L, 0.764R	VD	$VD = \dfrac{2 \times R \times L \times I}{1000}$ $VD = \dfrac{2 \times 0.764 \times 60 \times 40}{1000}$ $VD = \mathbf{3.67\ V}$
3φ	VD	$VD = \dfrac{2 \times R \times L \times I}{1000} \times 0.866$	208 V, 110 A, 75' L, 0.194 R, 0.866 multiplier	VD	$VD = \dfrac{2 \times R \times L \times I}{1000} \times 0.866$ $VD = \dfrac{2 \times 0.194 \times 75 \times 110}{1000} \times 0.866$ $VD = \mathbf{2.77\ V}$

VOLTAGE DROP VARIABLES

V = Voltage (in V) V$_s$ = Supply voltage (in V) I = Current (In A)	VD = Voltage drop (in V) K = Resistivity of conductor (in Ω) L = Length of conductor (in ft)	%VD = Percent voltage drop (in V) R = Resistance of conductor (in Ω/kft) 1000 = 1000' or less of conductor	V$_l$ = Voltage loss (in V) CM = Circular mils (in area) $0.866 = \dfrac{\sqrt{3}}{2}$

UNITS OF ENERGY

Energy	Btu	ft-lb	J	kcal	kWh
British thermal unit	1	777.9	1.056	0.252	2.930×10^{-4}
Foot-pound	1.285×10^{-3}	1	1.356	3.240×10^{-4}	3.766×10^{-7}
Joule	9.481×10^{-4}	0.7376	1	2.390×10^{-4}	2.778×10^{-7}
Kilocalorie	3.968	3.086	4.184	1	1.163×10^{-3}
Kilowatt-hour	3.413	2.655×10^{6}	3.6×10^{6}	860.2	1

VOLTAGE CONVERSIONS

To Convert	To	Multiply By
rms	Average	0.9
rms	Peak	1.414
Average	rms	1.111
Average	Peak	1.567
Peak	rms	0.707
Peak	Average	0.637
Peak	Peak-to-peak	2

DECIMAL EQUIVALENTS OF AN INCH

Fraction	Decimal	Fraction	Decimal	Fraction	Decimal	Fraction	Decimal
1/64	0.015625	17/64	0.265625	33/64	0.515625	49/64	0.765625
1/32	0.03125	9/32	0.28125	17/32	0.53125	25/32	0.78125
3/64	0.046875	19/64	0.296875	35/64	0.546875	51/64	0.796875
1/16	0.0625	5/16	0.3125	9/16	0.5625	13/16	0.8125
5/64	0.078125	21/64	0.328125	37/64	0.578125	53/64	0.828125
3/32	0.09375	11/32	0.34375	19/32	0.59375	27/32	0.84375
7/64	0.109375	23/64	0.359375	39/64	0.609375	55/64	0.859375
1/8	0.125	3/8	0.375	5/8	0.625	7/8	0.875
9/64	0.140625	25/64	0.390625	41/64	0.640625	57/64	0.890625
5/32	0.15625	13/32	0.40625	21/32	0.65625	29/32	0.90625
11/64	0.171875	27/64	0.421875	43/64	0.671875	59/64	0.921875
3/16	0.1875	7/16	0.4375	11/16	0.6875	15/16	0.9375
13/64	0.203125	29/64	0.453125	45/64	0.703125	61/64	0.953125
7/32	0.21875	15/32	0.46875	23/32	0.71875	31/32	0.96875
15/64	0.234375	31/64	0.484375	47/64	0.734375	63/64	0.984375
1/4	0.250	1/2	0.500	3/4	0.750	1	1.000

ENGLISH SYSTEM

		Unit	Abbr	Equivalents
LENGTH		mile	mi	5280', 320 rd, 1760 yd
		rod	rd	5.50 yd, 16.5'
		yard	yd	3', 36"
		foot	ft *or* '	12", 0.333 yd
		inch	in. *or* "	0.083', 0.028 yd
AREA		square mile	sq mi *or* mi^2	640 A, 102,400 sq rd
$A = l \times w$		acre	A	4840 sq yd, 43,560 sq ft
		square rod	sq rd *or* rd^2	30.25 sq yd, 0.00625 A
		square yard	sq yd *or* yd^2	1296 sq in., 9 sq ft
		square foot	sq ft *or* ft^2	144 sq in., 0.111 sq yd
		square inch	sq in. *or* in^2	0.0069 sq ft, 0.00077 sq yd
VOLUME		cubic yard	cu yd *or* yd^3	27 cu ft, 46,656 cu in.
$V = l \times w \times t$		cubic foot	cu ft *or* ft^3	1728 cu in., 0.0370 cu yd
		cubic inch	cu in. *or* in^3	0.00058 cu ft, 0.000021 cu yd
CAPACITY	*U.S. liquid measure*	gallon	gal.	4 qt (231 cu in.)
		quart	qt	2 pt (57.75 cu in.)
WATER, FUEL, ETC.		pint	pt	4 gi (28.875 cu in.)
		gill	gi	4 fl oz (7.219 cu in.)
		fluidounce	fl oz	8 fl dr (1.805 cu in.)
		fluidram	fl dr	60 min (0.226 cu in.)
		minim	min	1/6 fl dr (0.003760 cu in.)
	U.S. dry measure	bushel	bu	4 pk (2150.42 cu in.)
VEGETABLES, GRAIN, ETC.		peck	pk	8 qt (537.605 cu in.)
		quart	qt	2 pt (67.201 cu in.)
		pint	pt	1/2 qt (33.600 cu in.)
	British imperial liquid and dry measure	bushel	bu	4 pk (2219.36 cu in.)
		peck	pk	2 gal. (554.84 cu in.)
		gallon	gal.	4 qt (277.420 cu in.)
		quart	qt	2 pt (69.355 cu in.)
		pint	pt	4 gi (34.678 cu in.)
DRUGS		gill	gi	5 fl oz (8.669 cu in.)
		fluidounce	fl oz	8 fl dr (1.7339 cu in.)
		fluidram	fl dr	60 min (0.216734 cu in.)
		minim	min	1/60 fl dr (0.003612 cu in.)
MASS AND WEIGHT	*avoirdupois*	ton		2000 lb
		short ton	t	2000 lb
		long ton		2240 lb
COAL, GRAIN, ETC.		pound	lb *or* #	16 oz, 7000 gr
		ounce	oz	16 dr, 437.5 gr
		dram	dr	27.344 gr, 0.0625 oz
		grain	gr	0037 dr, 0.002286 oz
	troy	pound	lb	12 oz, 240 dwt, 5760 gr
GOLD, SILVER, ETC.		ounce	oz	20 dwt, 480 gr
		pennyweight	dwt *or* pwt	24 gr, 0.05 oz
		grain	gr	0.042 dwt, 0.002083 oz
	apothecaries'	pound	lb ap	12 oz, 5760 gr
		ounce	oz ap	8 dr ap, 480 gr
DRUGS		dram	dr ap	3 s ap, 60 gr
		scruple	s ap	20 gr, 0.333 dr ap
		grain	gr	.005 s, 0.002083 oz, 0.0166 dr ap

METRIC SYSTEM

	Unit	Abbr	Number of Base Units
LENGTH	kilometer	km	1000
	hectometer	hm	100
	dekameter	dam	10
	*meter	m	1
	decimeter	dm	0.1
	centimeter	cm	0.01
	millimeter	mm	0.001
AREA	square kilometer	sq km or km^2	1,000,000
	hectare	ha	10,000
	are	a	100
	square centimeter	sq cm or cm^2	0.0001
VOLUME	cubic centimeter	cu cm, cm^3, or cc	0.000001
	cubic decimeter	dm^3	0.001
	*cubic meter	m^3	1
CAPACITY	kiloliter	kl	1000
	hectoliter	hl	100
	dekaliter	dal	10
	*liter	l	1
	cubic decimeter	dm^3	1
	deciliter	dl	0.10
	centiliter	cl	0.01
	milliliter	ml	0.001
MASS AND WEIGHT	metric ton	t	1,000,000
	kilogram	kg	1000
	hectogram	hg	100
	dekagram	dag	10
	*gram	g	1
	decigram	dg	0.10
	centigram	cg	0.01
	milligram	mg	0.001

* base units

METRIC PREFIXES

Multiples and Submultiples	Prefixes	Symbols	Meaning
$1,000,000,000,000 = 10^{12}$	tera	T	trillion
$1,000,000,000 = 10^9$	giga	G	billion
$1,000,000 = 10^6$	mega	M	million
$1000 = 10^3$	kilo	k	thousand
$100 = 10^2$	hecto	h	hundred
$10 = 10^1$	deka	d	ten
Unit $1 = 10^0$			
$0.1 = 10^{-1}$	deci	d	tenth
$0.01 = 10^{-2}$	centi	c	hundredth
$0.001 = 10^{-3}$	milli	m	thousandth
$0.000001 = 10^{-6}$	micro	µ	millionth
$0.000000001 = 10^{-9}$	nano	n	billionth
$0.000000000001 = 10^{-12}$	pico	p	trillionth

ABBREVIATIONS . . .

A

abbreviation	ABBR
acoustic	ACST
acoustical tile	ACT. or AT.
adhesive	ADD. or ADH
adjustable	ADJ
adjustable-trip circuit breaker	ATCB
aggregate	AGGR
aileron	AIL
air conditioner	A/C
air handler	A/H
air tight	AT
alarm	ALM
alloy	ALY
alternating current	AC
aluminum	AI
ambient	AMB
Ambulatory Health Care Center	AHCC
American National Standards Institute	ANSI
American Wire Gauge	AWG
ammeter	A or AM
ampere	A or AMP
ampere interrupting rating	AIR.
amps	A
anchor bolt	AB
anode	A
antenna	ANT.
apartment	APT
appliance	APPL
approved	APPD or APVD
approximate	APPROX
approximately	APPROX
architectural	ARCH.
architecture	ARCH.
area	A
area drain	AD
armature	A or ARM.
asphalt	ASPH
asphalt tile	AT.
as required	AR
Assured Equipment Grounding Program	AEGP
astragal	A
Authority Having Jurisdiction	AHJ
automatic	AU or AUTO
automatic sprinkler	AS
auxiliary	AUX
avenue	AVE
azimuth	AZ

B

basement	BSMT
bathroom	B
bathtub	BT
battery (electric)	BAT.
beam	BM
bearing	BRG
bearing plate	BPL or BRG PL
bedroom	BR
benchmark	BM
black	BK
block	BLK
blocking	BLKG
blue	BL
board	BD
board foot	BF
bonding jumper	BJ
boulevard	BLVD
brake relay	BR
brass	BRS
brick	BRK
bridge	BRDG
bronze	BRZ
brown	BR
building	BL or BLDG
building line	BL
built-up roofing	BUR
bypass	BYP

C

cabinet	CAB.
cable	CA
Cable Antenna Television	CATV
cantilever	CANTIL
capacitor	CAP.
cased opening	CO
casement	CSMT
casing	CSG
cast iron	CI
cast-iron pipe	CIP
cast steel	CS
cast stone	CS or CST
catch basin	CB
cathode	K
ceiling	CLG
cellar	CEL
Celsius	°C
cement	CEM
cement floor	CF
center	CTR
centerline	CL
center-to-center	C to C
centigrade	C
central processing unit	CPU
ceramic	CER
ceramic tile	CT
ceramic-tile floor	CTF
channel	CHAN
chapter	CH
chimney	CHM
circuit	CIR or CKT
circuit breaker	CB
circuit interrupter	CI
circular mils	CM
cleanout	CO
clockwise	CW
closet	CLO
coarse	CRS
coated	CTD
coaxial	COAX
Code-Making Panel	CMP
cold air	CA
cold water	CW
column	COL
compacted	COMP
concrete	CONC
concrete block	CCB or CONC BLK
concrete floor	CCF
concrete pipe	CP
condenser	COND
conductor resistivity	K
conduit	C or CND
construction joint	CJ
continuous	CONT
contour	CTR
control	CONT
control joint	CJ or CLJ
control relay	CR
control relay master	CRM
copper	Cu
corner	COR
cornice	COR

corrugated	CORR
counterclockwise	CCW
counter electromotive force	CEMF
county	CO
cubic	CU
cubic foot	CU FT
cubic foot per minute	CFM
cubic foot per second	CFS
cubic inch	CU IN.
cubic yard	CU YD
current	I
current transformer	CT
cut out	CO
cycles per second	CPS

D

damper	DMPR
dampproofing	DP
dead load	DL
decibel	DB
deck	DK
demolition	DML
depth	DP
detail	DET or DTL
diagonal	DIAG
diagram	DIAG
diameter	D or DIA
dimension	DIM.
dimmer	DMR
dining room	DR
diode	DIO
direct current	DC
disconnect switch	DS
dishwasher	DW
distribution	DISTR
distribution panel	DPNL
division	DIV
door	DR
dormer	DRM
double-acting	DBL ACT
double hung window	DHW
double-pole	DP
double-pole double-throw	DPDT
double-pole double-throw switch	DPDT SW
double-pole single-throw	DPST
double-pole single-throw switch	DPST SW
double-pole switch	DP SW
double strength glass	DSG
double-throw	DT
down	D or DN
downspout	DS
drain	DR
drain tile	DT
drawing	DWG
drinking fountain	DF
drum switch	DS
dryer	D
drywall	DW
duplex	DX
dust tight	DT
dutch door	DD
duty cycle	DTY CY
dynamic braking contactor or relay	DB

E

each	EA
east	E
efficiency	Eff
ejector pump	EP
electric	ELEC

. . . ABBREVIATIONS . . .

electrical	ELEC	frequency	FREQ	inch	IN.
electrical metallic tubing	EMT	front view	FV	inch per second	IPS
electric panel	EP	full-load amps	FLA	inch-pound	IN. LB
electromechanical	ELMCH	full-load current	FLC	infrared	IR
electromotive force	EMF	full-load torque	FLT	inside diameter	ID
electronic	ELEK	furnace	FUR.	instantaneous overload	IOL
elevation	EL	furring	FUR.	instantaneous-trip circuit breaker	ITB
elevator	ELEV	fuse	FU	insulation	INSUL
enamel	ENAM	fuse block	FB	integrated circuit	IC
entrance	ENTR	fuse box	FUBX	interior	INT or INTR
equipment	EQPT	fuse holder	FUHLR	interlock	INTLK
equipment bonding jumper	EBJ	fusible	FSBL	intermediate	INT
equipment grounding conductor	EGC	future	FUT	intermediate metal conduit	IMC
equivalent	EQUIV			International Electrotechnical	
estimate	EST			Commission	IEC
excavate	EXC	**G**		interrupt	INT
exception	Ex.			inverse time breaker	ITB
exhaust	EXH	gallon per hour	GPH	inverse-time circuit breaker	ITCB
existing	EXIST. or EXST	gallon per minute	GPM	iron	I
expanded metal	EM	galvanized	GALV	iron pipe	IP
expansion joint	EXP JT	garage	GAR	isolated ground	IG
explosionproof	EP	gas	G		
exterior	EXT	gate	G		
exterior grade	EXT GR	gauge	GA	**J**	
		generator	GEN		
		glass	GL or GLS	jamb	JB or JMB
F		glass block	GLB	joint	JT
		glaze	GLZ	joist	J or JST
face brick	FB	gold	Au	junction	JCT
Fahrenheit	°F	grade	GR	junction box	JB
fast	F	grade line	GL		
field	F	gravel	GVL		
figure	FIG.	gray	GY	**K**	
fine print note	FPN	green	G or GR		
finish	FNSH	gross vehicle weight	GVW	key way	KWY
finish all over	FAO	ground	G, GND, GRD, or GRND	kick plate	KPL
finished floor	FNSH FL	grounded (outlet)	G	kiln-dried	KD
finished grade	FG or FIN GR	ground-fault circuit interrupter	GFCI	kilo (1000)	k
finish one side	F1S	ground fault protection of equipment	GFPE	1000 circular mils	kcmil
finish two sides	F2S	grounding electrode conductor	GEC	1000'	kFT
firebrick	FBCK	grounding electrode system	GES	kilovolt amps	kVA
fire door	FDR	gypsum	GYP	kilowatt	kW
fire extinguisher	FEXT	gypsum sheathing board	GSB	kilowatt-hour	kWh
fire hydrant	FHY			kitchen	K or KIT.
fireplace	FP			knife switch	KN SW
fireproof	FPRF	**H**		knockout	KO
fire wall	FW				
fixed window	FX WDW	hand-off-auto	HOA		
fixture	FXTR	handrail	HNDRL	**L**	
flammable	FLMB	hardboard	HBD		
flashing	FL	hardware	HDW	lamp	LT
flat	FL	hardwood	HDWD	lath	LTH
flexible metallic conduit	FMC	hazardous	HAZ	laundry	LAU
float switch	FS	header	HDR	laundry tray	LT
floor	FL	heating	HTG	lavatory	LAV
floor drain	FD	heating, air-conditioning, refrigeration	HACR	left	L
flooring	FLG or FLR	heating, ventilating, and air		left hand	LH
floor line	FL	conditioning	HVAC	less-flammable, liquid-insulated	LFLI
flow switch	FLS	heavy-duty	HD	library	LBRY or LIB
fluorescent	FLUOR or FLUR	hertz	Hz	light	LT
flush	FL	highway	HWY	lighting	LTG
flush mount	FLMT	hollow core	HC	lighting panel	LP
footing	FTG	hollow metal door	HMD	lights	LTS
foot per minute	FPM	horizontal	HOR	limit switch	LS
foot per second	FPS	horsepower	HP	line	L
foot switch	FTS	hose bibb	HB	linoleum	LINO or LINOL
forward	F or FWD	hot water	HW	lintel	LNTL
foundation	FDN	hot water heater	HWH	live load	LL
four-pole	4P	hours	HRS	living room	LR
four-pole double-throw switch	4PDT SW	hydraulic	HYDR	load	L
four-pole single-throw switch	4PST SW			location	LOC
four-pole switch	4PSW	**I**		locked-rotor ampacity	LRA
frame	FR			locked-rotor current	LRC
		immersion detection circuit interrupter	IDCI	louver	LVR or LV
				lumber	LBR

...ABBREVIATIONS...

M

magnetic brake	MB
main	MN
main bonding jumper	MBJ
main control center	MCC
manhole	MH
manual	MAN., MN, or MNL
manufacturer	MFR
marble	MRB
masonry	MSNRY
masonry opening	MO
material	MATL or MTL
maximum	MAX
maximum working pressure	MWP
mechanical	MECH
medicine cabinet	MC
medium	MED
memory	MEM
metal	MET or MTL
metal anchor	MA
metal door	METD
metal flashing	METF
metal jalousie	METJ
metal lath and plaster	MLP
metal threshold	MT
mezzanine	MEZZ
miles per gallon	MPG
miles per hour	MPH
minimum	MIN
mirror	MIR
miscellaneous	MISC
molding	MLDG
monolithic	ML
mortar	MOR
motor	M, MOT, or MTR
motor branch-circuit, short-circuit, ground-fault	MBCSCGF
motor circuit switch	MCS
motor control center	MCC
motor starter	M
motor switch	MS
mounted	MTD

N

nameplate	NPL
National Electrical Code®	NEC®
National Electrical Manufacturers Association	NEMA
National Electrical Safety Code	NESC
National Fire Protection Association	NFPA
negative	NEG
net weight	NTWT
neutral	N or NEUT
nominal	NOM
nonadjustable-trip circuit breaker	NATCB
non-time delay fuse	NTDF
normally closed	NC
normally open	NO
north	N
nosing	NOS
not to scale	NTS
number	NO.

O

Occupational Safety and Health Administration	OSHA
ohmmeter	OHM.
on center	OC
opaque	OPA

opening	OPNG
open web joist	OJ or OWJ
orange	O
ounces per inch	OZ/IN.
outlet	OUT.
outside diameter	OD
overall	OA
overcurrent	OC
overcurrent protection device	OCPD
overhang	OVHG
overhead	OH.
overload	OL
overload relay	OL

P

paint	PNT
painted	PTD
pair	PR
panel	PNL
pantry	PAN.
parallel	PRL
peak-to-peak	P-P
perpendicular	PERP
personal computer	PC
phase	PH
piece	PC
piling	PLG
pillar	PLR
pilot light	PL
piping	PP
pitch	P
plank	PLK
plaster	PLAS
plastic	PLSTC
plate	PL
plate glass	PLGL
plugging switch	PLS
plumbing	PLMB or PLBG
plywood	PLYWD
pneumatic	PNEU
point of beginning	POB
pole	P
polyvinyl chloride	PVC
porcelain	PORC
porch	P
positive	POS
pound(s)	LB
pounds per feet	LB-FT
pounds per inch	LB-IN.
pounds per square foot	PSF
pounds per square inch	PSI
power	P or PWR
power consumed	P
power factor	PF
precast	PRCST
prefabricated	PFB or PREFAB
prefinished	PFN
pressure switch	PS
primary switch	PRI
property line	PL
pull box	PB
pull switch	PS
pull-up torque	PUT
pushbutton	PB

Q

quadrant	QDRNT
quantity	QTY
quarry tile	QT

quarry-tile roof	QTR
quart	QT
quick-acting	QA

R

radius	R
raintight	RT
random	RDM or RNDM
range	R or RNG
receipt of comments	ROC
receipt of proposals	ROP
receptacle	RECEPT or RCPT
recess	REC
recessed	REC
rectifier	REC
red	R
reference	REF
refrigerator	REF
register	REG or RGTR
reinforce	RE
reinforced concrete	RC
reinforcing steel	RST
reinforcing steel bar	REBAR
required	REQD
resistance	R
resistor	R or RES
return	RTN
reverse	R or REV
reverse-acting	RACT
revision	REV
revolutions per minute	RPM
revolutions per second	RPS
rheostat	RH
ribbed	RIB
right	R
right hand	RH
rigid	RGD
riser	R
road	RD
roll roofing	RR
roof	RF
roof drain	RD
roofing	RFG
room	R or RM
root mean square	RMS
rotor	RTR
rough	RGH
rough opening	RO
rough sawn	RS

S

safety switch	SSW
sanitary	SAN
scale	SC
schedule	SCH or SCHED
screen	SCR
scuttle	S
secondary	SEC
section	SECT
selector switch	SS
series	S
service	SERV
service entrance	SE
service factor	SF
sewer	SEW.
shake	SHK
sheathing	SHTHG
sheet	SH or SHT
sheet metal	SM
shelf and rod	SH & RD

. . . ABBREVIATIONS

shelving	SHELV				
shingle	SHGL	**T**		**V**	
shower	SH	telephone	TEL	valley	VAL
shutter	SHTR	television	TV	valve	V
siding	SDG	temperature	TEMP	vapor seal	VS
silicon controlled rectifier	SCR	tempered	TEMP	vaportight	VT
sill cock	SC	terazzo	TER	vent	V
silver	Ag	terminal	T or TERM.	ventilation	VENT.
single-phase	1PH	terminal board	TB	vent pipe	VP
single-pole	SP	terra cotta	TC	vent stack	VS
single-pole circuit breaker	SPCB	thermal	THRM	vertical	V or VERT
single-pole double-throw	SPDT	thermally protected	TP	vinyl tile	VTILE or VT
single-pole double-throw switch	SPDT SW	thermostat	THERMO	violet	V
single-pole single-throw	SPST	thermostat switch	THS	volt	V
single-pole single-throw switch	SPST SW	three-phase	3PH	voltage	E or V
single-pole switch	SP SW	three-pole	3P	voltage drop	VD
single strength glass	SSG	three-pole double-throw	3PDT	volt amps	VA
sink	S or SK	three-pole single-throw	3PST	volts	V
skylight	SLT	three-way	3WAY	volts alternating current	VAC
slate	S, SL, or SLT	three-wire	3W	volts direct current	VDC
sliding door	SLD	threshold	TH	volume	VOL
slope	SLP	time	T		
slow	S	time delay	TD		
smoke detector	SD	time-delay fuse	TDF		
socket	SOC	time delay relay	TR	**W**	
soffit	SF	toilet	T	wainscot	WAIN
soil pipe	SP	tongue-and-groove	T & G	walk-in closet	WIC
solenoid	SOL	torque	T	warm air	WA
solid core	SC	transformer	T, TRANS, or XFMR	washing machine	WM
south	S	transformer, primary side	H	waste pipe	WP
spare	SP	transformer, secondary side	X	waste stack	WS
specification	SPEC	tread	TR	water	WTR
splash block	SB	triple-pole double-throw	3PDT	water closet	WC
square	SQ	triple-pole double-throw switch	3PDT SW	water heater	WH
square foot	SQ FT	triple-pole single-throw	3PST	water meter	WM
square inch	SQ IN.	triple-pole single-throw switch	3PST SW	waterproof	WP
square yard	SQ YD	triple-pole switch	3P SW	watt(s)	W
stack	STK	truss	TR	weatherproof	WP
stainless steel	SST	two-phase	2PH	welded	WLD
stairs	ST	two-pole	DP	welded wire fabric	WWF
standard	STD	two-pole double-throw	DPDT	west	W
standpipe	SP	two-pole single-throw	DPST	white	W
starter	START or STR	typical	TYP	wide flange	WF
steel	STL			wire gauge	WG
stone	STN	**U**		wire mesh	WM
storage	STOR			with	W/
street	ST	unclamp	UCL	without	W/O
structural glass	SG	underground	UGND	wood	WD
sump pump	SP	underground feeder	UF	wrought iron	WI
supply	SPLY	undervoltage	UV		
surface four sides	S4S	Underwriters Laboratories Inc.	UL		
surface one side	S1S	unexcavated	UNEXC	**Y**	
switch	S or SW	unfinished	UNFIN		
switched disconnect	SWD	up	U	yellow	Y
		utility room	U RM		

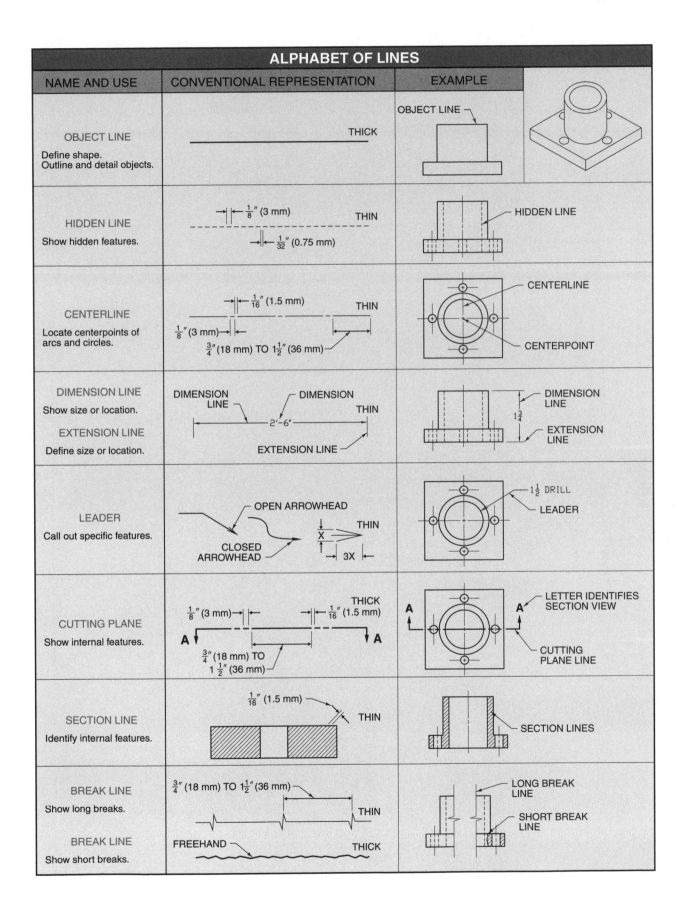

ALPHABET OF LINES

NAME AND USE	CONVENTIONAL REPRESENTATION	EXAMPLE	
OBJECT LINE — Define shape. Outline and detail objects.	THICK	OBJECT LINE	
HIDDEN LINE — Show hidden features.	$\frac{1}{8}$″ (3 mm) THIN — $\frac{1}{32}$″ (0.75 mm)	HIDDEN LINE	
CENTERLINE — Locate centerpoints of arcs and circles.	$\frac{1}{16}$″ (1.5 mm) THIN — $\frac{1}{8}$″ (3 mm) — $\frac{3}{4}$″ (18 mm) TO $1\frac{1}{2}$″ (36 mm)	CENTERLINE — CENTERPOINT	
DIMENSION LINE — Show size or location. **EXTENSION LINE** — Define size or location.	DIMENSION LINE — DIMENSION — 2′-6″ THIN — EXTENSION LINE	DIMENSION LINE — $1\frac{3}{4}$ — EXTENSION LINE	
LEADER — Call out specific features.	OPEN ARROWHEAD — CLOSED ARROWHEAD — X THIN 3X	$1\frac{1}{2}$ DRILL — LEADER	
CUTTING PLANE — Show internal features.	$\frac{1}{8}$″ (3 mm) $\frac{1}{16}$″ (1.5 mm) THICK A ↓ A ↓ $\frac{3}{4}$″ (18 mm) TO $1\frac{1}{2}$″ (36 mm)	LETTER IDENTIFIES SECTION VIEW A A — CUTTING PLANE LINE	
SECTION LINE — Identify internal features.	$\frac{1}{16}$″ (1.5 mm) THIN	SECTION LINES	
BREAK LINE — Show long breaks. **BREAK LINE** — Show short breaks.	$\frac{3}{4}$″ (18 mm) TO $1\frac{1}{2}$″ (36 mm) THIN — FREEHAND THICK	LONG BREAK LINE — SHORT BREAK LINE	

ARCHITECTURAL SYMBOLS . . .

Material	Elevation	Plan	Section
EARTH			
BRICK	WITH NOTE INDICATING TYPE OF BRICK (COMMON, FACE, ETC.)	COMMON OR FACE / FIREBRICK	SAME AS PLAN VIEWS
CONCRETE		LIGHTWEIGHT / STRUCTURAL	SAME AS PLAN VIEWS
CONCRETE MASONRY UNIT		OR	OR
STONE	CUT STONE RUBBLE	CUT STONE RUBBLE / CAST STONE (CONCRETE)	CUT STONE / CAST STONE (CONCRETE) RUBBLE OR CUT STONE
WOOD	SIDING PANEL	WOOD STUD / REMODELING / DISPLAY	ROUGH MEMBERS FINISHED MEMBERS PLYWOOD
PLASTER		WOOD STUD, LATH, AND PLASTER / METAL LATH AND PLASTER / SOLID PLASTER	LATH AND PLASTER
ROOFING	SHINGLES	SAME AS ELEVATION VIEW	
GLASS	OR / GLASS BLOCK	GLASS / GLASS BLOCK	SMALL SCALE LARGE SCALE

... ARCHITECTURAL SYMBOLS

Material	Elevation	Plan	Section
FACING TILE	CERAMIC TILE	FLOOR TILE	CERAMIC TILE LARGE SCALE / CERAMIC TILE SMALL SCALE
STRUCTURAL CLAY TILE			SAME AS PLAN VIEW
INSULATION		LOOSE FILL OR BATTS / RIGID / SPRAY FOAM	SAME AS PLAN VIEWS
SHEET METAL FLASHING		OCCASIONALLY INDICATED BY NOTE	
METALS OTHER THAN FLASHING	INDICATED BY NOTE OR DRAWN TO SCALE	SAME AS ELEVATION	SMALL SCALE / STEEL / CAST IRON / ALUMINUM / BRONZE OR BRASS
STRUCTURAL STEEL	INDICATED BY NOTE OR DRAWN TO SCALE	OR	REBARS / SMALL SCALE / LARGE SCALE / L-ANGLES, S-BEAMS, ETC.

PLOT PLAN SYMBOLS

NORTH	FIRE HYDRANT	WALK	ELECTRIC SERVICE
POINT OF BEGINNING (POB)	MAILBOX	IMPROVED ROAD	NATURAL GAS LINE
UTILITY METER OR VALVE	MANHOLE	UNIMPROVED ROAD	WATER LINE
POWER POLE AND GUY	TREE	BUILDING LINE	TELEPHONE LINE
LIGHT STANDARD	BUSH	PROPERTY LINE	NATURAL GRADE
TRAFFIC SIGNAL	HEDGE ROW	PROPERTY LINE	FINISH GRADE
STREET SIGN	FENCE	TOWNSHIP LINE	EXISTING ELEVATION + XX.00'

ELECTRICAL SYMBOLS . . .

LIGHTING OUTLETS

OUTLET BOX AND INCANDESCENT LIGHTING FIXTURE
CEILING   WALL

INCANDESCENT TRACK LIGHTING

BLANKED OUTLET   (B)   (B)

DROP CORD   (D)

EXIT LIGHT AND OUTLET BOX. SHADED AREAS DENOTE FACES.

OUTDOOR POLE-MOUNTED FIXTURES

JUNCTION BOX   (J)   (J)

LAMPHOLDER WITH PULL SWITCH   (L)$_{PS}$ (L)$_{PS}$

MULTIPLE FLOODLIGHT ASSEMBLY

EMERGENCY BATTERY PACK WITH CHARGER   B

INDIVIDUAL FLUORESCENT FIXTURE

OUTLET BOX AND FLUORESCENT LIGHTING TRACK FIXTURE

CONTINUOUS FLUORESCENT FIXTURE

SURFACE-MOUNTED FLUORESCENT FIXTURE

PANELBOARDS

FLUSH-MOUNTED PANELBOARD AND CABINET

SURFACE-MOUNTED PANELBOARD AND CABINET

CONVENIENCE OUTLETS

SINGLE RECEPTACLE OUTLET

DUPLEX RECEPTACLE OUTLET

TRIPLEX RECEPTACLE OUTLET

SPLIT-WIRED DUPLEX RECEPTACLE OUTLET

SPLIT-WIRED TRIPLEX RECEPTACLE OUTLET

SINGLE SPECIAL-PURPOSE RECEPTACLE OUTLET

DUPLEX SPECIAL-PURPOSE RECEPTACLE OUTLET

RANGE OUTLET   R

SPECIAL-PURPOSE CONNECTION   DW

CLOSED-CIRCUIT TELEVISION CAMERA

CLOCK HANGER RECEPTACLE   (C)

FAN HANGER RECEPTACLE   (F)

FLOOR SINGLE RECEPTACLE OUTLET

FLOOR DUPLEX RECEPTACLE OUTLET

FLOOR SPECIAL-PURPOSE OUTLET

UNDERFLOOR DUCT AND JUNCTION BOX FOR TRIPLE, DOUBLE, OR SINGLE DUCT SYSTEM AS INDICATED BY NUMBER OF PARALLEL LINES

BUSDUCTS AND WIREWAYS

SERVICE, FEEDER, OR PLUG-IN BUSWAY | B | B | B |

CABLE THROUGH LADDER OR CHANNEL | C | C | C |

WIREWAY | W | W | W |

SWITCH OUTLETS

SINGLE-POLE SWITCH   S

DOUBLE-POLE SWITCH   S$_2$

THREE-WAY SWITCH   S$_3$

FOUR-WAY SWITCH   S$_4$

AUTOMATIC DOOR SWITCH   S$_D$

KEY-OPERATED SWITCH   S$_K$

CIRCUIT BREAKER   S$_{CB}$

WEATHERPROOF CIRCUIT BREAKER   S$_{WCB}$

DIMMER   S$_{DM}$

REMOTE CONTROL SWITCH   S$_{RC}$

WEATHERPROOF SWITCH   S$_{WP}$

FUSED SWITCH   S$_F$

WEATHERPROOF FUSED SWITCH   S$_{WF}$

TIME SWITCH   S$_T$

CEILING PULL SWITCH   S

SWITCH AND SINGLE RECEPTACLE   S

SWITCH AND DOUBLE RECEPTACLE   S

A STANDARD SYMBOL WITH AN ADDED LOWERCASE SUBSCRIPT LETTER IS USED TO DESIGNATE A VARIATION IN STANDARD EQUIPMENT   a,b   a,b   S$_{a,b}$

... ELECTRICAL SYMBOLS

COMMERCIAL AND INDUSTRIAL SYSTEMS

PAGING SYSTEM DEVICE

FIRE ALARM SYSTEM DEVICE

COMPUTER DATA SYSTEM DEVICE

PRIVATE TELEPHONE SYSTEM DEVICE

SOUND SYSTEM

FIRE ALARM CONTROL PANEL — FACP

SIGNALING SYSTEM OUTLETS FOR RESIDENTIAL SYSTEMS

PUSHBUTTON

BUZZER

BELL

BELL AND BUZZER COMBINATION

COMPUTER DATA OUTLET

BELL RINGING TRANSFORMER — BT

ELECTRIC DOOR OPENER — D

CHIME — CH

TELEVISION OUTLET — TV

THERMOSTAT — T

UNDERGROUND ELECTRICAL DISTRIBUTION OR ELECTRICAL LIGHTING SYSTEMS

MANHOLE — M

HANDHOLE — H

TRANSFORMER-MANHOLE OR VAULT — TM

TRANSFORMER PAD — TP

UNDERGROUND DIRECT BURIAL CABLE

UNDERGROUND DUCT LINE

STREET LIGHT STANDARD FED FROM UNDERGROUND CIRCUIT

ABOVE-GROUND ELECTRICAL DISTRIBUTION OR LIGHTING SYSTEMS

POLE

STREET LIGHT AND BRACKET

PRIMARY CIRCUIT

SECONDARY CIRCUIT

DOWN GUY

HEAD GUY

SIDEWALK GUY

SERVICE WEATHERHEAD

PANEL CIRCUITS AND MISCELLANEOUS

LIGHTING PANEL

POWER PANEL

WIRING – CONCEALED IN CEILING OR WALL

WIRING – CONCEALED IN FLOOR

WIRING EXPOSED

HOME RUN TO PANEL BOARD
Indicate number of circuits by number of arrows. Any circuit without such designation indicates a two-wire circuit. For a greater number of wires indicate as follows: —///— (3 wires) —////— (4 wires), etc.

FEEDERS
Use heavy lines and designate by number corresponding to listing in feeder schedule

WIRING TURNED UP

WIRING TURNED DOWN

GENERATOR — G

MOTOR — M

INSTRUMENT (SPECIFY) — I

TRANSFORMER — T

CONTROLLER

EXTERNALLY-OPERATED DISCONNECT SWITCH

PULL BOX

PLUMBING SYMBOLS . . .

FIXTURES...	...FIXTURES	...PIPING

FIXTURES...

Fixture		Fixture		Piping	
STANDARD BATHTUB		LAUNDRY TRAY		CHILLED DRINKING WATER SUPPLY	—— DWS ——
OVAL BATHTUB		BUILT-IN SINK		CHILLED DRINKING WATER RETURN	—— DWR ——
WHIRLPOOL BATH		DOUBLE OR TRIPLE BUILT-IN SINK		HOT WATER	————
SHOWER STALL		COMMERCIAL KITCHEN SINK		HOT WATER RETURN	————
SHOWER HEAD		SERVICE SINK	SS	SANITIZING HOT WATER SUPPLY (180° F)	—/—/—
TANK-TYPE WATER CLOSET		CLINIC SERVICE SINK		SANITIZING HOT WATER RETURN (180° F)	—/—/—
WALL-MOUNTED WATER CLOSET		FLOOR-MOUNTED SERVICE SINK		DRY STANDPIPE	—— DSP ——
FLOOR-MOUNTED WATER CLOSET		DRINKING FOUNTAIN	DF	COMBINATION STANDPIPE	—— CSP ——
LOW-PROFILE WATER CLOSET		WATER COOLER		MAIN SUPPLIES SPRINKLER	—— S ——
BIDET		HOT WATER TANK	HWT	BRANCH AND HEAD SPRINKLER	—o——o—
WALL-MOUNTED URINAL		WATER HEATER	WH	GAS – LOW PRESSURE	—G——G—
FLOOR-MOUNTED URINAL		METER	M	GAS – MEDIUM PRESSURE	—— MG ——
TROUGH-TYPE URINAL		HOSE BIBB	HB	GAS – HIGH PRESSURE	—— HG ——
WALL-MOUNTED LAVATORY		GAS OUTLET	G	COMPRESSED AIR	—— A ——
PEDESTAL LAVATORY		GREASE SEPARATOR	G	OXYGEN	—— O ——
BUILT-IN LAVATORY		GARAGE DRAIN		NITROGEN	—— N ——
WHEELCHAIR LAVATORY		FLOOR DRAIN WITH BACKWATER VALVE		HYDROGEN	—— H ——
CORNER LAVATORY		**PIPING...**		HELIUM	—— HE ——
FLOOR DRAIN		SOIL, WASTE, OR LEADER – ABOVE GRADE	————	ARGON	—— AR ——
FLOOR SINK		SOIL, WASTE, OR LEADER – BELOW GRADE	— — —	LIQUID PETROLEUM GAS	—— LPG ——
		VENT	— — — —	INDUSTRIAL WASTE	—— INW ——
		COMBINATION WASTE AND VENT	—— SV ——	CAST IRON	—— CI ——
		STORM DRAIN	—— SD ——	CULVERT PIPE	—— CP ——
		COLD WATER	—·—·—	CLAY TILE	—— CT ——
				DUCTILE IRON	—— DI ——
				REINFORCED CONCRETE	—— RCP ——
				DRAIN – OPEN TILE OR AGRICULTURAL TILE	====

... PLUMBING SYMBOLS

PIPE FITTING AND VALVE SYMBOLS

	FLANGED	SCREWED	BELL & SPIGOT		FLANGED	SCREWED	BELL & SPIGOT		FLANGED	SCREWED	BELL & SPIGOT
BUSHING				REDUCING FLANGE				AUTOMATIC BY-PASS VALVE			
CAP				BULL PLUG				AUTOMATIC REDUCING VALVE			
REDUCING CROSS				PIPE PLUG				STRAIGHT CHECK VALVE			
STRAIGHT-SIZE CROSS				CONCENTRIC REDUCER				COCK			
CROSSOVER				ECCENTRIC REDUCER				DIAPHRAGM VALVE			
45° ELBOW				SLEEVE				FLOAT VALVE			
90° ELBOW				STRAIGHT-SIZE TEE				GATE VALVE			
ELBOW – TURNED DOWN				TEE – OUTLET UP				MOTOR-OPERATED GATE VALVE			
ELBOW – TURNED UP				TEE – OUTLET DOWN				GLOBE VALVE			
BASE ELBOW				DOUBLE-SWEEP TEE				MOTOR-OPERATED GLOBE VALVE			
DOUBLE-BRANCH ELBOW				REDUCING TEE				ANGLE HOSE VALVE			
LONG-RADIUS ELBOW				SINGLE-SWEEP TEE				GATE HOSE VALVE			
REDUCING ELBOW				SIDE OUTLET TEE – OUTLET DOWN				GLOBE HOSE VALVE			
SIDE OUTLET ELBOW – OUTLET DOWN				SIDE OUTLET TEE – OUTLET UP				LOCKSHIELD VALVE			
SIDE OUTLET ELBOW – OUTLET UP				UNION				QUICK-OPENING VALVE			
STREET ELBOW				ANGLE CHECK VALVE				SAFETY VALVE			
CONNECTING PIPE JOINT				ANGLE GATE VALVE – ELEVATION				GOVERNOR-OPERATED AUTOMATIC VALVE			
EXPANSION JOINT				ANGLE GATE VALVE – PLAN							
LATERAL				ANGLE GLOBE VALVE – ELEVATION							
ORIFICE FLANGE				ANGLE GLOBE VALVE – PLAN							

HVAC SYMBOLS

EQUIPMENT SYMBOLS

Symbol	Name
EXPOSED RADIATOR	
RECESSED RADIATOR	
FLUSH ENCLOSED RADIATOR	
PROJECTING ENCLOSED RADIATOR	
UNIT HEATER (PROPELLER) – PLAN	
UNIT HEATER (CENTRIFUGAL) – PLAN	
UNIT VENTILATOR – PLAN	
STEAM	
DUPLEX STRAINER	
PRESSURE-REDUCING VALVE	
AIR LINE VALVE	
STRAINER	
THERMOMETER	
PRESSURE GAUGE AND COCK	
RELIEF VALVE	
AUTOMATIC 3-WAY VALVE	
AUTOMATIC 2-WAY VALVE	
SOLENOID VALVE	S

DUCTWORK

Symbol	Name
DUCT (1ST FIGURE, WIDTH; 2ND FIGURE, DEPTH)	12 X 20
DIRECTION OF FLOW	
FLEXIBLE CONNECTION	
DUCTWORK WITH ACOUSTICAL LINING	
FIRE DAMPER WITH ACCESS DOOR	FD AD
MANUAL VOLUME DAMPER	— VD
AUTOMATIC VOLUME DAMPER	
EXHAUST, RETURN OR OUTSIDE AIR DUCT – SECTION	20 X 12
SUPPLY DUCT – SECTION	20 X 12
CEILING DIFFUSER SUPPLY OUTLET	20" DIA CD 1000 CFM
CEILING DIFFUSER SUPPLY OUTLET	20 X 12 CD 700 CFM
LINEAR DIFFUSER	96 X 6-LD 400 CFM
FLOOR REGISTER	20 X 12 FR 700 CFM
TURNING VANES	
FAN AND MOTOR WITH BELT GUARD	
LOUVER OPENING	20 X 12-L 700 CFM

HEATING PIPING

HIGH-PRESSURE STEAM	HPS
MEDIUM-PRESSURE STEAM	MPS
LOW-PRESSURE STEAM	LPS
HIGH-PRESSURE RETURN	HPR
MEDIUM-PRESSURE RETURN	MPR
LOW-PRESSURE RETURN	LPR
BOILER BLOW OFF	BD
CONDENSATE OR VACUUM PUMP DISCHARGE	VPD
FEEDWATER PUMP DISCHARGE	PPD
MAKEUP WATER	MU
AIR RELIEF LINE	V
FUEL OIL SUCTION	FOS
FUEL OIL RETURN	FOR
FUEL OIL VENT	FOV
COMPRESSED AIR	A
HOT WATER HEATING SUPPLY	HW
HOT WATER HEATING RETURN	HWR

AIR CONDITIONING PIPING

REFRIGERANT LIQUID	RL
REFRIGERANT DISCHARGE	RD
REFRIGERANT SUCTION	RS
CONDENSER WATER SUPPLY	CWS
CONDENSER WATER RETURN	CWR
CHILLED WATER SUPPLY	CHWS
CHILLED WATER RETURN	CHWR
MAKEUP WATER	MU
HUMIDIFICATION LINE	H
DRAIN	D

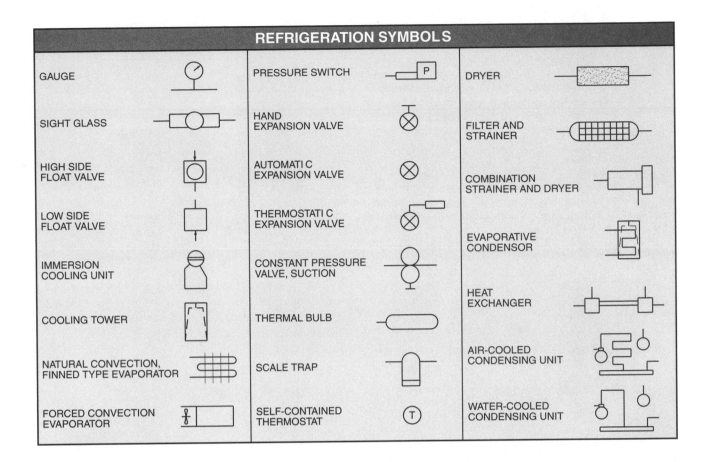

REFRIGERATION SYMBOLS

GAUGE		PRESSURE SWITCH		DRYER	
SIGHT GLASS		HAND EXPANSION VALVE		FILTER AND STRAINER	
HIGH SIDE FLOAT VALVE		AUTOMATIC EXPANSION VALVE		COMBINATION STRAINER AND DRYER	
LOW SIDE FLOAT VALVE		THERMOSTATIC EXPANSION VALVE		EVAPORATIVE CONDENSOR	
IMMERSION COOLING UNIT		CONSTANT PRESSURE VALVE, SUCTION		HEAT EXCHANGER	
COOLING TOWER		THERMAL BULB		AIR-COOLED CONDENSING UNIT	
NATURAL CONVECTION, FINNED TYPE EVAPORATOR		SCALE TRAP		WATER-COOLED CONDENSING UNIT	
FORCED CONVECTION EVAPORATOR		SELF-CONTAINED THERMOSTAT			

HAZARDOUS LOCATIONS

Class	Division	Group	Material
I	1 or 2	A	Acetylene
	1 or 2	B	Hydrogen, butadiene, ethylene oxide, propylene oxide
	1 or 2	C	Carbon monoxide, ether, ethylene, hydrogen sulfide, morpholine, cyclopropane
	1 or 2	D	Gasoline, benzene, butane, propane, alcohol, acetone, ammonia, vinyl chloride
II	1 or 2	E	Metal dusts
	1 or 2	F	Carbon black, coke dust, coal
	1 or 2	G	Grain dust, flour, starch, sugar, plastics
III	1 or 2	No groups	Wood chips, cotton, flax, nylon

ENCLOSURES . . .

An enclosure is a case or housing of apparatus to prevent personnel from accidentally contacting energized parts, or to protect the equipment from physical damage. See 100. Enclosures for hazardous (classified) locations are marked with a ✓.

Type 1 Enclosure

Type 1 enclosures are intended for indoor use primarily to provide a degree of protection against limited amounts of falling dirt. NEMA Standards Publication 250.2008.

General Electric Company

General duty safety switches from General Electric are available in Type 1 and 3R enclosures that are designed for residential or high commercial applications where duty is not severe.

Type 2 Enclosure

Type 2 enclosures are intended for indoor use primarily to provide a degree of protection against limited amounts of falling water and dirt. NEMA Standards Publication 250.2008.

Type 3 Enclosure

Type 3 enclosures are intended for indoor or outdoor use primarily to provide a degree of protection against falling dirt, windblown dust, rain, sleet, snow, and external ice formation. NEMA Standards Publication 250.2008.

Type 3R Enclosure

Type 3R enclosures are intended for outdoor use primarily to provide a degree of protection against falling dirt, rain, sleet, snow, and external ice formation. NEMA Standards Publication 250.2008.

Type 3S Enclosure

Type 3S enclosures are intended for indoor or outdoor use primarily to provide a degree of protection against falling dirt, windblown dust, rain, sleet, snow, and to provide for operation of external mechanisms when ice laden. NEMA Standards Publication 250.2008.

Square D Company

Heavy-duty safety switches from Square D Company have visible blades and are available in NEMA Type 3R rainproof enclosures.

Type 3X Enclosure

Type 3X enclosures are intended for indoor or outdoor use primarily to provide a degree of protection against falling dirt, windblown dust, rain, sleet, snow, corrosion, and external ice formation. NEMA Standards Publication 250.2008.

Type 3RX Enclosure

Type 3RX enclosures are intended for indoor or outdoor use primarily to provide a degree of protection against falling dirt, rain, sleet, snow, corrosion, and external ice formation. NEMA Standards Publication 250.2008.

Type 3SX Enclosure

Type 3SX enclosures are intended for indoor or outdoor use primarily to provide a degree of protection against falling dirt, windblown dust, rain, sleet, snow, corrosion, and to provide for operation of external mechanisms when ice laden. NEMA Standards Publication 250.2008.

Type 4 Enclosure

Type 4 enclosures are intended for indoor or outdoor use primarily to provide a degree of protection against windblown dust and rain, splashing water, hose-directed water, and damage from external ice formation. NEMA Standards 250.2008.

Type 4X Enclosure

Type 4X enclosures are intended for indoor or outdoor use primarily to provide a degree of protection against corrosion, windblown dust and rain, splashing water, hose-directed water, and damage from external ice formation. NEMA Standards Publication 250.2008.

. . . ENCLOSURES

Crouse-Hinds Division, Cooper Industries, Inc.

Crouse-Hinds manufactures enclosures made from high-impact strength fiberglass-reinforced polyester material that is corrosion-resistant, dust tight, watertight, and weatherproof, and rated as NEMA Type 3, 4X, and 12.

Type 5 Enclosure

Type 5 enclosures are intended for indoor use primarily to provide a degree of protection against settling airborne dust, falling dirt, and dripping noncorrosive liquids. NEMA Standards Publication 250.2008.

Type 6 Enclosure

Type 6 enclosures are intended for indoor or outdoor use primarily to provide a degree of protection against hose-directed water, the entry of water during occasional temporary submersion at a limited depth, and damage from external ice formation. NEMA Standards Publication 250.2008.

Type 6P Enclosure

Type 6P enclosures are intended for indoor or outdoor use primarily to provide a degree of protection against hose-directed water, the entry of water during prolonged submersion at a limited depth, and damage from external ice formation. NEMA Standards Publication 250.2008.

Type 7 Enclosure ✓

Type 7 enclosures are intended for indoor use in locations classified as Class I, Groups A, B, C, or D, as defined in the National Electrical Code. NEMA Standards Publication 250.2008.

Rockwell Automation, Allen-Bradley Company, Inc.

Bulletin 609U manual starting switches from Allen-Bradley are available with undervoltage protection in Type 3R, 7, and 9 enclosures.

Type 8 Enclosure ✓

Type 8 enclosures are intended for indoor or outdoor use in locations classified as Class I, Groups A, B, C, and D, as defined in the National Electrical Code. NEMA Standards Publication 250.2008.

Type 9 Enclosure ✓

Type 9 enclosures are for use in indoor locations classified as Class II, Groups E, F, or G, as defined in the National Electrical Code. NEMA Standards Publication 250.2008.

Type 10 Enclosure ✓

Type 10 enclosures are constructed to meet the applicable requirements of the Mine Safety and Health Administration. NEMA Standards Publication 250.2008.

Type 12 Enclosure

Type 12 enclosures are intended for indoor use primarily to provide a degree of protection against circulating dust, falling dirt, and dripping noncorrosive liquids. NEMA Standards Publication 250.2008.

Rockwell Automation, Allen-Bradley Company, Inc.

Bulletin 100 molded enclosed starters from Allen-Bradley are available in Type 1, 4/4X, and 12 construction.

Type 12K Enclosure

Type 12K enclosures with knockouts are intended for indoor use primarily to provide a degree of protection against circulating dust, falling dirt, and dripping noncorrosive liquids. NEMA Standards Publication 250.2008.

Type 13 Enclosure

Type 13 enclosures are intended for indoor use primarily to provide a degree of protection against dust, spraying of water, oil, and noncorrosive coolant. NEMA Standards Publication 250.2008.

ENCLOSURE TESTING

Enclosures for electrical equipment up to 1000 V maximum are designed and tested per NEMA 250-1991, *Enclosures for Electrical Equipment (1000 V Maximum)*. Section 4 of NEMA 250-1991 includes descriptions, applications, features, and test criteria for nonhazardous location enclosures (Type 1, 2, 3, 3R, 3S, 4, 4X, 5, 6, 6P, 12, 12K, and 13). Section 5 of NEMA 250-1991 includes descriptions, applications, features, and test criteria for hazardous (classified) location enclosures (Type 7, 8, 9, and 10).

Nonhazardous location enclosures are designed according to 11 design tests. See Enclosures. The design tests are given in Section 6 of NEMA 250-1991. For example, the external icing test is used to test Type 3, 3R, 3S, 4, 4X, 6, and 6P enclosures. The enclosures must be manually operated by one person while ice-laden and must be undamaged after the ice is melted. See External Icing Test. For additional design tests, see NEMA 250-1991, Section 6.

Design Test	1	2	3	3R	3S	3X	3RX	3SX	4	4X	5	6	6P	12	12K	13
Rod Entry	•	•		•												
Drip		•									•			•	•	
Rain			•	•	•	•	•	•								
Dust			•		•	•		•			•			•	•	
External Icing			•	•	•	•	•		•	•		•	•			
Hosedown									•	•		•	•			
Rust Resistance	•	•									•			•	•	•
Corrosion Protection			•	•	•	•	•	•	•	•		•	•			
Submersion												•	•			
Air Pressure																
Oil Exclusion																•

ENCLOSURES

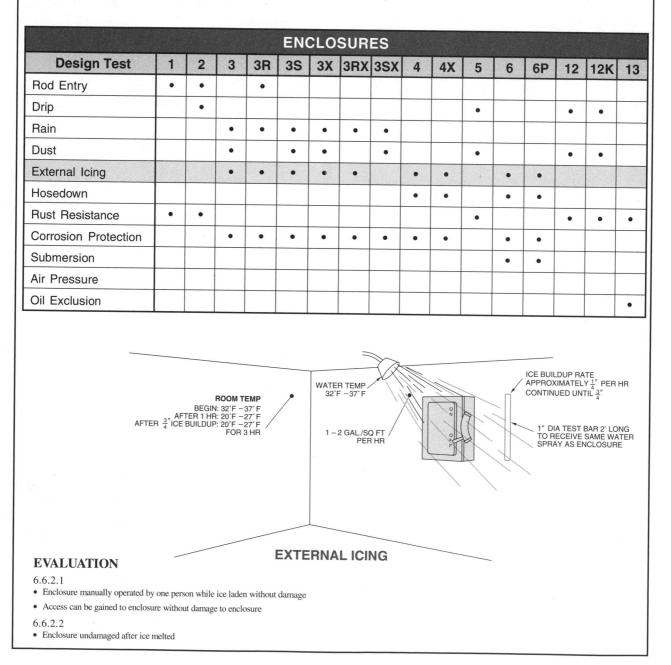

ROOM TEMP
BEGIN: 32°F – 37°F
AFTER 1 HR: 20°F – 27°F
AFTER 3/4" ICE BUILDUP: 20°F – 27°F FOR 3 HR

WATER TEMP 32°F – 37°F

1 – 2 GAL./SQ FT PER HR

ICE BUILDUP RATE APPROXIMATELY 1/4" PER HR CONTINUED UNTIL 3/4"

1" DIA TEST BAR 2' LONG TO RECEIVE SAME WATER SPRAY AS ENCLOSURE

EXTERNAL ICING

EVALUATION

6.6.2.1
• Enclosure manually operated by one person while ice laden without damage
• Access can be gained to enclosure without damage to enclosure

6.6.2.2
• Enclosure undamaged after ice melted

LOCKING WIRING DEVICES

2-POLE, 3-WIRE

WIRING DIAGRAM	NEMA ANSI	RECEPTACLE CONFIGURATION	RATING
	ML2 C73.44		15 A 125 V
	L5-15 C73.42		15 A 125 V
	L5-20 C73.72		20 A 125 V
	L6-15 C73.74		15 A 250 V
	L6-20 C73.75		20 A 250 V
	L6-30 C73.76		30 A 250 V
	L7-15 C73.43		15 A 277 V
	L7-20 C73.77		20 A 277 V
	L8-20 C73.79		20 A 480 V
	L9-20 C73.81		20 A 600 V

3-POLE, 3-WIRE

WIRING DIAGRAM	NEMA ANSI	RECEPTACLE CONFIGURATION	RATING
	ML3 C73.30		15 A 125/250 V
	L10-20 C73.96		20 A 125/250 V
	L10-30 C73.97		30 A 125/250 V
	L11-15 C73.98		15 A 3φ 250 V
	L11-20 C73.99		20 A 3φ 250 V
	L12-20 C73.101		20 A 3φ 480 V
	L12-30 C73.102		30 A 3φ 480 V
	L13-30 C73.103		30 A 3φ 600 V

3-POLE, 4-WIRE

WIRING DIAGRAM	NEMA ANSI	RECEPTACLE CONFIGURATION	RATING
	L14-20 C73.83		20 A 125/250 V
	L14-30 C73.84		30 A 125/250 V
	L15-20 C73.85		20 A 3φ 250 V
	L15-30 C73.86		30 A 3φ 250 V
	L16-20 C73.87		20 A 3φ 480 V
	L16-30 C73.88		30 A 3φ 480 V
	L17-30 C73.89		30 A 3φ 600 V

4-POLE, 4-WIRE

WIRING DIAGRAM	NEMA ANSI	RECEPTACLE CONFIGURATION	RATING
	L18-20 C73.104		20 A 3φ Y 120/208 V
	L18-30 C73.105		30 A 3φ Y 120/208 V
	L19-20 C73.106		20 A 3φ Y 277/480 V
	L20-20 C73.108		20 A 3φ Y 347/600 V

4-POLE, 5-WIRE

WIRING DIAGRAM	NEMA ANSI	RECEPTACLE CONFIGURATION	RATING
	L21-20 C73.90		20 A 3φ Y 120/208 V
	L22-20 C73.92		20 A 3φ Y 277/480 V
	L23-20 C73.94		20 A 3φ Y 347/600 V

NON-LOCKING WIRING DEVICES

2-POLE, 3-WIRE

WIRING DIAGRAM	NEMA ANSI	RECEPTACLE CONFIGURATION	RATING
	5-15 C73.11		15 A 125 V
	5-20 C73.12		20 A 125 V
	5-30 C73.45		30 A 125 V
	5-50 C73.46		50 A 125 V
	6-15 C73.20		15 A 250 V
	6-20 C73.51		20 A 250 V
	6-30 C73.52		30 A 250 V
	6-50 C73.53		50 A 250 V
	7-15 C73.28		15 A 277 V
	7-20 C73.63		20 A 277 V
	7-30 C73.64		30 A 277 V
	7-50 C73.65		50 A 277 V

4-POLE, 4-WIRE

WIRING DIAGRAM	NEMA ANSI	RECEPTACLE CONFIGURATION	RATING
	18-15 C73.15		15 A 3φ Y 120/208 V
	18-20 C73.26		20 A 3φ Y 120/208 V
	18-30 C73.47		30 A 3φ Y 120/208 V
	18-50 C73.48		50 A 3φ Y 120/208 V
	18-60 C73.27		60 A 3φ Y 120/208 V

3-POLE, 3-WIRE

WIRING DIAGRAM	NEMA ANSI	RECEPTACLE CONFIGURATION	RATING
	10-20 C73.23		20 A 125/250 V
	10-30 C73.24		30 A 125/250 V
	10-50 C73.25		50 A 125/250 V
	11-15 C73.54		15 A 3φ 250 V
	11-20 C73.55		20 A 3φ 250 V
	11-30 C73.56		30 A 3φ 250 V
	11-50 C73.57		50 A 3φ 250 V

3-POLE, 4-WIRE

WIRING DIAGRAM	NEMA ANSI	RECEPTACLE CONFIGURATION	RATING
	14-15 C73.49		15 A 125/250 V
	14-20 C73.50		20 A 125/250 V
	14-30 C73.16		30 A 125/250 V
	14-50 C73.17		50 A 125/250 V
	14-60 C73.18		60 A 125/250 V
	15-15 C73.58		15 A 3φ 250 V
	15-20 C73.59		20 A 3φ 250 V
	15-30 C73.60		30 A 3φ 250 V
	15-50 C73.61		50 A 3φ 250 V
	15-60 C73.62		60 A 3φ 250 V

RACEWAYS

EMT	Electrical Metallic Tubing
ENT	Electrical Nonmetallic Tubing
FMC	Flexible Metal Conduit
FMT	Flexible Metallic Tubing
HDPE	High Density Polyethylene Conduit
IMC	Intermediate Metal Conduit
LFMC	Liquidtight Flexible Metal Conduit
LFNC	Liquidtight Flexible Nonmetallic Conduit
NUCC	Nonmetallic Underexposed Conduit with Conductors
PVC	Rigid Polyvinyl Chloride Conduit
RNC	Rigid Nonmetallic Conduit
RTRC	Reinforced Thermosetting Resin Conduit

CABLES

AC	Armored Cable
BX	Tradename for AC
FCC	Flat Conductor Cable
IGS	Integrated Gas Spacer Cable
MC	Metal-Clad Cable
MI	Mineral-Insulated, Metal Sheathed Cable
MV	Medium Voltage
NM	Nonmetallic-Sheathed Cable (dry)
NMC	Nonmetallic-Sheathed Cable (dry or damp)
NMS	Nonmetallic-Sheathed Cable (dry)
SE	Service-Entrance Cable (dry)
TC	Tray Cable
UF	Underground Feeder Cable
USE	Underground Service-Entrance Cable

COMMON ELECTRICAL INSULATIONS

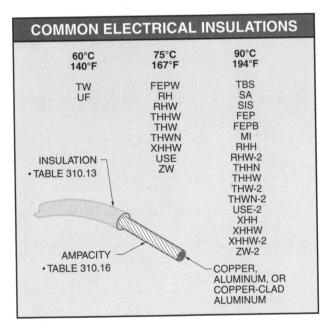

60°C 140°F	75°C 167°F	90°C 194°F
TW	FEPW	TBS
UF	RH	SA
	RHW	SIS
	THHW	FEP
	THW	FEPB
	THWN	MI
	XHHW	RHH
	USE	RHW-2
	ZW	THHN
		THHW
		THW-2
		THWN-2
		USE-2
		XHH
		XHHW
		XHHW-2
		ZW-2

INSULATION —
• TABLE 310.13

AMPACITY —
• TABLE 310.16

COPPER,
ALUMINUM, OR
COPPER-CLAD
ALUMINUM

FUSES AND ITCBs

Increase	Standard Ampere Ratings
5	15, 20, 25, 30, 35, 40, 45
10	50, 60, 70, 80, 90, 100, 110
25	125, 150, 175, 200, 225
50	250, 300, 350, 400, 450
100	500, 600, 700, 800
200	1000, 1200
400	1600, 2000
500	2500
1000	3000, 4000, 5000, 6000

1 A, 3 A, 6 A, 10 A, and 601 A are additional standard ratings for fuses.

MOTOR TORQUE

Torque	Starting Torque	Nominal Torque Rating
$T = \dfrac{HP \times 5252}{rpm}$ where T = torque HP = horsepower 5252 = constant $\left(\dfrac{33,000 \text{ lb-ft}}{\pi \times 2} = 5252\right)$ rpm = revolutions per minute	$T = \dfrac{HP \times 5252}{rpm} \times \%$ where HP = horsepower 5252 = constant $\left(\dfrac{33,000 \text{ lb-ft}}{\pi \times 2} = 5252\right)$ rpm = revolutions per minute $\%$ = motor class percentage	$T = \dfrac{HP \times 63,000}{rpm}$ where T = nominal torque rating (in lb-in) $63,000$ = constant HP = horsepower rpm = revolutions per minute

AC MOTOR CHARACTERISTICS

Motor Type 1φ	Typical Voltage	Starting Ability (Torque)	Size (HP)	Speed Range (rpm)	Cost*	Typical Uses
Shaded-pole	115 V, 230 V	Very low 50% to 100% of full load	Fractional ½ HP to ⅓ HP	Fixed 900, 1200, 1800, 3600	Very low 75% to 85%	Light-duty applications such as small fans, hair dryers, blowers, and computers
Split-phase	115 V, 230 V	Low 75% to 200% of full load	Fractional ⅓ HP or less	Fixed 900, 1200, 1800, 3600	Low 85% to 95%	Low-torque applications such as pumps, blowers, fans, and machine tools
Capacitor-start	115 V, 230 V	High 200% to 350% of full load	Fractional to 3 HP	Fixed 900, 1200, 1800	Low 90% to 110%	Hard-to-start loads such as refrigerators, air compressors, and power tools
Capacitor-run	115 V, 230 V	Very low 50% to 100% of full load	Fractional to 5 HP	Fixed 900, 1200, 1800	Low 90% to 110%	Applications that require a high running torque such as pumps and conveyors
Capacitor-start-and-run	115 V, 230 V	Very high 350% to 450% of full load	Fractional to 10 HP	Fixed 900, 1200, 1800	Low 100% to 115%	Applications that require both a high starting and running torque such as loaded conveyors
3φ Induction	230 V, 460 V	Low 100% to 175% of full load	Fractional to over 500 HP	Fixed 900, 1200, 3600	Low 100%	Most industrial applications
Wound rotor	230 V, 460 V	High 200% to 300% of full load	½ HP to 200 HP	Varies by changing resistance in rotor	Very high 250% to 350%	Applications that require high torque at different speeds such as cranes and elevators
Synchronous	230 V, 460 V	Very low 40% to 100% of full load	Fractional to 250 HP	Exact constant speed	High 200% to 250%	Applications that require very slow speeds and correct power factors

* based on standard 3φ induction motor

DC AND UNIVERSAL MOTOR CHARACTERISTICS

Motor Type	Typical Voltage	Starting Ability (Torque)	Size (HP)	Speed Range (rpm)	Cost*	Typical Uses
DC Series	12 V, 90 V, 120 V, 180 V	Very high 400% to 450% of full load	Fractional to 100 HP	Varies 0 to full speed	High 175% to 225%	Applications that require very high torque such as hoists and bridges
Shunt	12 V, 90 V, 120 V, 180 V	Low 125% to 250% of full load	Fractional to 100 HP	Fixed or adjustable below full speed	High 175% to 225%	Applications that require better speed control than a series motor such as woodworking machines
Compound	12 V, 90 V, 120 V, 180 V	High 300% to 400% of full load	Fractional to 100 HP	Fixed or adjustable	High 175% to 225%	Applications that require high torque and speed control such as printing presses, conveyors, and hoists
Permanent-magnet	12 V, 24 V, 36 V, 120 V	Low 100% to 200% of full load	Fractional	Varies from 0 to full speed	High 150% to 200%	Applications that require small DC-operated equipment such as automobile power windows, seats, and sun roofs
Stepping	5 V, 12 V, 24 V	Very low† .5 to 5000 oz/in.	Size rating is given as holding torque and number of steps	Rated in number of steps per sec (maximum)	Varies based on number of steps and rated torque	Applications that require low torque and precise control such as indexing tables and printers
AC/DC Universal	115 VAC, 230 VAC, 12 VDC, 24 VDC, 36 VDC, 120 VDC	High 300% to 400% of full load	Fractional	Varies 0 to full speed	High 175% to 225%	Most portable tools such as drills, routers, mixers, and vacuum cleaners

* based on standard 3φ induction motor
† torque is rated as holding torque

OVERCURRENT PROTECTION DEVICES

Motor Type	Code Letter	Motor Size	FLC (%) TDF	NTDF	ITB	ITCB
AC*	—	—	175	300	150	700
AC*	A	—	150	150	150	700
AC*	B–E	—	175	250	200	700
AC*	F–V	—	175	300	250	700
DC	—	⅛ to 50 HP	150	150	150	150
DC	—	Over 50 HP	150	150	150	175

* full-voltage and resistor starting

LOCKED ROTOR CURRENT

Apparent, 1φ	Apparent, 3φ	True, 1φ	True, 3φ
$LRC = \dfrac{1000 \times HP \times kVA/HP}{V}$	$LRC = \dfrac{1000 \times HP \times kVA/HP}{V \times \sqrt{3}}$	$LRC = \dfrac{1000 \times HP \times kVA/HP}{V \times PF \times E_{ff}}$	$LRC = \dfrac{1000 \times HP \times kVA/HP}{V \times \sqrt{3} \times PF \times E_{ff}}$

where
LRC = locked rotor current (in amps)
1000 = multiplier for kilo
HP = horsepower
kVA/HP = kilovolt amps per horsepower
V = volts

(3φ columns add: $\sqrt{3} = 1.73$)

True, 1φ also: PF = power factor, E_{ff} = motor efficiency

MAXIMUM OCPD

$OCPD = FLC \times R_M$
where
FLC = full-load current (from motor nameplate or NEC® Table 430.150)
R_M = maximum rating of OCPD

Motor Type	Code Letter	Motor Size	TDF	NTDF	ITB	ITCB
AC*	—	—	175	300	150	700
AC*	A	—	150	150	150	700
AC*	B–E	—	175	250	200	700
AC*	F–V	—	175	300	250	700
DC	—	1/8 to 50 HP	150	150	150	250
DC	—	Over 50 HP	150	150	150	175

FLC (%)

EFFICIENCY

Input and Output Power Known
$E_{ff} = \dfrac{P_{out}}{P_{in}}$
where
E_{ff} = efficiency (%)
P_{out} = output power (W)
P_{in} = input power (W)

Horsepower and Power Loss Known
$E_{ff} = \dfrac{746 \times HP}{746 \times HP + W_l}$
where
E_{ff} = efficiency (%)
746 = constant
HP = horsepower
W_l = watts lost

VOLTAGE UNBALANCE

$V_u = \dfrac{V_d}{V_a} \times 100$
where
V_u = voltage unbalance (%)
V_d = voltage deviation (V)
V_a = voltage average (V)
100 = constant

POWER

Power Consumed
$P = \dfrac{HP \times 746}{E_{ff}}$
where
P = power consumed (W)
HP = horsepower
746 = constant
E_{ff} = efficiency (%)

Operating Cost
$C_{/hr} = \dfrac{P_{/hr} \times C_{/kWh}}{1000}$
where
$C_{/hr}$ = operating cost per hour
$P_{/hr}$ = power consumed per hour
$C_{/kWh}$ = cost per kilowatt hour
1000 = constant to remove kilo

Annual Savings
$S_{Ann} = C_{Ann\,Std} - C_{Ann\,Eff}$
S_{Ann} = annual cost savings
$C_{Ann\,Std}$ = annual operating cost for standard motor
$C_{Ann\,Eff}$ = annual operating cost for energy-efficient motor

FULL-LOAD CURRENTS — DC MOTORS

Motor rating (HP)	Current (A)	
	120 V	240 V
¼	3.1	1.6
⅓	4.1	2.0
½	5.4	2.7
¾	7.6	3.8
1	9.5	4.7
1½	13.2	6.6
2	17	8.5
3	25	12.2
5	40	20
7½	48	29
10	76	38

FULL-LOAD CURRENTS — 1φ, AC MOTORS

Motor rating (HP)	Current (A)	
	115 V	230 V
⅙	4.4	2.2
¼	5.8	2.9
⅓	7.2	3.6
½	9.8	4.9
¾	13.8	6.9
1	6	8
1½	20	10
2	4	12
3	4	17
5	6	28
7½	80	40

FULL-LOAD CURRENTS — 3φ, AC INDUCTION MOTORS

Motor rating (HP)	Current (A)			
	208 V	230 V	460 V	575 V
¼	1.11	.96	.48	.38
⅓	1.34	1.18	.59	.47
½	2.2	2.0	1.0	.8
¾	3.1	2.8	1.4	1.1
1	4.0	3.6	1.8	1.4
1½	5.7	5.2	2.6	2.1
2	7.5	6.8	3.4	2.7
3	10.6	9.6	4.8	3.9
5	16.7	15.2	7.6	6.1
7½	24.0	22.0	11.0	9.0
10	31.0	28.0	14.0	11.0
15	46.0	42.0	21.0	17.0
20	59	54	27	22
25	75	68	34	27
30	88	80	40	32
40	114	104	52	41
50	143	130	65	52
60	169	154	77	62
75	211	192	96	77
100	273	248	124	99
125	343	312	156	125
150	396	360	180	144
200	—	480	240	192
250	—	602	301	242
300	—	—	362	288
350	—	—	413	337
400	—	—	477	382
500	—	—	590	472

TYPICAL MOTOR EFFICIENCIES

HP	Standard Motor (%)	Energy-Efficient Motor (%)	HP	Standard Motor (%)	Energy-Efficient Motor (%)
1	76.5	84.0	30	88.1	93.1
1.5	78.5	85.5	40	89.3	93.6
2	79.9	86.5	50	90.4	93.7
3	80.8	88.5	75	90.8	95.0
5	83.1	88.6	100	91.6	95.4
7.5	83.8	90.2	125	91.8	95.8
10	85.0	90.3	150	92.3	96.0
15	86.5	91.7	200	93.3	96.1
20	87.5	92.4	250	93.6	96.2
25	88.0	93.0	300	93.8	96.5

STANDARD CALCULATION: ONE-FAMILY DWELLING

1. GENERAL LIGHTING: *Table 220.12*

_____ sq ft × 3 VA = _____ V₄

Small appliances: *220.52(A)*

_____ VA × _____ circuits = _____ VA

Laundry: *220.52(B)*

_____ VA × 1 = _____ VA

_____ VA

Applying Demand Factors: *Table 220.42.*

First 3000 VA × 100% = 3000 VA

Next _____ VA × 35% = _____ VA **PHASES**   **NEUTRAL**

Remaining _____ VA × 25% = _____ VA

Total _____ VA _____ VA _____ VA

2. FIXED APPLIANCES: *220.53*

Dishwasher = _____ VA

Disposer = _____ VA

Compactor = _____ VA

Water heater = _____ VA

_____ = _____ VA

_____ = _____ VA

_____ = _____ VA (120 V Loads × 75%)

Total _____ VA × 75% = _____ VA _____ VA _____ VA

3. DRYER: *220.54; Table 220.54*

_____ VA × _____ % = _____ VA _____ VA × 70% = _____ VA

4. COOKING EQUIPMENT: *Table 220.55; Notes*

Col A _____ VA × _____ % = _____ VA

Col B _____ VA × _____ % = _____ VA

Col C _____ VA × _____ % = _____ VA

Total _____ VA _____ VA × 70% = _____ VA

5. HEATING or A/C: *220.60*

Heating unit = _____ VA × 100% = _____ VA

A/C unit = _____ VA × 100% = _____ VA

Heat pump = _____ VA × 100% = _____ VA

Largest Load _____ VA _____ VA _____ VA

6. LARGEST MOTOR: *220.14(C) (220.50)*

φ _____ VA × 25% = _____ VA _____ VA

N _____ VA × 25% = _____ VA _____ VA

1φ service: PHASES $I = \dfrac{VA}{V} =$ _____ A

NEUTRAL $I = \dfrac{VA}{V} =$ _____ A _____ VA _____ VA

220.61(B) First 200 A × 100% = 200 A

Remaining _____ A × 70% = _____ A

Total _____ A

OPTIONAL CALCULATION: ONE-FAMILY DWELLING

1. HEATING or A/C: *220.82(C)(1 – 6)*

Heating units (3 or less) = _____ VA × 65% = _____ VA

Heating units (4 or more) = _____ VA × 40% = _____ VA

A/C unit = _____ VA × 100% = _____ VA

Heat pump = _____ VA × 100% = _____ VA

Largest Load _____ VA

Total _____ VA   _____ VA

PHASES

2. GENERAL LOADS: *220.82(B)(1 – 4)*

General lighting: *220.82(B)(1)*

_____ sq ft × 3 VA   _____ VA

Small appliance and laundry loads: *220.82(B)(2)*

_____ VA × _____ circuits =   _____ VA

Special loads: *220.82(B)(3 – 4)*

Dishwasher = _____ VA

Disposer = _____ VA

Compactor = _____ VA

Water heater = _____ VA

_____ = _____ VA

_____ = _____ VA

_____ = _____ VA

_____ = _____ VA

_____ = _____ VA

_____ VA   _____ VA

Total _____ VA

Applying Demand Factors: *220.82(B)*

First 10,000 VA × 100% =   10,000 VA

Remaining _____ VA × 40% =   _____ VA

Total _____ VA   _____ VA

NEUTRAL (Loads from Standard Calculation)

1. General lighting = _____ VA
2. Fixed appliances = _____ VA
3. Dryer = _____ VA
4. Cooking equipment = _____ VA
5. Heating or A/C = _____ VA
6. Largest motor = _____ VA

Total ☐ VA

1φ service: PHASES $I = \dfrac{\text{_____ VA}}{\text{V}} = $ _____ A

NEUTRAL $I = \dfrac{\text{_____ VA}}{\text{V}} = $ _____ A   ☐ VA

Glossary

A

abbreviation: Shortened form of a word or phrase.

accessible: Equipment admits close approach and is not guarded by locked doors, elevation, etc.

adjustable-trip circuit breakers (ATCBs): Circuit breakers whose trip setting can be changed by adjusting the ampere setpoint, trip time characteristics, or both, within a particular range.

ambient temperature: The temperature of air around a piece of equipment.

ampacity: The current that a conductor can carry continuously, under the conditions of use.

appliance: Any utilization equipment which performs one or more functions, such as clothes washing, air conditioning, cooking, etc.

appliance branch circuit: A branch circuit that supplies energy to one or more outlets to which appliances are to be connected.

approved: Acceptable to the AHJ.

armored cable (AC): A factory assembly that contains the conductors within a jacket made of a spiral wrap of steel.

array: An assembly of modules or panels mounted on a support structure to form a direct current, power-producing unit.

askarel: A group of nonflammable synthetic chlorinated hydrocarbons that were once used where nonflammable insulating oils were required.

autotransformers: Single-winding transformers that share a common winding between the primary and secondary circuits.

auxiliary gutter: A metal or nonmetallic enclosure equipped with hinged or removable covers that is used to supplement wiring space.

B

bare conductor: A conductor with no insulation or covering of any type.

bathroom: An area with a basin and one or more of a toilet, tub, or shower.

bend: Any change in direction of a raceway.

bonding: Joining metal parts to form a continuous path to conduct safely any current that is commonly imposed.

box: A metallic or nonmetallic electrical enclosure used for equipment, devices, and pulling or terminating conductors.

branch circuit: The portion of the electrical circuit between the last overcurrent device (fuse or CB) and the outlets or utilization equipment.

branch-circuit conductor: The circuit conductor(s) between the final overcurrent device protecting the circuit and the outlet(s).

branch-circuit rating: The ampere rating or setting of the overcurrent device protecting the conductors.

building: A structure that stands alone that is cutoff from adjoining structures by fire walls with all openings therein protected by approved fire doors.

bulk storage plant: A location where liquids are received by tank vessel, pipelines, or tank cars, and are stored or blended in bulk for the purpose of distribution to similar vehicles or containers.

bushing: A fitting placed on the end of a conduit to protect the conductor's insulation from abrasion.

busway: A sheet metal enclosure that contains factory-assembled aluminum or copper busbars which are supported on insulators.

C

cable: A factory assembly with two or more conductors and an overall covering.

cable assembly: A flexible assembly containing multiconductors with a protective outer sheath.

cablebus: An assembly of insulated conductors and terminations in an enclosed, ventilated protective metal housing.

cable tray system (CTS): An assembly of sections and associated fittings which form a rigid structural system used to support cables and raceways.

cadwelding: A welding process used to make electrical connections of copper to copper or copper to steel in which no outside source of heat or power is required.

cartridge fuse: A fuse constructed of a metallic link(s) which is designed to open at predetermined current levels to protect circuit conductors and equipment.

cell: A single, enclosed, tubular space in a precast cellular concrete slab floor with the direction of the cell being parallel to the direction of the floor member.

cellular metal floor raceway: The hollow spaces of cellular metal floors, with fittings, approved as enclosures for electric conductors.

circuit breaker (CB): A device which opens and closes circuits by nonautomatic means and opens circuits automatically when a predetermined overcurrent exists.

circular mil: A measurement used to determine the cross-sectional area of a conductor.

Class I location: A hazardous location in which sufficient quantities of flammable gases and vapors are present in the air to cause an explosion or ignite the hazardous materials.

Class II location: A hazardous location in which sufficient quantities of combustible dust are present in the air to cause an explosion or ignite the hazardous materials.

Class III location: A hazardous location in which easily-ignitable fibers or flyings are present in the air but not in a sufficient quantity to cause an explosion or ignite the hazardous materials.

conductor: A slender rod or wire that is used to control the flow of electrons in an electrical circuit.

conduit body: A conduit fitting that provides access to the raceway system through a removable cover at a junction or termination point.

conduit seal: A fitting which is inserted into runs of conduit to isolate certain electrical apparatus from atmospheric hazards.

continuous load: A load in which the maximum current is expected to continue for three hours or more.

controller: The device in a motor circuit which turns the motor ON or OFF.

cover: The shortest distance measured between a point on the top surface of any direct-buried conductor, cable, conduit, or other raceway and the top surface of finished grade, concrete, or similar cover.

covered conductor: A conductor not encased in a material recognized by the NEC®.

current-limiting fuses: Fuses that open a circuit in less than $\frac{1}{2}$ of a cycle to protect the circuit components from damaging short-circuit currents.

current transformer: A transformer that creates a constant ratio of primary to secondary current instead of attempting to maintain a constant ratio of primary to secondary voltage.

D

damp location: A partially protected area subject to some moisture.

dead front: A cover required for the operation of a plug or connector.

demand: The amount of electricity required at a given time.

demand factor: The ratio of the maximum demand of a system, or part of a system, to the total connected load of a system or the part of the system under consideration.

detail: Part of a plan, elevation, or section view drawn at a larger scale.

device box: A box which houses an electrical device.

devices: Electrical components, such as receptacles and switches, that are designed to carry electricity.

disconnecting means: A device or group of devices that separate or isolate the conductors of a circuit from their source of supply.

division: The classification assigned to each Class based upon the likelihood of the presence of the hazardous substance in the atmosphere.

Division 1 location: A hazardous location in which the hazardous substance is normally present in the air in sufficient quantities to cause an explosion or ignite the hazardous materials.

Division 2 location: A hazardous location in which the hazardous substance is not normally present in the air in sufficient quantities to cause an explosion or ignite the hazardous materials.

dry location: A location which is not normally damp or wet.

dry-niche lighting fixture: A lighting fixture installed in the pool wall in a niche sealed against pool water entry.

dry-type transformer: A transformer which provides air circulation based on the principle of heat transfer.

dust-ignitionproof: Enclosed in a manner which prevents the entrance of dusts and does not permit arcs, sparks, or excessive temperature to cause ignition of exterior accumulations of specified dust.

dustproof: Construction in which dust does not interfere with the successful operation of equipment.

dusttight: Construction that does not permit dust to enter the enclosing case under specified test conditions.

dwelling: A structure that contains eating, living, and sleeping space, and permanent provisions for cooking and sanitation.

dwelling unit: One or more rooms used by one or more persons for housekeeping with space for eating, living, and sleeping and having permanent provisions for cooking and sanitation.

dynamic load: A load that produces a small but constant vibration.

E

easement: Space set aside by agreement with property owners for use by utility companies to install and service their equipment.

Edison-base fuse: A plug fuse that incorporates a screw configuration which is interchangeable with fuses or other ampere ratings.

effectively grounded: Grounded with sufficient low impedance and current-carrying capacity to prevent hazardous voltage buildups.

electrical metallic tubing (EMT): A lightweight tubular steel raceway without threads on the ends.

electric-discharge lighting fixture: A lighting fixture that utilizes a ballast for the operation of the lamp.

elevation: An orthographic drawing showing vertical planes of a building

enclosure: The case or housing of equipment or other apparatus which provides protection from live or energized parts.

equipment: Any material, device, fixture, apparatus, appliance, etc. used in conjunction with electrical installations.

equipment bonding jumper (EBJ): A conductor that connects two or more parts of the EGC.

equipment grounding conductor (EGC): An electrical conductor that provides a low-impedance path between electrical equipment and enclosures and the system grounded conductor and GEC.

equipotential plane: An area in which all conductive elements are bonded or otherwise connected together in a manner which prevents a difference of potential from developing within the plane.

explosionproof apparatus: Equipment which is enclosed in a case that is capable of withstanding any explosion that may occur within it, without permitting the ignition of flammable gases or vapors on the outside of the enclosure.

exposed: As applied to wiring methods, is on a surface or behind panels which allow access.

F

FCC System: A complete wiring system (including FCC, connectors, terminators, adapters, boxes, and receptacles) designed for installation under carpet squares.

feeder: All circuit conductors between the service equipment or the source of a separately derived system and the final branch-circuit overcurrent device.

feeder neutral load: The maximum unbalance between any of the ungrounded conductors and the grounded conductor.

fitting: An electrical system accessory that performs a mechanical function.

flash point (fire point): The temperature at which liquids give off vapor sufficient to form an ignitable mixture with the air near the surface of the liquid.

flat conductor cable (FCC): A cable with three or more flat copper conductors edge-to-edge and separated and enclosed by an insulating material.

flexible cable: An assembly of one or more insulated conductors, with or without braids, contained within an overall outer covering and used for the connection of equipment to a power source.

flexible cord: An assembly of two or more insulated conductors, with or without braids, contained within an overall outer covering and used for the connection of equipment to a power source.

flexible metal conduit (FMC): A raceway of metal strips which are formed into a circular cross-sectional raceway.

floor plan: An orhtographic drawing of a building as though cutting planes were made through it horizontally.

forming shell: A structure mounted in a pool and designed to support a wet-niche lighting fixture assembly.

fuse: An overcurrent protection device with a fusible link that melts and opens the circuit when an overload condition or short circuit occurs.

G

general-purpose branch circuit: A branch circuit that supplies a number of outlets for lighting and appliances.

general-use snap switch: A form of general-use switch constructed so that it can be installed in device boxes, etc.

general-use switch: A switch for use in general distribution and branch circuits. The ampere-rated switch is capable of interrupting its rated current at its rated voltage.

generator: A device that is used to convert mechanical power to electrical power.

grade: The level or elevation of the earth on a job site.

ground: The earth.

grounded: Connected to the earth or a conductive body connected to the earth.

grounded conductor: A conductor that has been intentionally grounded.

ground fault: An unintentional connection between an ungrounded conductor and any grounded raceway, box, enclosure, fitting, etc.

ground-fault circuit interrupter (GFCI): A device whose function is to de-energize a circuit within established time when current to ground exceeds a predetermined value which is less than that required to operate the protective device of a supply circuit.

grounding conductor: The conductor that connects electrical equipment or the grounded conductor to the grounding electrode.

grounding electrode conductor (GEC): The conductor that connects the grounding electrode(s) to the system grounded conductor and/or the EGC.

grounding receptacles: Receptacles which include a grounding terminal connected to a grounding slot in the receptacle configuration.

group: An atmosphere containing flammable gases or vapors or combustible dust.

H

hazardous location: A location where there is an increased risk of fire or explosion due to the presence of flammable gases, vapors, liquids, combustible dusts, or easily-ignitable fibers or flyings.

header: A transverse metal raceway for electric conductors with access to predetermined cells, permitting the installation of electric conductors from a distribution center to the floor cells.

health care facility: A location, either a building or a portion of a building, which contains occupancies such as, hospitals, nursing homes, limited or supervisory care facilities, clinics, medical and dental offices, and either movable or permanent ambulatory facilities.

heating panel: A complete assembly with a junction box or a length of flexible conduit for connection.

heating panel set: A rigid or nonrigid assembly with nonheating leads or a terminal junction suitable for connection.

heating system: A complete system of components such as heating elements, fasteners, nonheating circuit wiring, leads, temperature controllers, safety signs, junction boxes, raceways, and fittings.

hermetic refrigerant motor-compressor: A combination of a compressor and motor enclosed in the same housing, having no external shaft or shaft seals, with the motor operating in the refrigerant.

high-intensity discharge (HID) lighting fixture: A lighting fixture that generates light from an arc lamp contained within an outer tube.

hydromassage bathtub: A permanently installed bathtub with a recirculating pump designed to accept, circulate, and discharge water for each use.

I

identified: Recognized as suitable for the use, purpose, etc.

immersion detection circuit interrupter (IDCI): Circuit interrupter designed to provide protection against shock when appliances fall into a sink or bathtub.

impedance: The total opposition to the flow of current in a circuit.

impedance heating system: A system in which heat is generated by pipe(s) and rod(s) by an electrical source.

indicating device: A pilot light, buzzer, horn, or other type of alarm. Often, the wiring of a motor control circuit is very elaborate and requires ten times the amount of wiring as the motor power circuit.

individual branch circuit: A branch circuit that supplies only one piece of utilization equipment.

in sight from: Visible and not more than 50′ away.

instantaneous-trip circuit breakers (ITBs): Circuit breakers with no delay between the fault or overload sensing element and the tripping action of the device.

instrument transformer: A transformer used to reduce higher voltage and current ratings to safer and more suitable levels for the purposes of control and measurement.

insulated conductor: A conductor covered with a material classified as electrical insulation.

interactive system: A system that operates in parallel with and is permitted to deliver power to a normal utility service connected to the same load.

intermittent duty motor: Motor that operates for alternate intervals of load and no load; load and rest; or load, no load, and rest.

intermittent load: A load in which the maximum current does not continue for three hours.

interrupting rating: The maximum amount of current that an OCPD can clear safely.

intrinsically safe system: A system with an assembly of intrinsically safe apparatus and associated apparatus which is interconnected and used in hazardous locations to supply equipment.

inverse-time circuit breakers (ITCBs): Circuit breakers with an intentional delay between the time when the fault or overload is sensed and the time when the circuit breaker operates.

inverter: Equipment used to change voltage level or waveform, or both, of electrical energy.

isolated-ground receptacles: Receptacles in which the grounding terminal is isolated from the device yoke or strap.

isolating switch: A switch designed to isolate an electric circuit from the power source. It has no interrupting rating and is operated only after the circuit has been opened by other means.

isolation transformer: A transformer that utilizes a shield between the primary and secondary windings and a transformer ratio of 1:1 to ensure that the load is separated from the power source.

J

junction box: A box in which splices, taps, or terminations are made.

K

K factor: The resistance of a circular-mil foot of wire at a set temperature.

kick: A single bend in a raceway.

L

labeled: Equipment acceptable to the AHJ and to which a label has been attached.

lampholders: Devices designed to accommodate a lamp for the purpose of illumination.

less-flammable liquid: An insulating oil which is flammable but has reduced flammable characteristics and a higher fire point.

lighting and appliance branch-circuit panelboard: A panelboard with more than 10% of its branch-circuit fuses or CBs rated at 30 A or less (15 A, 20 A, 25 A, or 30 A).

lighting outlet: An outlet intended for the direct connection of a lampholder, a lighting fixture, or pendant cord terminating in a lampholder.

lighting outlets: Outlets that provide power for lighting fixtures.

lighting track: An assembly consisting of an energized track and lighting fixture heads which can be positioned in any location along the track.

line surge: A temporary increase in the circuit or system voltage or current that may occur as a result of fluctuations in the electrical distribution system.

liquid-filled transformer: A transformer that utilizes some form of insulating liquid to immerse the core and windings of the transformer to aid in the removal of heat generated by the transformer windings.

liquidtight flexible metal conduit (LFMC): A raceway of circular cross section with an outer liquidtight, nonmetallic, sunlight-resistant jacket over an inner helically-wound metal strip.

liquidtight flexible nonmetallic conduit (LFNC): A raceway of circular cross-section with an outer jacket which is resistant to oil, water, sunlight, corrosion, etc. The inner core varies based on intended use.

listed: Equipment or material approved by the AHJ in a list.

longitudinal section: A section taken lengthwise.

luminaire: A complete lighting unit consisting of a lamp or lamps together with the parts designed to distribute the light, to position and protect the lamps and ballast (where applicable), and to connect the lamps to the power supply.

M

main bonding jumper (MBJ): The connection at the service equipment that ties together the EGC, the grounded conductor, and the GEC.

manufactured building: A building of closed construction, normally factory-made or assembled for installation or assembly and installation on building site. This does not include mobile homes, recreational vehicles, park trailers, or manufactured homes.

metal-clad cable (MC): A factory assembly of one or more conductors with or without fiber-optic members, enclosed in a metallic armor.

metal wireway: A sheet metal trough with a hinged or removable cover that houses and protects wires and cables laid in place after the wireway has been installed.

mil: 0.001″.

milliampere (mA): $\frac{1}{1000}$ of an ampere (1000 mA = 1 A).

mineral oil: A chemically untreated insulating oil that is distilled from petroleum.

mobile home: A transportable factory-assembled structure or structures constructed on a permanent chassis for use as a dwelling. A mobile home is not constructed on a permanent foundation but is connected to the required utilities. The term "mobile home" includes manufactured homes.

module: The smallest complete, environmentally-protected assembly of solar cells, exclusive of tracking, designed to generate DC power under sunlight.

motor branch circuit: The point from the last fuse or CB in the motor circuit out to the motor.

motor control center (MCC): An assembly of one or more enclosed sections with a common power bus and primarily containing motor control units.

motor control circuit: The circuit of a control apparatus or system which carries electric signals directing the performance of the controller, but does not carry the main power current.

multifamily dwelling: A dwelling with three or more dwelling units.

multiple receptacle: A single device with two or more receptacles.

multioutlet assembly: A surface, flush, or freestanding raceway that contains conductors and receptacles

multiwire branch circuit: A branch circuit with two or more ungrounded conductors having a potential difference between them, and is connected to the neutral or grounded conductor of the system.

N

nipple: A short piece of conduit or tubing that does not exceed 24″ in length.

nominal voltage: Any voltages within an acceptable range.

nonadjustable-trip CBs (NATCBs): Fixed CBs designed without provisions for adjusting either the ampere trip setpoint or the time-trip setpoint.

noncoincident loads: Loads that are are unlikely to be in use at the same time.

nonflammable liquid: A liquid that is noncombustible and does not burn when exposed to air.

nongrounding receptacles: Receptacles with two wiring slots for branch-circuit wiring systems that do not provide an equipment grounding conductor.

no-niche lighting fixture: A lighting fixture installed above or below the water without a niche.

nonlinear load: A load where the wave shape of the steady-state current does not follow the wave shape of the applied voltage.

nonmetallic extensions: Assemblies of two insulated conductors in a nonmetallic jacket or an extruded thermoplastic covering. Nonmetallic extensions are surface extensions intended for mounting directly on the surface of walls or ceilings.

nonmetallic-sheathed cable (NM): A factory assembly of two or more insulated conductors having an outer sheath of moisture-resistant, flame-retardant, nonmetallic material.

nonmetallic wireway: A flame-retardant nonmetallic trough with a removable cover that houses and protects wires and cables laid in place after the wireway has been installed.

non-selective system: A fault on an individual branch circuit not only opens the branch-circuit overcurrent protection device, but also opens the feeder overcurrent protection device.

non-time delay fuses (NTDFs): Fuses that may detect an overcurrent and open the circuit almost instantly.

O

offset: A double bend in a raceway, each containing the same number of degrees.

one-family dwelling: A building consisting solely of one dwelling unit.

outlet: Any point in the electrical system where current supplies utilization equipment.

outlet box: A box which houses a piece of utilization equipment.

overcurrent: Any current in excess of that for which the conductor or equipment is rated.

overlamping: Installing a lamp of a higher wattage than for which the fixture is designed.

overload: A small-magnitude overcurrent, that over a period of time, leads to an overcurrent which may operate the overcurrent protection device (fuse or CB).

oxidation: The process by which oxygen mixes with other elements and forms a rust-like material.

oxide: A thin, but highly resistive coating that forms on metal when exposed to the air.

P

panel: A collection of modules, secured together, wired, and designed to provide a field-installable unit.

panelboard: A single panel or group of assembled panels with buses and overcurrent devices, which may have switches to control light, heat, or power circuits.

panic hardware: Door hardware designed to open easily in an emergency situation.

parallel conductors: Two or more conductors that are electrically connected at both ends to form a single conductor.

pendants: Hanging light fixtures that use flexible cords to support the lampholder.

permanently-connected appliance: A hard-wired appliance that is not cord-and-plug connected.

permanently installed swimming pool: A pool constructed in ground or partially above ground and designed to hold over 42″ of water, and all indoor pools regardless of depth.

phase converter: An electrical device that converts 1φ power to 3φ power.

phase-to-ground voltage: The difference of potential between a phase conductor and ground.

phase-to-phase voltage: The maximum voltage between any two phases of an electrical distribution system.

photovoltaic output circuit: Conductors between photovoltaic source circuits and the inverter or DC utilization equipment.

photovoltaic power source: One or more arrays which generate DC power at system voltage and current.

photovoltaic source circuit: Conductors between modules and from modules to common junction point(s) of the DC system.

pilot device: A sensing device that controls the motor controller.

plot plan: Orthographic drawing that shows the location and orientation of a structure on a lot and the size of the lot.

place of assembly: A building, structure, or portion of a building designed or intended for use by 100 or more persons.

plug fuse: A fuse that uses a metallic strip which melts when a predetermined amount of current flows through it.

point of beginning: A fixed point from which all measurements, vertical and horizontal, are made to show the building lot.

potential transformer: A transformer which steps down higher voltages while allowing the voltage of the secondary to remain fairly constant from no-load to full-load conditions.

power panelboard: A panelboard with more than 10% of its branch-circuit fuses or CBs rated over 30 A or more.

premises wiring: Basically all interior and exterior wiring installed on the load side of the service point or the source of a separately derived system.

print: Detailed plan of a building drawn orthographically.

pull box: A box used as a point to pull or feed electrical conductors into the raceway system.

R

raceway: A metal or nonmetallic enclosed channel for conductors.

raceway system: An enclosed channel of metal or nonmetallic materials used to contain the wires or cables of an electrical system.

readily accessible: Capable of being reached quickly.

receptacle outlets: Outlets that provide power for cord-and-plug connected equipment.

receptacles: Contact devices installed at outlets for the connection of cord-connected electrical equipment.

recreational vehicle: A vehicular type unit which is self-propelled or is mounted on or pulled by another vehicle and is primarily designed as temporary living quarters for camping, travel, or recreational use.

redundant grounding: Grounding with two separate grounding paths.

resistance heating element: A specific element to generate heat that is embedded in or fastened to the surface.

rigid metal conduit (RMC): A threadable raceway generally made of steel or aluminum designed for physical protection of conductors and cables.

rigid nonmetallic conduit (RNC): A cross-sectional raceway of suitable nonmetallic material that is resistant to moisture and chemical atmospheres.

S

section: An orthographic drawing created by passing a cutting plane (either horizontal or vertical) through a portion of a building.

selective system: A fault on an individual branch circuit opens only the branch-circuit overcurrent protective device and does not affect the feeder overcurrent protection device.

self-grounding receptacles: Grounding type receptacles which utilize a pressure clip around the 6 – 32 mounting screw to ensure good electrical contact between the receptacle yoke and the outlet box.

separately derived system: A system that supplies premises with electrical power derived or taken from storage batteries, solar photovoltaic systems, generators, transformers, or converter windings.

service: The electrical supply, in the form of conductors and equipment, that provides electrical power to the building or structure.

service conductors: The conductors from the service point or other source of power to the service disconnecting means.

service drop: The conductors that extend from the overhead utility supply system to the service-entrance conductors at the building or structure.

service-entrance cable (SE): A single or multiconductor assembly with or without an overall covering.

service-entrance conductors – overhead systems: Conductors that connect the service equipment for the building or structure with the electrical utility supply conductors.

service-entrance conductors – underground systems: Conductors that connect the service equipment with the service lateral.

service equipment: All of the necessary equipment to control the supply of electrical power to a building or a structure.

service lateral: The underground service conductors that connect the utility's electrical distribution system with the service-entrance conductors.

service mast: An assembly consisting of a service raceway, guy wires or braces, service head, and any fittings necessary for the support of service-drop conductors.

service point: The connection point between the utility and the premises wiring.

short circuit: The unintentional connection of two ungrounded conductors that have a potential difference between them. The condition that occurs when two ungrounded conductors (hot wires), or an ungrounded and a grounded conductor of a 1ϕ circuit, come in contact with each other.

single receptacle: A single contact device with no other contact device on the same yoke.

skeleton tubing: Neon tubing that is the sign of outline lighting. It does not contain, nor is attached to a sign body or enclosure.

solar cell: Basic photovoltaic device that generates electricity when exposed to light.

solar photovoltaic system: The system used to convert solar energy into electrical energy suitable for connection to a utilization load.

spa (hot tub): An indoor or outdoor hydromassage pool or tub that is not designed to have the water discharged after each use.

special permission: The written approval of the authority having jurisdiction.

specifications: Sheets of paper gathered together into a pamphlet (or into several volumes for a large building) covering a number of subjects in detail.

step-down transformer: A transformer with more windings in the primary winding, which results in a load voltage which is less than the applied voltage.

step-up transformer: A transformer with more windings in the secondary winding, which results in a load voltage which is greater than the applied voltage.

storable swimming pool: A pool constructed on or above ground and designed to hold less than 42″ of water, or a pool with nonmetallic, molded polymeric walls or inflatable fabric walls regardless of size.

strut-type channel raceway: A surface raceway formed of moisture-resistant and corrosion-resistant metal.

supervised installation: An electrical installation in which the conditions of maintenance are such that only qualified persons monitor or service the electrical equipment.

surface raceway: An enclosed channel for conductors which is attached to a surface.

switch: A device, with a current and voltage rating, used to open or close an electrical circuit.

switchboard: A single panel or group of assembled panels with buses, overcurrent devices, and instruments.

T

temperature rise: The amount of heat that an electrical component produces above the ambient temperature.

thermally protected: Designed with an internal thermal protective device which senses excessive operating temperatures and opens the supply circuit to the fixtures.

time delay fuses (TDFs): Fuses that may detect and remove a short circuit almost instantly, but allow small overloads to exist for a short period of time.

torque: A turning or twisting force, typically measured in foot-pounds (ft-lb).

transformer: A device that converts or "transforms" electrical power at one voltage or current to another voltage or current.

transformers: Electrical devices that contain no moving parts and are used primarily to convert electrical power at one voltage and current rating to another voltage and current rating.

transverse section: A section taken crosswise.

travel trailer: A vehicle that is mounted on wheels, has a trailer area less than 320 sq ft (excluding wardrobes, closets, cabinets, kitchen units, fixtures, etc.), is of such size and weight that a special highway use permit is not required, and is designed as temporary living quarters while camping or traveling.

two-family dwelling: A dwelling with two dwelling units.

Type I building: A building in which all structural members (walls, columns, beams, girders, trusses, arches, floors, and roofs) are constructed of approved noncombustible or limited-combustible materials.

Type II building: A building that does not qualify as Type I construction in which the structural members (walls, columns, beams, girders, trusses, arches, floors, roofs, etc.) are constructed of approved noncombustible or limited-combustible materials.

Type S fuse: A plug fuse that incorporates a screw and adapter configuration which is not interchangeable with fuses of another ampere rating.

U

unfinished basement: The portion of area of a basement which is not intended as a habitable room, but is limited to storage areas, work areas, etc.

unit switch: A switch designed to control a specific unit load.

USE: Underground service-entrance cable with a moisture-resistant covering that is not flame-resistant.

utilization equipment: Any electrical equipment which uses electrical energy for electronic, electromechanical, chemical, heating, lighting, etc. purposes.

V

voltage drop: Voltage that is lost due to the resistance of conductors.

voltage-to-ground: The difference of potential between a given conductor and ground.

W

wet location: Any location in which a conductor is subjected to excessive moisture or saturation from any type of liquids or water.

wet-niche lighting fixture: A lighting fixture installed in a forming shell and completely surrounded by water.

wireway: A metallic or nonmetallic trough with a hinged or removable cover designed to house and protect conductors and cables.

within sight: Visible and not more than 50′ away.

Index

A

abbreviations, 12, 34
AC voltage measurement, 98
adapters, 55
addition to existing building, 94
adjustable trip circuit breaker, 53
agricultural building, 143
air conditioning equipment, 97
aircraft hangars, 206
air plenum wiring methods, 134
aluminum conductor, 80, 118
aluminum wire and conduit, 117
amateur transmitting and receiving
    station, 217
ampacity of fixture wires, 80
amplifiers, 212
angle pulls, 138
antenna systems, 217
apartment
  details and elevations, 84–85
  floor plan, 81–82
  load calculations, 81–83
  service calculations, 83–84
appliance, 70
  branch circuit protection, 95
  panelboard, 55, 130
applying the NEC®, 4–5
area, 33
armored cable, 60
array, 144
askarel transformers, 179
authority having jurisdiction, 4
autotransformer, 179
  dimmer, 211
auxiliary gutter, 138
  recessed, 212
available fault current, 128

B

bathroom, 31
bonding jumpers, 133
bonding swimming pools, 189

box
  angle pull, 138
  device, 51, 52
  junction, 138
  metal, 51
  outlet, 51, 52
  pull, 138
box fill, 51
bracket fixture wiring, 211
braided-covered insulated
    conductors, 183
branch circuit, 100
  calculations, 34, 36
  conductor, 142
  conductor identification, 100
  load calculation, 142
  loads, 100
  protection, 95
  tap rules, 133
  taps, 142
building, 123
  addition to existing, 94
  opening service drop clearances,
    41
bulk storage plant, 207
bushing, 60
busways, 166
  no plug-in, 166
  plug-in, 166
  trolley, 166
BX, 60

C

cable
  aerial, 91
  connectors, 211
  Type SE, 41
  Type UF, 67
cablebus, 167
cable fished into masonry walls, 67
cable limiter, 124
cable trays, 210

calculations, 31–39
capacitors
  motor circuits, 183
  over 600 V, 183
  600 V or less, 183
cartridge fuses, 56–57
  classes, 57
  dual-element, time-delay, 56
  dual-element, time-delay
    current-limiting, 56
  single-element, current-limiting,
    56
cell, 136
cellular concrete floor raceway, 136
cellular metal floor raceway, 136
circuit breaker, 53, 128
circuits, 66
  Class 1, 214
  Class 2, 214
  Class 3, 214
  combining Class 2 and Class 3,
    214
  communication, 216
  less than 50 volts, 213
Class I, Division 1 locations, 202
Class I, Division 2 locations, 203
Class II, Division 1 locations, 204
Class II, Division 2 locations, 204
Class III, Division 1 locations, 205
Class III, Division 2 locations, 205
clearances for service drops, 40
clearances for service drops from
    building openings, 41
clearances for service laterals, 40
coating processes, 208
Code-Making Panel, 2
combustible, 200
commercial garages, 206
commercial locations, 109
  electrical plans, 110–118
  lighting system, 117
  service calculation, 115–116
  service conduit size, 116–117